The Cell
in Medical Science

Volume 1:

The Cell and its Organelles

Freeze-etched preparation of a portion of a mesenchymal cel. The cytoplasm contains mitochondria (M), tracts of endoplasmic reticulum (ER), and numerous vesicles (V). Some mitochondria show internal cristae (arrows). A pinocytic invagination (Pi) of the plasma membrane is also seen. A portion of the nucleus (N) is seen surrounded by the two membranes (NE) of the nuclear envelope. × 39 000. Courtesy of Dr. E. Katchburian, The London Hospital Medical College.

The Cell
in Medical Science

Volume 1:
The Cell and its Organelles

edited by

F. BECK, *Department of Anatomy, University of Leicester*
and
J. B. LLOYD, *Research Unit in Biochemistry, Keele University*

1974

Academic Press
London New York

A Subsidiary of Harcourt Brace Jovanovich, Publishers

611.018
C393
v. 1

ACADEMIC PRESS INC. (LONDON) LTD.
24/28 Oval Road,
London NW1

United States Edition published by
ACADEMIC PRESS INC.
111 Fifth Avenue
New York. New York 10003

Library of Congress Catalog Card Number: 78-172366
ISBN: 0 12 084201 7

PRINTED IN GREAT BRITAIN BY
WILLMER BROTHERS LIMITED, BIRKENHEAD

Contributors to Volume 1

A. C. ALLISON, *Clinical Research Centre, Harrow, Middlesex, HA1 3UJ.*

H. BAUM, *Department of Biochemistry, Chelsea College of Science and Technology, London, SW3 6LX.*

F. BECK, *Department of Anatomy, University of Leicester, University Road, Leicester, LE1 7RH.*

P. N. CAMPBELL, *Department of Biochemistry, University of Leeds, Leeds, LS2 9LS.*

A. VON DER DECKEN, *The Wenner-Gren Institute for Experimental Biology, Stockholm, Sweden.*

J. B. LLOYD, *The Research Unit in Biochemistry, University of Keele, Keele, Staffordshire, ST5 5BG.*

J. A. LUCY, *Department of Biochemistry, Royal Free Hospital School of Medicine, University of London, London, WC1N 1BP.*

A. P. MATHIAS, *Wellcome Laboratories, Department of Biochemistry, University College, London, Gower Street, London, WC1E 6BT.*

G. R. MOORES, *Department of Cell Biology, University of Glasgow, Glasgow, G11 6NU.*

T. A. PARTRIDGE, *Experimental Pathology Unit, Charing Cross Hospital Medical School, London.*

E. N. WILLMER, *Yew Garth, Grantchester, Cambridge.*

Preface

The increasing importance of cell biology in Medical Science is becoming clear to clinicians and laboratory scientists alike. It is the meeting ground of many traditional disciplines and forms a central theme for many others. Its impact on subjects as diverse as immunology and neurobiology is already very great and one cannot but appreciate the potential that the application of its techniques and discipline must have for the future.

Many excellent introductory books of cell biology are available but beyond these one has in general to pass either to reviews or to original articles in order to probe more deeply. The present volumes are designed for readers who already have an elementary knowledge of cell biology; they present various aspects of the subject in depth and try to indicate some of the directions in which contemporary cell biology is moving and the methods it uses. No attempt is made to provide a comprehensive cover of cell biology, but the topics are chosen so as to produce a coherent work rather than a series of unconnected essays. We are greatly indebted to our contributors for their willingness to work within a number of constraints and for their patience with a multitude of editorial requests made to help us achieve our aims.

Volume 1 of this book is concerned with the generalized cell unit, and its chapters deal with the biophysical and biochemical basis of the structure and function of the chief subcellular organelles. Volume 2 contains a series of chapters on the relationship of the cell to developmental processes both within the cell itself and in the organism as a whole. There follow chapters on specific cellular specializations, particularly within the neuromuscular system. Here morphological adaptation for specific functional purposes is described in detail. This theme is again followed in Volume 3 where connective tissues, various endocrines, absorptive and secretory cells are dealt with. Volume 4 begins with three chapters on metabolic control mechanisms, and these are followed by chapters on the relationship between the cell and its environment in various pathological states; immunological processes, inflammation, wound healing and carcinogenesis are treated within this framework.

We believe this book will be of value to senior undergraduate students and to research workers looking for summaries on a variety of related topics concerned with cell structure and function. The contributors have been asked to provide only brief bibliographies which enable the reader to develop his own interest; the chapters do not attempt to include an extensive review of the literature.

Like many before us, we are indebted to Academic Press for the patience, forbearance and unfailing courtesy of numerous members of their staff.

F. BECK *February 1974*
University of Leicester
J. B. LLOYD
University of Keele

Contents of Volume 1

The Cell and its Organelles

1. The Cell as a Unit
E. N. WILLMER

2. Membranes and Trans-Membrane Transport
J. A. LUCY

3. The Cell Surface
G. R. MOORES and T. A. PARTRIDGE

4. The Nucleus
A. P. MATHIAS

5. Cytomembranes and Ribosomes
P. N. CAMPBELL and A. VON DER DECKEN

6. Mitochondria and Peroxisomes
H. BAUM

7. Lysosomes
J. B. LLOYD and F. BECK

8. Subcellular Pathology
A. C. ALLISON

Contents of Volume 2
Cellular Genetics and Development

Cellular Specializations

Contents of Volume 3
Cellular Specializations (continued)

Contents of Volume 4
Cellular Control Mechanisms

Cellular Responses to Environment

*Deceased

The Cell and its Organelles

1.   The Cell as a Unit

E. N. WILLMER

Emeritus Professor of Histology, University of Cambridge, England

I. INTRODUCTION

Once upon a time cells in the body used to be compared with the bricks in a building and, just as there may be different types of bricks for different purposes, so each cell was regarded as something rather rigid and determinate : for example, there were muscle cells, liver cells, nerve cells and so on. This view, however, is now as dead as the Dodo. A more apt, though still inadequate, analogy is sometimes drawn between the cells in the body and the individuals in a human community. This comparison is at least more vital.

Cells were first described as essentially micro-anatomical units and their size was such that, at the time, investigation of their physiology and biochemistry was well-nigh impossible. However, subsequent progress in micro-techniques has so emphasized the diversity, adaptability, individuality, versatility and plasticity of cells that the concept of the cell as a rigid morphological unit is now quite untenable and the present concept is very different from that prevalent in the days of Schleiden and Schwann (1838–9). Even to describe cells as physiological units is sometimes fallacious in view of the fact that the func-

B

tional unit is often a combination of cells. Cardiac muscle is an obvious example, and the partnership between nerve and neuroglia (Chapter 19) is another less obvious though equally important one.

II. Evolutionary Aspects

It may be helpful first to review the concept of the cell in relation to evolution and differentiation, as this may emphasize the dynamic and adaptive qualities of cells. The cells of the metazoa can, with considerable justification and in spite of essays to the contrary, be compared in many respects with the protozoa, and it is now recognized that with only a few relatively minor exceptions the free-living protozoa (and indeed probably the bacteria too) have developed all the essential metabolic pathways and enzyme systems that are known to be necessary to sustain the life of metazoan cells. In other words, they possess in their genomes all the necessary DNA sequences for the construction of the normal metabolic enzymes and proteins for cellular activity, though not necessarily those required for all the special products like keratin, collagen, or the various proteins of the blood. With this in mind, the problems of human cytology are seen to be largely those that arise as a consequence of grouping, or possibly subdividing, the original protozoan units into colonial units, and of the subsequent evolution of these colonial forms: i.e. to develop organs and tissues which have adapted them to life among their competitors and in special environments. Human cytology can be no more studied without reference to the ancestral past than human behaviour can be considered as independent of human history and pre-history.

Within certain defined limits the protozoon is an adaptable form. It may or may not possess polarity, i.e. an anterior and a posterior end, but each individual is normally adapted to living in an environment which, compared with itself, is large and uniform and which therefore affects all parts of the cell surface more or less equally. Many different environments have been successfully colonized by the protozoa and this has led to a great diversity in form and function among them; each type is suited to the particular niche that it occupies, whether that niche be the ocean, the fresh-water pond, the soil or the cytoplasm of an human erythrocyte. Protozoa are essentially cells whose responses are dependent upon the interplay between their own inherited characters and their immediate environment. Occasionally they are able to change from one form into another. For example, *Naegleria gruberi* (Fig. 1) can be either amoeboid or flagellate, and it changes form according to the salt concentration of the environment. In some protozoa, a similar change of form follows a cell division, as in the production of flagellate gametes, but *Naegleria* is peculiar in that the change of form (which is comparable with some processes of differentiation in higher animals) can occur directly, without involving mitosis (Willmer, 1956). Changes in the ionic content of the surrounding fluid cause almost immediate changes of form, even inducing or eliminating polarity. *Naegleria* is certainly an exceptional organism, but nevertheless it has its lessons for the human cytologist, in showing how quickly and extensively protoplasmic units can respond to their immediate environment, to say nothing of the more subtle and unseen biochemical adaptations that may also be occurring.

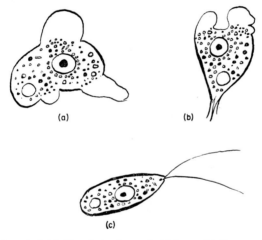

Fig. 1. *Naegleria gruberi* (a) Unpolarized amoeboid form. (b) Intermediate, polarized, amoeboid form. (c) Free-swimming flagellate form.

A. COLONIES AND THE DEVELOPMENT OF
MICROENVIRONMENTS

When, however, protozoan cells (?) first united into colonies, or multinucleate protozoa became multicellular organisms (and it matters not for the present discussion how these colonies or organisms originated) then the cells of each colony or organism immediately created new and different environments for themselves and for all the other constituent cells. The uniform external environment of the protozoa was for ever lost, so far as the individual cells were concerned. For successful colony formation therefore, the cells had to be able to adapt themselves to these new microenvironments, without at the same time producing conditions inimical to the other cells of the colony. Moreover, the stability of the colony as a whole had to be preserved. The availability of oxygen, the elimination of toxic products and all such-like requirements had to be met. Similarly, if the colony took the form of a hollow blastula-like organism (Fig. 2), as is certainly suggested by the widespread occurrence of this form in the embryo-genesis of invertebrates, all the cells had to work together to maintain its coherence, shape, size and probably also its buoyancy. Furthermore,

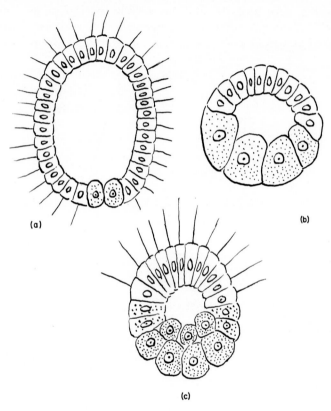

(a)

(b)

(c)

Fig. 2. Various forms of blastula. (a) and (b) Blastulae with sharp division between two types of cells. (c) Blastula with a gradation of cells.

if in the interest of stability and coherence the cells formed "tight junctions" between themselves then the contained fluid could well differ materially from the fluid of the surroundings. In that case the cells inevitably become polarized, with one surface facing the outside world and the opposite surface facing the fluid of the "blastocoel" cavity. In such an organism, toxic substances have to be excreted and hydrostatic and ionic equilibria have to be preserved. It is possible that all this could be done by a uniform population of cells, but a comparative study of the many different organisms of this kind, as they occur in the form of the blastulae of invertebrates, suggests that the conditions may be best satisfied by the combination of cells of two or more classes. In calcareous sponges, for example, the blastulae are composed partly of flagellated cells and partly of more phagocytic cells without flagella. In nemertine worms, the cells of the animal pole of the blastula differ from those of the vegetal pole. It has been shown that when the blastula is bisected equatorially the cells of the "animal" half reunite and form a hollow sphere, while those of the "vegetal" half cohere to form a solid mass or "morula". Obviously the two groups of cells have different means of coherence or deal with their surrounding fluids quite differently, and it seems probable that in the normal blastula they combine together to maintain the equilibrium of the embryo, because the normal embryo neither shrinks nor swells unduly, nor does it suffer from the accumulation of toxic products. In many embryos there is no sharp division between the "animal" and "vegetal" cells, but a more or less gradual transition from one to the other, i.e. an axial gradient, as shown in Figure 2c.

Again, if the early gastrula of the frog is divided latitudinally into three segments, an animal pole region, an equatorial region, and a vegetal pole region, the equatorial part alone can survive and reform into a viable embryo. It presumably contains an adequate mixture of the "animal" and "vegetal" cells. It may be noted that this survival does not depend on the presence of the dorsal lip, since this was contained in the vegetal third (Paterson, 1957).

B. CONTROL OF THE CELLULAR ENVIRONMENT

Symbiotic Pairs

These observations thus point to the very early evolution of a kind of symbiosis between cells of different sorts. A symbiosis by which equilibrium can, presumably, be more effectively conserved than it could be by one type of cell only, whose activities if left unbalanced and unchecked could easily lead to the fatal deterioration of the immediate cellular environment. If such a symbiosis exists, even in relatively primitive and simple organisms like the embryos of invertebrates and amphibia, it is likely that similar associations continue to exist between the cells of higher organisms, and that they have developed into even more elaborate patterns among the adult tissues of higher vertebrates.

Thus it may be often misleading, even wrong, to consider cells as separate and independent units. Probably they more often exist as members of partnerships. It is certainly notable that practically every epithelium in the body contains at least two visibly distinct classes of cells. Serous and mucous cells, ciliated and goblet cells, and many other pairs spring readily to mind as examples. Clear and specific physiological functions can usually be ascribed to each of these partners. For example, there is an obvious functional partnership between the epithelial cells of the trachea in that they work together to produce a current of mucous fluid to cleanse the surface of foreign particles. But the origin of the duality or multiplicity is probably far more deep-seated. The skins of many primitive aquatic invertebrates have exactly comparable mechanisms and the ionic regulation of such animals is often dependent on the skin. There must also be interesting ionic problems for the human respiratory epithelia to solve when their owner flies from the arid African desert and lands in the saturated atmosphere of a London fog. In neither case is the concentration or dilution of the fluid in the respiratory passages allowed to alter the concentration of the blood.

The Units of Epithelia

All epithelia separate two environments. Usually these are, as they originally were, the outside world and "the internal environment". As we have seen, the original internal environment was probably the fluid contained in some blastocoel-like cavity in the most primitive metazoa. So the primary function of an epithelium, with which the specialist activities of its constituent cells must still be consistent, is the preservation of the constancy of the internal environment, and this important function probably precedes and determines any later specializations. An example from the tissue culture of the skin of the chick may make this clear. The skin normally consists of an apparently uniform germinative layer from which the daughter cells ultimately keratinize. This is a specialization that causes it to become impervious to water because its superficial layers act as a protective barrier, a development which incidentally allowed animals to leave their initially aquatic environment and to colonize the land and the air. The basal layer meanwhile retains its reproductive character and it can be shown to consist of cells with at least two different potentialities. When the epithelium is treated in tissue culture with a medium rich in vitamin A, the keratinizing layers tend to be sloughed off and the underlying layer then produces cells that differentiate along quite different paths. They no longer produce keratin but either become ciliated or develop into goblet cells that secrete mucus (Fig. 3). If the vitamin A content is then lowered, these differentiated cells can again be sloughed away and the basal cells once more produce a layer of keratinizing cells (Fell and Mellanby, 1953). Thus the cells of the basal layer have at

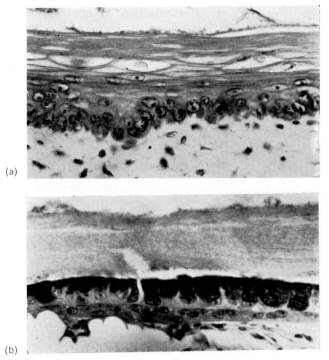

Fig. 3. Effects of vitamin A on chick skin tissue culture. (a) Normal skin with keratinizing layers. (b) Skin in medium rich in vitamin A. Note the layer of mucous secreting cells (With permission from Fell, 1960).

least three potentialities, though once they have assumed one of the forms of behaviour the change seems then to be irreversible; sooner or later the differentiated cells die. It is certain that the basal cells can produce ciliated cells, mucoid cells or keratinizing cells, but it is not clear whether all the cells of the basal layer are initially the same, or whether there are some that can only differentiate into ciliated cells and others that are determined as mucoid cells. This, in fact, may only be a matter of the timing of the differentiation, i.e. an initially uniform population of cells eventually divides into three different sorts of cells. Nevertheless, the underlying duality of the epithelium, when it is not keratinizing, is a matter of great interest in relation to the "boundary function" of the skin. It suggests that two types of cells are always potentially present and perhaps are still involved in maintaining the equilibrium between the two sides. We must remember that the skins of many aquatic animals are built on the same general plan as the respiratory epithelia of higher vertebrates and the presence of ciliated and goblet cells in an epithelium antedates the necessity for cleansing the tracheal surface by many millions of years, though cleansing must also be one of the essential properties of skin at a very early stage of evolution. Conversely it is interesting to note that one of the symptoms of vitamin A deficiency in rats is the keratin- ization of the respiratory epithelium (Fig. 4).

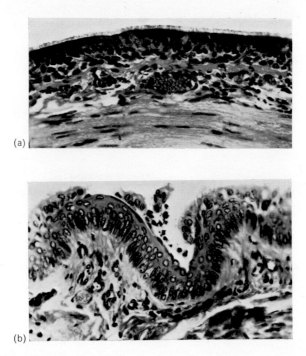

Fig. 4. Effects of vitamin A deficiency on respiratory epithelium. (a) Normal epithelium. (b) Keratinizing epithelium in vitamin A deficiency. (Preparation by Wolbach. With permission from Fell, 1960).

Probably the most notable example of symbolic pairs of cells in the body, where the function of the complete mechanism depends on both partners, is, as already mentioned, in the nervous system. Morphologically, nerve cells always exist in close association with supporting cells: for example satellite cells in ganglia, Schwann cells on peripheral nerve fibres, oligodendroglia and astrocytes in the central nervous system. Nerve cells always seem to require these asso- ciates and, probably, neural function will only be successfully un- ravelled when the nature of these relationships is thoroughly under- stood (see Chapter 19 for further details).

III. THE UNITS OF "THE INNER MASS"
AND THE ORIGIN OF CELL FAMILIES

To return again to the immediate consequences of colonial organization, the "internal environment" that was initially formed by the enclosure of a "blastocoel" provides a potential new environment for cells, an environment which may be isolated and different from the outside world. Thus sooner or later it becomes capable of being controlled and determined by the activities of the cells of its surrounding epithelium. Within it cells could live, differentiate independently of the outside world and contribute new activities to the colony as a whole, providing, for example, a skeleton, a circulatory system and so on. The cells that could enter this environment and so form the cells of "the inner mass" (not to be confused with the "inner cell mass" in mammalian development) must initially have been epithelial cells. But, in making their entry, their epithelial properties of cohesion and polarization could well be lost and the way would be opened for specializations in many new and different directions; determined by the multiplicity of micro-environments that could now be created and controlled.

These considerations can probably provide a rational explanation for the well-known phenomenon that, when tissues from almost any metazoan animal from sponges to man are known in tissue culture, and their cells are allowed to migrate out into the medium on to a flat surface that they can wet, i.e. they are allowed to undergo "unorganized" growth, they show one of three general forms of behaviour. They either behave as an epithelium (Fig. 5a) whose cells (epitheliocytes) emerge as a sheet, each cell attached continuously to its neighbours; or they migrate out more or less radially, forming a loose network of mechanocytes (fibroblast-like cells) (Fig. 5b); or they emerge by apparently random amoeboid movement as isolated phagocytic amoebocytes (macrophage-like cells) (Fig. 5c). In vertebrates the epithelial behaviour is characteristic of skin, intestinal mucosa, kidney tubules and in fact of any epithelium normally in contact with the outside world. It is notable that cells of ectodermal (e.g. skin), endodermal (e.g. intestine), or mesodermal (e.g. kidney) origin may all behave in this manner. On the other hand, the mechanocytes and amoebocytes are essentially derived from the cells of the "inner mass" (i.e. from the invaders of the primordial blastocoel). While the germ layers are certainly useful descriptive terms in morphological development, they have little significance from the physiological point of view. Indeed, they are probably far less useful than the subdivisions just made between the initially epithelial cells, and the cells derived from the inner mass, i.e. the amoebocytes and the mechanocytes.

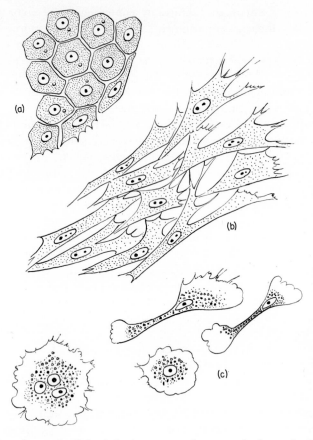

Fig. 5. Three forms of cellular behaviour as seen in unorganized growth of tissues in culture. (a) Epitheliocytes (epithelial cells). (b) Mechanocytes (fibroblasts, etc.). (c) Amoebocytes (macrophages, etc.).

By their ability to produce collagen and mucopolysaccharides in varying proportions the mechanocytes have in course of time become directly responsible for laying down the skeletogenous elements of the body, cartilage, bone, tendon, fibrous tissue etc. It is notable that in many of these tissues mechanocytes are accompanied by amoebocytes, the cells that among other activities are responsible for the phagocytic scavenging systems of the body: i.e. the macrophages, reticulo-endothelial system, polymorphonuclear leucocytes etc. The one situation in which mechanocytes appear to exist by themselves in almost pure form is in the formation of cartilage, and perhaps this affords an example of the consequences of grouping together cells that are all of one type; within this tissue, away from the influence of "opposing" cells, the accumulation of ground substance proceeds relatively unchecked. In bone, the mechanocytic osteoblasts similarly produce a collagenous ground substance in large amounts, but the process is opposed by the activities of the osteoclasts and these are cells of amoebocytic type. Once again it is perhaps significant that vitamin A deficiency upsets this balance. Proteolytic enzymes no longer attack the matrix and osteoclasts are inactive. Thus there is probably a sort of reciprocal relationship in all the skeletal and connective tissues between the activities of the mechanocytes and the amoebocytes whereby the growth of the skeleton and supporting tissues is organized and controlled. The activities of the units of the one family of cells are opposed by the activities of units of the other.

These three families of cells, as exemplified by the three forms of behaviour of cells in culture, are undoubtedly archaic and fundamental. They are found to include or characterize practically all the cells of the body, with the possible exceptions of the neural tissues and to a lesser extent, the muscle tissues which may behave somewhat differently. Moreover, they are not just a product of the tissue culture technique for they are also seen in the process of wound-healing, and are again reflected in the broad sub-division of malignant growths into carcinomata, sarcomata and leukaemias. Neural tissues do not readily grow *in vitro* as unorganized tissues and their behaviour in culture is largely restricted to reparative processes and re-differentiation into organized neural "tissues". The fibres of skeletal muscle tend to split up, presumably into their original constituent myoblasts, and it is notable that cultures of muscle may show both mechanocytes and amoebocytes. How these cells relate to the original multinucleate fibres is still problematical and needs further investigation, particularly with reference to the various types of muscle fibres, fast, slow, red, white etc. From the evolutionary point of view it should be emphasized that muscle is not a simple tissue. In different phyla, muscle originates in many different ways and it performs many different functions; it is structurally very diverse. It also shows some remarkable instances of convergence. Even in man it has many different embryological origins. There are many types of plain muscle (e.g. vascular, ciliary, pupillary, intestinal, uterine etc.), and many different sources and kinds of striated muscle (e.g. mandibular muscles, trunk muscles, limb muscles, nuclear bag and nuclear chain fibres in muscle spindles). Chapters 20, 21 and 22 deal with muscle in detail.

IV. Modification of Cellular Function

The behaviour of bone and cartilage cells *in vitro* is particularly illuminating in relation to the concept of cells as units. From cultures of endosteal bone there emerge mechanocytic osteoblasts and amoebocytic osteoclasts. When the former are encouraged to grow, they can produce large colonies that are predominantly composed of mechanocytes, and the original explant can be removed, so that only the emigrated cells remain. If these are then treated with a medium that does not encourage their growth (e.g. pure plasma rather than plasma and embryo-tissue-juice), the colonies start to differentiate again in their centres and generally lay down osteoid tissue, if not actual bone (Fig. 6). Sometimes they produce cartilage (Fig. 7), and the evidence now indicates that bone formation and/or cartilage formation are both possible for mechanocytic cells derived from skeletogenous tissues, and that local conditions including mechanical pressure and tension, oxygen tension, and probably the relative numbers of mechanoctyes and amoebocytes, determine what sort of ground-substance shall be produced. Thus, neither bone cells nor cartilage cells are rigidly determined entities, but both depend on local conditions for their characteristic functions (Fell, 1933; Hall, 1970).

A. DIFFERENTIAL CELL DIVISIONS

Recent work, particularly on the development of striated muscle, suggests that, during development, cells tend to undergo periods of growth and cell division in which they remain in a more or less undifferentiated form, e.g. as mechanocytes etc. and that they may maintain this multiplication for many generations. During this phase the cells do not produce those substances that characterize the adult tissues, e.g. myosin in the case of muscle. Sooner or later a mitosis occurs that sets one or both of the daughter cells on to another path of activity: e.g. as a myoblast that can now produce myosin and is capable of fusing with other myoblasts instead of retaining the independence of the original mechanocytes (Okazaki and Holtzer, 1966).

Similarly, at a certain stage chondroblasts begin to produce mucopolysaccharides, and osteoblasts produce collagen, though in these cases the change is probably quantitative rather than qualitative. It is probable that local conditions can determine the onset of such "quantal" or differentiating mitoses and also to some extent at least, the type of differentiation that follows them, though how this is achieved is still problematical. Furthermore, it is important to remember that *Naegleria* (see p. 5) produces flagella without the intervention of mitosis, so one should not automatically assume that all differentiation depends on "quantal" or differentiating mitoses.

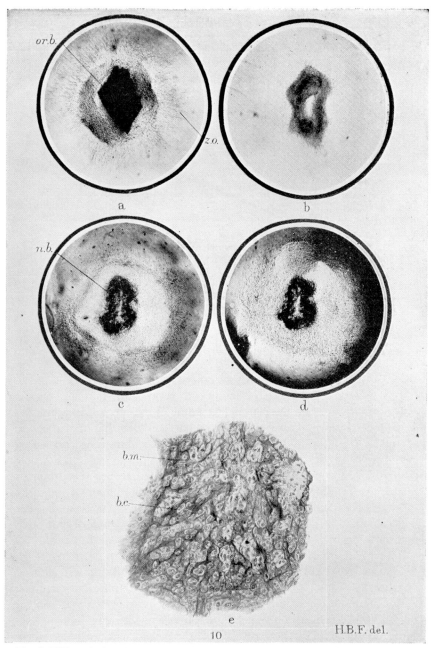

Fig. 6. Differentiation of bone *in vitro* from endosteum. (a) 5 day old culture of piece of endosteum from chick tibia, showing outgrowth predominantly of "mechanocyte" cells. (b) Same culture after removal of original explant. (c) Same culture 5 days later, showing new bone in the centre of the tissue. (d) Same culture after 7 days. (e) Stained section of the 7-day-old culture.

b.c. bone cell; b.m. bone matrix; n.b. new bone; or.b. original bone; z.o. zone of outgrowing cells. (With permission from Fell, 1932).

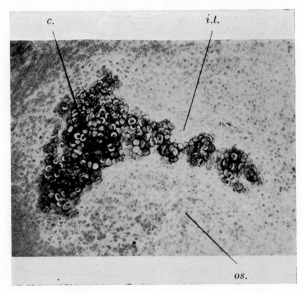

Fig. 7. Development of cartilage in a culture of endosteal bone similar to that illustrated in Fig. 6. (c) Cartilage; i.t. intermediate tissue; o.s. osteoid tissue. (With permission from Fell, 1933).

B. THE GENOME, AND THE CONTROL OF ITS ACTIVITY

It is now recognized as probably a general truth that all somatic cells normally contain the complete genome of the species. In amphibia it has been shown (Gurdon and Woodland, 1968) that the injection of nuclei from brain or from intestinal epithelium into enucleated eggs results in the development of these composite cells into whole embryos. Thus the nuclei of these differentiated tissues (i.e. brain or intestinal epithelium) must still possess all the genetic material required for the formation of a whole individual. Moreover, these and other experiments have made it fairly clear that it is the cytoplasm that often determines which parts of the nuclear material shall function.

The main problem in the study of differentiation is therefore to determine by what means only certain parts of the genome are allowed or caused to function at any one time : i.e. to determine, for example, why a mammalian ovum does not normally produce haemoglobin or myosin for which it must carry the necessary genetic information, since certain of the descendents of the ovum can produce these substances. As seen in this way, it is clear that the ovum is just as much a differentiated cell as is the muscle cell, the blood cell or any of the cells of the somatic tissues. Cells move from one state of differentiation to another and there is no stage when the whole of the genome is active. Some pathways of differentiation lead to the production of special proteins and these paths are often irreversible (keratinization for instance), but tissue culture studies have indicated that differentiation is far more flexible than was at one time supposed.

During the normal course of embryonic differentiation cells, probably by virtue of their immediate local environments acting on their cytoplasm and then the cytoplasm acting on the nucleus, become, after a period of multiplication, divided into families with greater potentialities in some directions than in others. Cell divisions of the quantal type, rather than of the proliferative type, probably play a determining role in this phenomenon, though what factors are required for the initiation of the quantal divisions still remain unknown. Chapters 14 and 15 deal with the problems of cell differentiation, growth and morphogenesis in detail.

Nuclear and Cytoplasmic Interactions

One of the most histologically obvious, but at the same time least investigated features of adult tissues is the characteristic pattern of their nuclei (Fig. 8). The nucleus of a nerve cell differs grossly from that of a muscle cell and both differ from the nuclei of leucocytes or pancreatic cells and so on. If examined closely enough the nuclear pattern of each cell type would presumably be found to be specific,

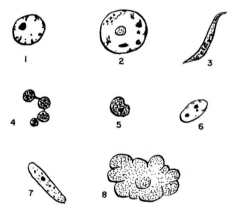

Fig. 8. Various forms of nuclei in cells of vertebrates. (1) Pancreatic cell. (2) Nerve cell. (3) Schwann cell. (4) Neutrophil leucocyte. (5) Lymphocyte. (6) Fibroblast. (7) Muscle cell. (8) Megacaryocyte.

though at the histological level some nuclei belonging to different classes of cell may appear to be very similar. The unique nature of the nucleus in each cell type reflects the specific interactions between nuclear and cytoplasmic materials that characterize each type of cell, and behind this pattern must lie the key to the locking up or unlocking of the parts of the DNA code that are to be used or not used by the cells in question. So far, the key has not been found. In parenthesis, it may be suggested that since the DNA is very tightly coiled during the actual mitosis, when the barrier of the nuclear membrane is broken down, it may be the period when the threads are uncoiling in the new cytoplasm of the daughter cell (i.e. during the reconstruction of the telophase nucleus) that is the most potent for the nuclear-cytoplasmic interaction. Obviously however there must be some interchange possible at other times also. Hormones change the activities of cells without involving mitosis, and these changes often involve the production of proteins or polypeptides.

Cell Division and Adaptation

In tissue cultures, whenever cell division is restricted, organized growth is more likely to occur. When, however, cell division is stimulated, as by the addition of embryo juice, the cells rapidly dedifferentiate. This is a bad way of describing their behaviour,

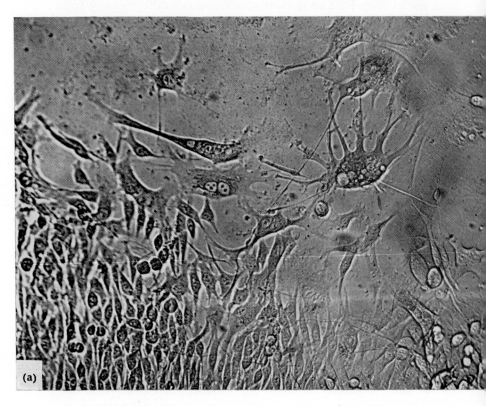

Fig. 9. Morphology of cells in culture. (a) Culture of fresh "fibroblasts".

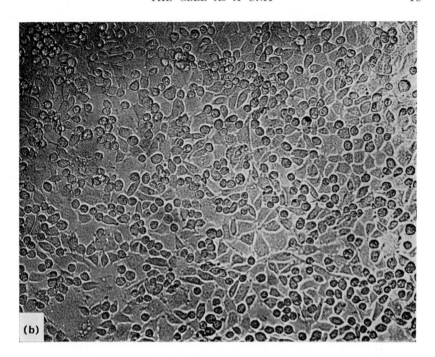

Fig. 9. Morphology of cells in culture. (b) Fibroblasts after prolonged culture *in vitro* (With permission from Earle, 1961).

because under these conditions they are really not de-differentiating in the sense of returning to a more embryonic condition, but are adapting themselves for survival as independent units under the special conditions of culture. At first, the cells that are thus encouraged to grow from developing long-bones for example, may still be recognizably osteoblasts, chondroblasts, etc. and be capable of reviving such characteristic activities comparatively easily if given the appropriate conditions. After more prolonged growth, however, these special tendencies become progressively lost and the cells become "generalized", first as mechanocytes and later as "tissue culture cells" (Fig. 9). Thus, strains of epithelial cells from liver, after they have been kept growing in culture for long enough, have been found to become almost indistinguishable from strains of periosteal fibroblasts kept under similar conditions (Fig. 10). In any case, it should be noted that liver cells and periosteal cells emerging from their respective explants and creeping on the glass are creating for themselves entirely novel and unnatural environments, so that immediate adaptation is necessary for their survival. When growth in a culture is suppressed, the cells, if

A*

they have not "adapted" too far, may begin to reconstitute the tissues from which they were derived and, by recreating the same local environments, some semblance of the normal histology and physiology may be maintained or reacquired. This only seems to happen in the early stages of culture. Presumably after a time the adaptive changes are so great that there is no turning back. The previous local environments can no longer be re-established and under these conditions the cells must progressively adapt to life *in vitro* or perish in the attempt.

These considerations make it quite clear that there is nothing absolute about a liver cell, a bone cell or any other tissue cell. They are not immutable units. They are what they are because of where they are and by virtue of their past history both ontogenetic and phylogenetic (see Willmer, 1970). When their local environment changes, then their activities may also change accordingly. If the cell is to survive, the changes must of course always be in the direction of restoration of the equilibrium between the cell and its surroundings.

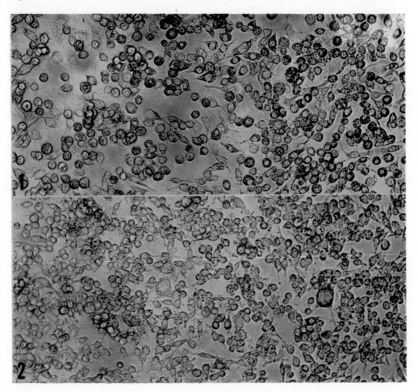

Fig. 10. Two strains of mouse liver cells after prolonged culture *in vitro*. (With permission from Evans *et al.*, 1958).

During the course of embryonic and later development a cell, like a human being, passes through a series of environments and these gradually mould its final qualities and behaviour. The changes during development are all relatively irreversible, since all experiences leave their mark and help to make each cell a special case with an unique history. The more special the experiences, the more special is the resulting cell, and the more difficult does reversibility become. Just as a man changes during his development and cannot be entirely freed from his past, so the cells of an organism become determined. Nevertheless, as the result of the process of "brain–washing" on the one hand, or by the activation of growth and cell division on the other, the determination and differentiation is not now believed to be as absolute and irreversible as it was once thought to be. Provided that the environmental changes in tissue cultures are such that the cells can remain capable of adapting themselves to them, they could presumably lead the cell in almost any direction. In practice most changes in the environment of tissue culture cells are inevitably crude and they generally prove fatal to the cells.

Determining Factors in Cellular Function

The behaviour of a cell can now be seen to be determined by the interaction of at least five factors or systems. There is first of all the "genome". This means the sum total of all the genetic material, i.e. DNA, characteristic of the species. By means of this material it is theoretically possible that the cells could make every protein or polypeptide for which coding exists in the inherited pattern of DNA. Moreover, in addition to the information actually used by the individual, it is probable that the genome of any species also contains information that is no longer used, but which was at one time important in ancestral history and which has as yet not been eliminated or irretrievably modified. In practice all these possible proteins are not made, because of what can be loosely described as the repressor system. Only a fraction of the genome is apparently allowed to function at any one time. How the repression and selection of active genes is brought about is still under investigation. It could be by repressor genes, by histones or by other proteins or substances limiting the activity of the DNA sequences. It could occur either in the transcription of the code into messenger RNA or at some later step in the synthesis of the protein. However this may be, the second factor is this repressor system, and it is that which determines the functional part of the genome. DNA threads and their repressors do not, however, exist *in vacuo*. They exist in the special nuclear and cytoplasmic complex that they inherited in the ovum and that they have continu-

ously modified ever since. This immediate (nucleoplasmic) environment of the DNA system at least partly determines the activity
of the genome and the extent of its repression. But, once again, this
nuclear and cytoplasmic organization or complex does not exist *in
vacuo,* but that of each cell has a very special local environment.
For example, a liver cell may have surfaces in contact with Kupffer
cells, bile ducts, fibroblasts, reticulin fibres and with blood at a particular position in the sinusoid : i.e. with particular O_2 and CO_2 tensions
and a general composition characteristic of and probably peculiar
to that position. This special local environment of the nucleo–
cytoplasmic complex is the fourth factor in the system. Thus there
are really five parts of this constantly interacting system, the genome,
which is divisible into a repressed part and an active part, the repressors, the nucleo-cytoplasmic complex, and the local environment.

If a hormone starts to circulate in the blood and so enters the
environment of a cell, it could theoretically act on any one of these
parts of the system, and if it did so, the consequences of its action
would almost inevitably involve all the rest. If it acted first on the
cell membrane, this could so disturb the whole balance of the cell that
immediate repair processes would become necessary. These repairs
could require new proteins or new enzymes and thus an altered
distribution of repressors. This would mean that different parts of the
DNA would be "exposed" and become active. Alternatively, it is
possible that the hormone might penetrate all the way to the nucleus
and genome, and there directly initiate modifications leading to the
formation of new proteins and hence a new cellular organization.
Thus it is to some extent an academic question as to where a hormone
acts, because the system is such that alteration in one part inevitably
produces changes throughout the whole system. Moreover, the cell is
a dynamic system, constantly in a state of flux and perpetually
engaged in adapting itself to its immediate surroundings. The presence
of a hormone is just one more factor. The surface membrane of a
cell, except in those like the blood cells that float freely in a relatively
uniform environment, cannot be considered as a single entity, though
it is of course continuous. It probably differs from place to place
and from time to time as the cell moves and probes the surroundings
with its processes. It changes, too, whenever the activities of the cells
in the immediate vicinity impinge upon its various parts. (see also
Chapters 2 and 3).

It is clear that a transitory action on a cell or some part of a cell
may lead to far-reaching changes of a relatively permanent character.
For example a hormone or, more generally, a secretion or bi-product
from a neighbouring cell may so alter the surface constitution that

the cell may never again be able to reconstruct an exactly similar membrane. The greased surface of a mechanical bearing, when momentarily acted upon by a detergent ceases to function properly and it can only be made to do so by recoating it with an appropriate grease. Moreover, the detergent after producing its "inhibiting" action may not remain in the bearing. Thus in seeking for information on how various hormones, evocators and other directive influences act on cells, the possibility of their transitory action should be borne in mind. Furthermore, the place where the hormone or other activator finally comes to rest and is detectable by histochemical or biochemical methods may not necessarily be the site of its primary physiological action, any more than the sewage plant is the place of action for the detergent that cleansed the dinner plates.

V. SUB-CELLULAR UNITS

It must be abundantly clear from all that has been said that there is no such thing as a typical animal cell, though there are certain basic features that every cell possesses to a greater or less extent (Fig. 11). Normally there is a nucleus, though most mammalian erythrocytes carry out their physiological functions for many weeks without either nucleus or mitochondria. Moreover and as noted earlier, nuclei are very different structures in different classes of cells within any one animal. Similarly the organelles that cells may contain differ in number and form from cell to cell, and sometimes from moment to moment. They should not be regarded as constant entities, but rather as types of structural organization that regularly recur, often with their own distinctive features, in the make-up of most cells. Mitochondria, for example, may be numerous or almost absent; they may be spheroidal or tubular. Their cristae can assume many different forms, from lamelliform to tubular. Mitochondria may subdivide and re-unite, and indeed the ciné camera shows them to be always in a state of flux. Their numbers can increase rapidly and they can as readily decrease again. Some mitochondria at least contain DNA and their membranes are so different in kind from other membranes in the cell, that it has been suggested that the membrane that forms the cristae is in fact the surface membrane of a distinct organism that now symbiotically inhabits the cells of plants and animals. "Mitochondrion" is really a generic term, for a special group of cell organelles. "The mitochondrion" is a dangerous over–simplification. Mitochondria are discussed in detail in Chapter 6.

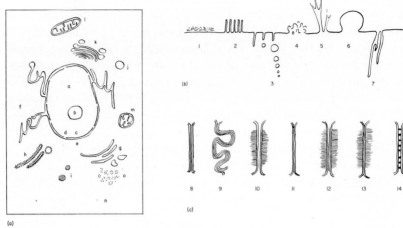

Fig. 11. Basic features of a cell.

(a) Essential structures. a, nucleus. b, nucleolus. c, inner nuclear membrane. d, pore in the membrane. e, outer nuclear membrane. f, endoplasmic reticulum. g, cisternae and membranes of smooth endoplasmic reticulum. h, rough endoplasmic reticulum. i, liposomes. j, lysosome. k, Golgi lamellae and vacuoles. l, mitochondrion longitudinal section. m, mitochondrion transverse section. n, plasma membrane. o, polysomes.

(b) Variations of the cell membrane. 1, Glycocalyx of attached glycoprotein. 2, Microvilli. 3, Pinocytotic channels and vesicles. 4, "Ruffles". 5, Pointed pseudopodia. 6, Lobose pseudopodium. 7, Invaginations.

(c) Types of attachment to neighbouring cells. 8, Simple adhesion. 9, Interlocking surfaces. 10, Adherence with intracellular fibrils. 11, Tight junction (nexus). 12, Desmosome (macula adherens). 13, Tight junction with intracellular fibrils (zona occludens). 14, Septate desmosome.

A. THE IMPORTANCE OF PHOSPHOLIPIDS

Cell membranes, including mitochondrial membranes, are almost certainly dependent on their phospholipid layers and micelles as skeletons for any associated proteins or polysaccharides that may endow them with their own special properties (see also Chapter 2). It is thus relevant to the understanding of cellular activity to observe the behaviour of such phospholipid systems. They are almost unique in their molecular behaviour on account of their peculiar combination and arrangement of hydrophilic and hydrophobic moieties, Alcoholic extracts of neural tissue are very rich in phospholipids and when they are allowed to evaporate and the ensuing deposit is immersed in sodium chloride solutions the lipids show the most remarkable spontaneous formations of membranes (Fig. 12) and continuous movements of "myelin figures". These figures result from the phospholipid molecules jockeying for position with their bodies in the lipid and their feet in the aqueous phase. Such myelin figures are extremely sensitive to local conditions and continue to organize and reorganize themselves for very long periods, perhaps indefinitely and certainly until molecular

equilibrium is reached. Although the system described is entirely artificial, the behaviour of these transitory and self-organizing structures
is probably directly comparable with the behaviour of the similar
lipoprotein complexes in living cells. The realization and appreciation
of this capacity of phospholipids for perpetual re-organization is essential to the proper interpretation of cellular activity. Moreover, since
proteins also have lipophilic and hydrophilic groups they readily form
structural complexes with phospholipids and constitute molecular
groups with very special properties.

Phospholipid molecules in water, automatically become non-random
in their arrangement. The consequent non-random character of the
distribution of any other molecules that may become associated with
phospholipids in living systems is therefore a fundamental property
of those systems, and must be fully appreciated if cells are to be understood.

Not only are the phospholipids and proteins themselves responsible for this non-randomness, but the almost infinitely small sub-
division of endoplasmic reticula, vacuoles etc. often means that the
distribution of molecules within the cytoplasm is very uneven. On a
somewhat larger scale, for example in the fermentations of sugars by
yeast, certain by-products appear in such small quantities that, if they
were to be distributed evenly among all the yeast cells that produce
them, any one cell might contain only a few molecules. The laws
of mass action and others based on statistical probability cannot therefore be satisfactorily applied to such small systems. How much more
true therefore does this concept become when applied to the distribution of molecules within the minute granules, vacuoles and membrane–
systems of most cells.

There are several classes of naturally occurring phospholipids and
the manner and proportions in which they may pack together is almost
infinite. Still more variety is introduced by the other hydrophobic
compounds, for example cholesterol and the steroids, neutral fats
and parts of proteins, that may pack with them. Finally phospholipid
complexes may be entirely changed by changing the ionic constitution
of the aqueous phase.

B. VACUOLES AND LYSOSOMES

The cytoplasm of cells generally contains fat droplets and vacuoles
of various forms and kinds, including the lysosomes that are characterized by the possession within them of a battery of lytic enzymes:
for example nucleases, proteases, carbohydratases, lipases, phosphatases
etc. Vacuoles, which are probably as variable in constitution as the
cells that possess them may originate *de novo* from endoplasmic

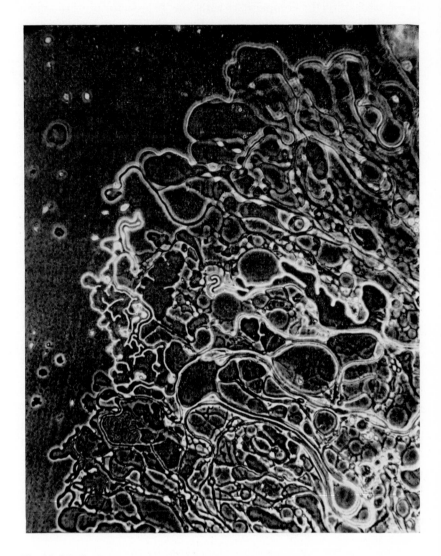

Fig. 12. Myelin figures. 0·06% NaCl solution added to lipids deposited on a microscope slide after evaporation of an alcoholic extract of brain tissue. (a) After 1 min.

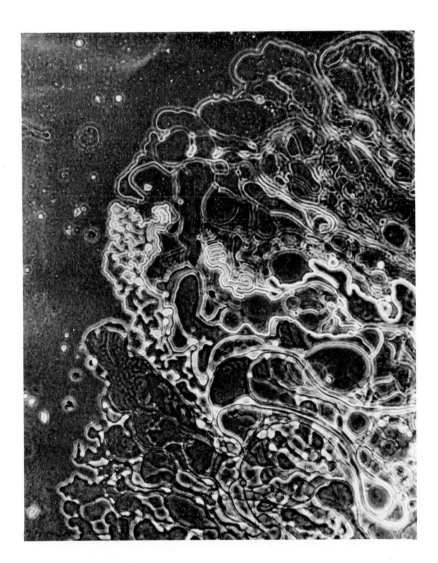

Fig. 12. (b) After 5 min.

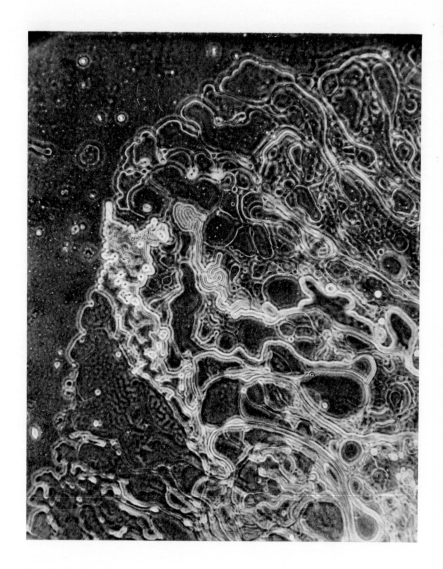

Fig. 12. (c) After 10 min.

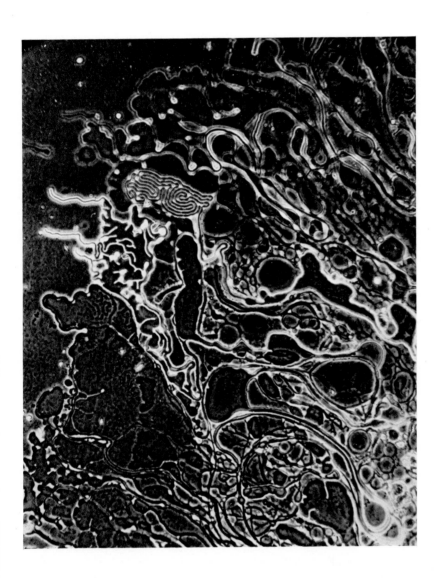

Fig. 12. (d) After 20 min.

reticulum, from the Golgi complex, or they may be inclusions of the outside world, such as pinocytic and phagocytic vacuoles. They may contain fat, or secretions of one sort or another, the latter being proteinaceous when they are derived from the "rough" endoplasmic reticulum with its associated ribosomes. Frequently the protein becomes associated with a polysaccharide after passage through the Golgi complex. It must be remembered that the electron-microscope, which is often needed to reveal the presence of vacuoles, tells very little of their chemical and molecular constitution and nothing of their variation with time. Ciné photography, though not capable of such high magnification, certainly accentuates the versatility and impermanence of these cell structures.

An elaborate system of names has necessarily been proposed for the various classes of vacuoles (see for example Chapter 7), but in the present state of our ignorance it is probably better to be somewhat cautious in its use, because the giving of names almost certainly oversimplifies the position and discourages further study, just as the naming of a disease may satisfy the patient but tends to stop further investigation of its causes and aetiology.

C. SURFACE CONTACTS

An important feature of the cell as a unit is the nature of its relationship with its neighbours (Chapter 3).

Cell surfaces all have their special characters (Fig. 11b). Some cells are "sticky" and cohere with their neighbours, for example, epithelial cells. Others, like macrophages, keep a respectful distance from their neighbours. Myoblasts can divide and remain loosely attached to each other, but when they become myocytes their surfaces change and they can then fuse with each other. Neutrophil leucocytes can adhere to glass and move on its surface: neither erythrocytes nor lymphocytes can do so, though the latter can move and become amoeboid on agar.

Cells that cohere, like epithelial cells, may do so through various structures classified as tight junctions, desmosomes, or interdigitations (Fig. 11c). Moreover the properties of the intercellular membranes probably differ from those of the membranes on the free surfaces of the same cells.

The surface of pinocytosing and phagocytosing cells seems to be very labile and, at least *in vitro,* numerous microvilli are often a prominent feature as they wave slowly about in the external medium though they are themselves transitory structures. In tissue cultures their production often alternates with the development of "ruffles". Some cells of this class show pinocytosis, some micropinocytosis, some phagocytosis, while others may show none of these activities. Many cells have specific polysaccharide or glycoprotein coatings—the so-called glycocalyx whose importance in determining cellular behaviour is only now beginning to be appreciated.

D. CILIA ETC.

A surprising number of cells, other than those in typical ciliated epithelia may possess more or less well developed (or degenerate) cilia or flagella. Such cells have been recorded in many unexpected places, such as the pancreas, anterior pituitary, parathyroids and even include fibroblasts in connective tissue (Fig. 13). Cilia and flagella are remarkably uniform in ultrastructure wherever they occur, from protozoa to man. Most of them possess the characteristic nine pairs of peripheral fibrils and a centre pair (Fig. 14), though the central pair is absent from some "sensory" cilia.

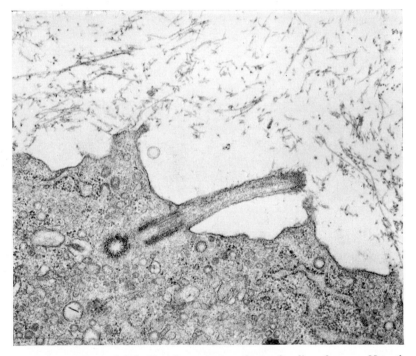

Fig. 13. Cilium on a chondroblast in an organ culture of radius of mouse. Note the basal bodies. (With permission from Scherft and Daems, 1967).

Cilia tend to have a straight active stroke and a bent recovery stroke; the beat is directional and is usually coordinated with that of neighbouring cilia. Flagella, which in general are relatively longer than cilia, pass waves of bending along their length either from base

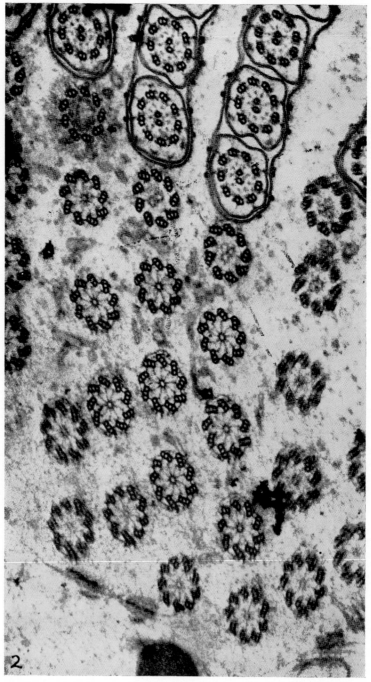

Fig. 14. Transverse sections through basal bodies and flagella of *Pseudotrichonympha*. Free flagella in grooves in the cytoplasm appear at the top, and the proximal ends of the basal bodies appear at the bottom. (With permission from Gibbons and Grimstone, 1960).

to tip or from tip to base. The bend may travel as a spiral along the flagellum.

Motile cilia generally have an elaborate system of spurs and root-fibres at their base where the centriole-like basal body or granule is also a constant feature. Transverse fibres sometimes connect the root of one cilium to that of the next in rather specific formations. The direction of the spurs and the plane of the central fibrils can be used to determine the direction of the active stroke of the cilium. In spite of this relative uniformity of structure, the beat of cilia can take many different forms, and, again, within the general label of cilium, there may be grouped together a whole host of different functional entities. The relationship between the structure and the mode of its action is not yet fully elucidated though a sliding system between the fibrils, akin to that for muscle (Chapter 20) has been suggested.

E. ELECTRON-MICROSCOPIC INFORMATION

Much of this discussion about cell organelles has resulted from the study of electron-microscope photographs. Let it never be forgotten therefore that: (1) living matter exists in an essentially aqueous medium, though its fats and phospholipids tend to escape from it if they can: (2) proteins have hydrophilic and hydrophobic parts: (3) fixation for electron-microscopy renders some proteins insoluble, and treatment by heavy metals attaches metal ions to the proteins and other molecules and it is largely those metal ions that are "seen": (4) embedding in araldite or other matrix necessitates the removal of all the water (what happens to a solution of gelatin when the water is replaced by air or araldite?): (5) the structures relevant to the functions of a living cell that can be revealed by the electron-microscope are as limited as those that a fossil can reveal of the functions of the living animal: (6) even electron-microscope sections have thickness, and moiré patterns (Fig. 15) are frequently caused by overlapping regular systems: (7) the more instantaneous the process of fixation the more defined the picture; but, the more defined the picture, the less easy does it become to interpret it as merely a phase in a continuously changing system (cf. the snap-shot of a high-jump or a horse race): (8) patterns seen in electron-microscope photographs of cells probably have more of the character of the patterns of soldiers on a parade ground or of formation dancers than of the permanent walls shown on architectural plans. The patterns are more or less transient but because of their molecular constitution they are always reforming, and thus they give an illusion of permanence and rigidity. Thus the interpretation of electron-microscope pictures in terms of the aqueous living system is no easy task, and in spite of their universal

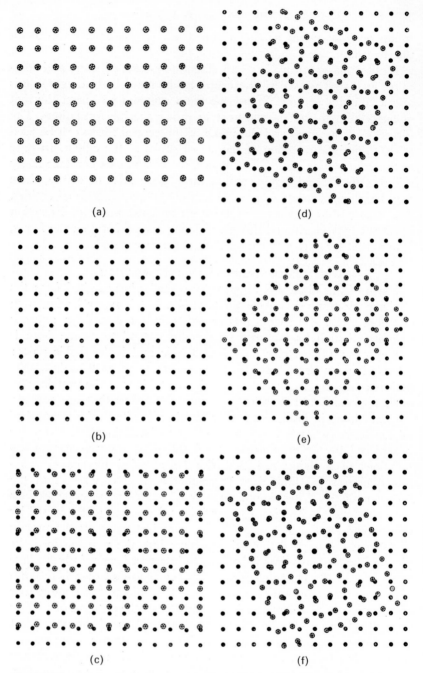

Fig. 15. If two sheets of cellophane each with a pattern of dots as shown in (a and b) are superimposed, the pattern seen in (c) is obtained; but, if the two sheets are rotated with respect to each other but kept in the same plane, then patterns like those in (d), (e) and (f) develop.

use such pictures should be interpreted with great caution and compared with the information revealed by ciné photography of living cells with the highest powers of the light microscope.

Many of these strictures on the use of the electron-microscope apply

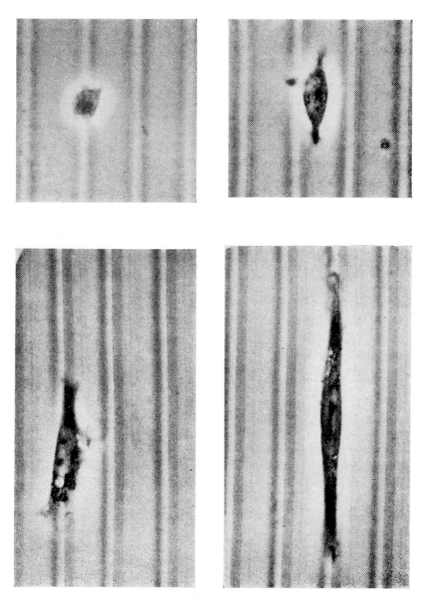

Fig. 16. Four stages in the orientation of a cell along the groove of scored glass. (Photographs by P. Weiss. From Willmer, 1970).

B*

also to histochemical localizations and tracer studies whenever these depend on fixation and the use of high magnifications. For all that, the value of carefully interpreted time-dependent studies and cytochemical techniques is beyond doubt.

As emphasized earlier, many cells have a polarity, probably imposed upon them initially by virtue of their position in a bounding epithelium; others, like macrophages, appear to have little, if any, polarity. The polarity is to be distinguished from orientation and directional movement. For example, if cells in culture are provided with fibres on which to creep or with a glasss surface etched with parallel lines, they will first orientate themselves parallel to the fibres or the lines and then they will creep along them (Fig. 16). They may move in either direction. A polarized cell on the other hand, has a front end and a back end, as in ciliated cells, secretory cells, and particularly nerve cells. In such cells the products of activity manifest themselves only in certain regions, at one of the poles for example, as does acetylcholine at a synapse. Such polarities and orientations are obviously of extreme importance in the creation of local environments for neighbouring cells and have not been given the attention that they deserve.

VI. DYNAMIC CHARACTER OF CELLS

Thus, although there is no doubt that the cell can usefully be considered as a unit, and that the body is constructed from such units, the variation of form and function within the framework of "the cell" is almost as infinite as are the characters, functions and ways of life of the individuals in a human community. It is clearly a mistake to think of liver cells, muscle cells etc. except when they are in their normal context in the living body. It is even more erroneous to consider the functions of any given organ or tissue without reference to its cellular constitution and condition at the time of observation. For example, "membrane permeability" has in the past been investigated by stretching portions of toad bladder across an aperture between two solutions whose compositions could be varied and measured. It has not always been appreciated, however, that such "membranes" are composed of several sorts of epithelial cells that are more or less polarized, several other sorts of less polarized cells (e.g. fibroblasts, macrophages etc) as well as "inanimate" though by no means unchangeable "basement membranes". In the body all these cells and tissues would normally exist in a special relationship with their blood vessels, tissue fluids etc. or with the contents of the bladder. Thus, although clear and repeatable measurements may be made on the permeability or electrical properties of the system as a whole by suspending the "membrane" between "suitable" solutions, these measurements provide no information about the permeability of any individual cell membrane as it actually exists in the normal tissue. Moreover, the solutions used in experiments of this type have not always even been compatible with the continued life of the cells! "Normal" histological and cytological appearance of the cells after the observations have been made is a first requirement even for the interpretation of the functions of the membrane as a whole.

Similarly when liver function is under investigation, it must be remembered that the liver is composed partly of hepatocytes (and it is by no means certain that these are all identical) partly of Kupffer cells, fibroblasts, cells of the bile ducts, etc. and the function of the whole depends on the interactions of all these with themselves, with the blood perfusing the organ and with the bile. Thus the attribution of particular functions to particular cells is not always an easy matter, and partnerships in function may be difficult to detect, though none the less important.

More than a century after the original description of cells was given it is still relevant to consider the cell as a unit, but the concept of the nature of that unit has certainly changed. Accent now has to be thrown on adaptability, partnerships, antagonisms and mutual interactions of the units on the one hand and of the subdivision of the cell into still smaller units on the other. For the cell physiologist, "$\pi\alpha\nu\tau\alpha$ $\overset{\varsigma}{\rho}\epsilon\iota$, $o\dot{\upsilon}\delta\epsilon\nu$ $\mu\epsilon\nu\epsilon\iota$".

REFERENCES

Earle, W. R. (1961). *Nat. Cancer Inst. Monogr.* **7**, 213–36.
Evans, V. J., Hawkins, N. M., Westfall, B. B. and Earle, W. R. (1958). *Cancer Res.* **18**, 261–6.
Fell, H. B. (1932). *J. Anat.* **66**, 157–80.
Fell, H. B. (1933). *Proc. Roy. Soc. Ser. B.* **112**, 417–27.
Fell, H. B. (1960). *Proc. Nutr. Soc.* **19**, 50–4.
Fell, H. B. and Mellanby, E. (1953). *J. Physiol. London* **119**, 470–88.
Gibbons, I. R. and Grimstone, A. V. (1960). *J. Biophys. Biochem. Cytol.* **7**, 697–716.
Gurdon, J. B. and Woodland, H. R. (1968). *Biol. Rev.* **43**, 233–66.
Hall, B. K. (1970). *Biol. Rev.* **45**, 455–84.
Okazaki, K. and Holtzer, H. (1966). *Proc. Nat. Acad. Sci. U.S.A.* **56**, 1484–90.
Paterson, M. C. (1957). *J. Exp. Zool.* **134**, 183–205.
Scherft, J. P. and Daems, W. T. (1967). *J. Ultrastruct. Res.* **19**, 546–55.
Willmer, E. N. (1956). *J. Exp. Biol.* **33**, 583–603.
Willmer, E. N. (1970). *Cytology and Evolution.* Academic Press, New York.

2. Membranes and Trans-Membrane Transport

J. A. LUCY
Department of Biochemistry, Royal Free Hospital
School of Medicine, University of London, England.

I. GENERAL INTRODUCTION

Without lipoprotein membranes, cells, as we know them, would not exist. The plasma or boundary membrane of a cell enables it to maintain a controlled environment within which complex processes, such as protein synthesis, can proceed relatively independently of the surrounding milieu. A number of suggestions have been made concerning the possible origin of the cell boundary. One of these is based on the physical properties of phospholipids which, with proteins and glycoproteins, are the chief constituents of biological membranes. Phospholipid molecules being amphipathic, i.e. containing chemical groups some of which have a high and others a low affinity for water, will orientate spontaneously at an air–water interface to form a monomolecular layer. It has been proposed that the action of atmospheric wind currents on lipid monolayers at the surface of a primeval sea may have led to the occlusion within bimolecular leaflets (bilayers of lipid) of assemblies of molecules which subsequently evolved into a primitive cytoplasm. A lipid bilayer is a more stable structure in water than either a monolayer or a thick membrane that is bounded on each side by a layer of phospholipid and contains large amounts of relatively non-polar material, such as tri-glyceride, in its interior. As a result, phospholipid molecules spontaneously form bilayers with properties that closely resemble those of biological membranes. It has been suggested that the bilayer may be regarded as a natural organization for phospholipid molecules in the same sense that the α-helix of protein and the double helix of DNA are natural units of structure.

Membrane integrity is of prime importance for the cell. Any significant failure of the barrier properties of membranes is likely to lead to pathological disturbances in cell and tissue functions. For example, damage to the plasma membranes of erythrocytes by lysolecithin released from lecithin by the action of phospholipases present in snake venom quickly leads to intravascular haemolysis. Similarly serious consequences may follow a loss of integrity in the boundary membranes of subcellular organelles such as lysosomes and mitochondria. Thus the structure of membranes is important in medicine and so is considered in the first part of this chapter. It is also important, however, to be aware of the more complex ways in which membranes control the ingress and egress of metabolites and ions by the various processes of permeability and transport. The second part of the present chapter therefore deals with trans-membrane transport.

Many complex enzyme reactions that are not necessarily concerned with transport are also located in or on biological membranes. The fact that membranes provide sites for integrated enzyme reactions is an additional reason for the study of membranes in medical science, but this aspect of membrane biochemistry is considered in more detail in the chapters concerned with, for example, mitochondria than here.

II. Structures of Membranes

A. INTRODUCTION

The investigation of membranes has attracted an increasing number of workers during the past ten years. As a result, there has been a marked increase both in the number of research papers and journals concerned with membranes and also in the variety and complexity of physical techniques used in experimental studies on membranes. There are now many different models (see Table 3) which purport to describe membrane structure and the interactions between the lipid and protein moieties of membranes. However it is clear from the multiplicity of these theories that, although we possess much more information about membranes, we are very far from fully understanding their structure, let alone the intimate relationships between structure and function. Both the extent of the data now available and the difficulty of attempting to provide a short but balanced account of current concepts are illustrated by a recent review article (Stoeckenius and Engleman, 1969) which ran to approximately twice the length of the present chapter and referred to 290 publications.

B. CHEMICAL COMPOSITION

Lipids

All animal cell membranes contain phospholipids. Among the most commonly found are phosphatidyl choline (lecithin) and phosphatidyl ethanolamine (Table 1). However, each molecular species in fact represents a class of substances and not a single substance. Thus many different lecithins are present in the membrane of human erythrocytes. The lecithin molecules differ not only in the lengths of their fatty acid chains, which may range from C_{12} to C_{22} with C_{16} and C_{18} fatty acids being predominant, but also in their degree of unsaturation. From this it can be seen that a membrane such as the erythrocyte membrane probably contains 150–200 chemically different lipid molecules (Van Deenen, 1969). Some membranes contain sterols and sterol esters: the sterol being almost exclusively cholesterol in vertebrates. In addition, some membranes contain glycolipids. Among vertebrates, there seems to be relatively little species variation in the phospholipid composition of individual membranes from the same organ or cell type—except for variations in plasma membranes which can differ widely. For any given cell, however, the lipoprotein membranes of its constituent organelles differ in lipid composition in varying degrees, both from each other and from the plasma membrane.

Current studies of the phospholipid composition of membranes may contribute to an understanding of some of the complex interactions between membranes that occur *in vivo*. In the secretion of proteins, newly synthesized protein passes from within the cisternae of the endoplasmic reticulum, via Golgi vesicles, to the cell exterior. Thus, membrane material derived from the endoplasmic reticulum may be transformed in the Golgi region into vesicle membranes which resemble plasma membranes and are capable of fusing with the plasma membrane. It is interesting therefore that the phospholipid composition of the Golgi apparatus has recently been found to be intermediate between that of endoplasmic reticulum and the plasma

TABLE 1. A comparison of the lipid composition of a sub-cellular fraction enriched in Golgi apparatus with other fractions from rat liver. (Reproduced with permission from T. W. Keenan and D. J. Morré (1970) *Biochemistry*, **9**, 19–25)

Compound	% of Total fraction		
	Endoplasmic reticulum	Golgi apparatus	Plasma membrane
Total phospholipids*	84·9	53·9	61·9
Sphingomyelin†	3·7 ± 1·1	12·3 ± 2·5	18·9 ± 2·3
Phosphatidylcholine	60·9 ± 2·2	45·3 ± 2·7	39·9 ± 2·8
Phosphatidylserine	3·3 ± 2·2	4·2 ± 1·1	3·5 ± 1·8
Phosphatidylinositol	8·9 ± 2·3	8·7 ± 2·8	7·5 ± 1·3
Phosphatidylethanolamine	18·6 ± 1·1	17·0 ± 1·9	17·8 ± 1·5
Lysophosphatidylcholine	4·7 ± 3·4	5·9 ± 0·4	6·7 ± 0·7
Lysophosphatidylethanolamine	‡	6·3 ± 1·0	5·7 ± 2·7
Total neutral lipids§	15·1	46·1	38·1
Cholesterol	24·6	16·5	34·5
Free fatty acids	40·6	38·9	35·1
Triglycerides	24·7	35·1	22·4
Cholesterol esters	10·1	9·6	8·0

* Mean value of three determinations, phospholipid = phosphorus value × 25.
† Individual phospholipids expressed as percentage of total lipid phosphorus. Values are mean plus and minus standard deviation of four to six determinations.
‡ Not detected.
§ Mean value of three to four determinations. Individual neutral lipids expressed as percentage of total neutral lipid fraction.

membrane with respect to levels of phosphatidyl choline and sphingomyelin in rat liver (see Table 1).

Proteins

Studies on membrane proteins have been rendered difficult by the insoluble character of these proteins and by their tendency to polymerize (Kaplan and Criddle, 1971). In a recently published paper, Marchesi and his colleagues have drawn attention to three important questions that need to be answered to advance the understanding of membrane proteins. (1) The number and size of the proteins in any given membrane; (2) the characteristics of these proteins which make them especially suited to a structural role; and (3) the particular arrangement of the proteins which, with associated lipid molecules, gives a membrane its characteristic stability, insolubility and selective permeability.

Methods reported for the preparation of soluble erythrocyte membrane proteins include treatment of the membrane with organic solvents (to extract the lipids), followed by detergents, urea or other denaturants, and in some cases by sonication, to solubilize the proteins These methods are not designed to distinguish different membrane proteins from one another so much as to separate lipid and protein and to solubilize all of the membrane proteins. The proteins obtained by such procedures are highly insoluble in aqueous solvents and they are extremely heterogeneous on ultracentrifugal and electrophoretic analyses. Soluble preparations of protein from enrythrocyte membranes have been obtained, however, by lysis of erythrocytes in water containing a non-ionic detergent in the presence of a mixed–bed ion–exchange resin whose purpose is to remove ions, released from the cells as they are lysed. Another procedure involves treating red cell ghosts with cold, aqueous pyridine. Marchesi *et al* have selectively solubilized a membrane protein, termed spectrin, from human red cells by the use of aqueous solutions of low ionic strength containing 1mM ethylenediamine-tetraacetate. In neutral salt solutions, this purified protein polymerizes into two major species: in 6M guanidine the aggregates are dissociated into a single monomeric unit with a molecular weight of approximately, 140 000. The yield of spectrin obtained represented less than a quarter of the total membrane protein. Other workers have found proteins of fourteen different molecular weight classes in human red cell membranes by gel electrophoresis in the presence of sodium dodecyl sulphate. Four proteins with molecular weights between 86 000 and 255 000 accounted for 60–65% of the membrane protein.

Glycoproteins

Those proteins containing carbohydrates (attached by covalent link-ages to the polypeptide moiety) which are made up of neutral sugars (e.g. D-galactose, D-mannose, L-fucose), amino sugars (e.g. D-galacto-samine and D-glucosamine which are usually acetylated), and amino sugar acids (sialic acid), but contain no uronic acids or sulphate groups are termed glycoproteins. These substances are present as a "cell coat" on plasma membranes where they confer both charge properties on the cell and antigenic specificity on the plasma membrane. They are also present on the membranes of sub-cellular organelles. Their presence has been demonstrated by the use of cell electrophoresis, immunological investigations and electron micro-scope studies. Glycoproteins in membranes have been reviewed by Cook (1968) who has commented that the possibility that surface glycoproteins may act as sites of interaction and recognition between cells could, perhaps, explain why cells have devised a biosynthetic mechanism for the production of complex and varied membrane glycoproteins from a small number of monosaccharides. (The role of glycoproteins at the cell surface is discussed further in Chapter 3).

C. ORIGIN OF THE BILAYER CONCEPT

It would appear that the earliest suggestion that the lipids of membranes might be organized in the form of a bilayer of lipid molecules arranged back-to-back was made by Gorter and Grendel in 1925. They obtained an estimate of the average surface area of red cells and compared it with the area occupied by a monolayer, at an air-water interface, of lipids extracted from a known number of erythrocytes. Since these two measurements were related by a factor of two, they concluded that the erythrocyte is bounded by a bimolecular layer or bilayer of lipid molecules. Recent experiments undertaken on closely similar lines have indicated that their conclusion was actually based on unreliable experimental data since the early workers used too small a value for the surface area of erythrocytes: this error was compensated for by incomplete acetone extraction of the membrane lipids.

In general, membranes behave as lipoid barriers: i.e. those substances that penetrate into cells with ease have a relatively high solubility in lipid solvents. Conversely, unless special diffusion and transport mechanisms are involved, water-soluble substances enter cells much less readily than lipid-soluble substances since they cannot "dissolve" in the membrane and hence cannot cross it. On the basis of findings like this, Danielli proposed the bimolecular leaflet as a general model for the structure of cell membranes since the characteristic permeability properties of living cells could be explained if the typical surface (plasma membrane) of the cell were lipoidal and only a few molecules thick. In the Danielli model, protein was postulated to occur on each side of the membrane to explain the low interfacial tension observed at the surfaces of living cells. More recent studies on "black" lipid membranes (see the section on electrical properties, p. 45) have shown, however, that phospholipid molecules themselves, when arranged in a bilayer, have a very low interfacial tension with the surrounding aqueous phase. It was also originally thought that the first layer of protein adsorbed to the lipid bilayer might be either extended or denatured because proteins are known to unfold at an interface between water and non-polar lipids. However, if protein is adjacent to the polar groups of the membrane phospholipids, the protein need not necessarily be unfolded. Equally, if the protein is in the interior of the membrane and interacts with the lipid moieties of membranes by hydrophobic bonding (see Table 3), the protein may retain a folded configuration.

The rates at which water and other small polar molecules enter cells is apparently too rapid to be explicable in terms of the boundary membrane being composed of an unbroken bilayer of lipid molecules. This anomaly has been interpreted to mean that small polar solutes in transit through the membrane interact minimally or not at all with the hydrocarbon chains of membrane lipids. Thus, in later modifications of the Danielli model, water-filled "pores" extend across the thickness of the membrane from one side to the other (c.f. Table 3).

D. ELECTRICAL PROPERTIES : LIPID BILAYER MEMBRANES

The electrical resistance of biological membranes is generally very high, and values in the range of 10^3 to $10^6 \Omega cm^2$ are commonly observed. At first sight these values would seem to be consistent with a continuous bimolecular leaflet structure for biological membranes, as bilayers of lipid would be anticipated to have resistances of these orders of magnitude. Recent experimental studies have, however, revealed an interesting point in this connection. Advantage can now be taken of newly developed techniques for the formation and study of artificial lipid bilayers: these methods allow electrical measurements to be made on lipid films of known composition and thickness (Henn and Thompson, 1969). Films can be made from solutions of purified phospholipid in a non-polar hydrocarbon such as decane : no protein is required to obtain stable films. Optical and other measurements have shown that lipid films may be prepared with lecithin of a thickness of approximately 7·5 nm. From the known physical properties of phospholipids and from thermodynamic considerations it is highly likely that these films are, in fact, simple lipid bilayers. But one can see from Table 2 that the d.c. electrical resistances of such model membranes (10^6 to $10^9 \Omega cm^2$) are actually several orders of magnitude greater than those of most biological membranes. Nevertheless the electrical resistance of black lipid films can be reduced, either transiently or permanently, and brought within the range of values found for natural membranes by the addition of a variety of organic substances including lysolecithin, surfactants, ionophorous antibiotics (see the section on passive permeability to ions, p. 68) and some proteins. Such agents might cause the membrane phospholipids to adopt a more permeable arrangement (such as localized "micellar" structures, c.f. Table 3) than the bilayer, or they could, like cyclic ionophorus antibiotics, either provide pores through the membrane or behave as ion carriers. These findings indicate that a phospholipid bilayer may be the basic structure for many biological membranes but it would also seem that perturbations of the bilayer structure occur in natural membranes and that these perturbations reduce the electrical resistance below that expected for an unbroken bilayer. Calculations have been made which indicate that the effective area of "pores" in erythrocyte membranes lies between 0·01% and 1% of the total surface area.

The electrical capacitances of black lipid films approximate to those to be expected for a lipid leaflet of bimolecular thickness: the values obtained are also similar to those reported for biological membranes which range from $0 \cdot 5$ to $1 \cdot 3$ $\mu F/cm^2$ (Table 2). The capacitance of a black film depends on its composition. Thus, when cholesterol is present, the capacitance is increased up to a value of $0 \cdot 6$ $\mu F/cm^2$ as compared with $0 \cdot 38$ $\mu F/cm^2$ for a directly comparable membrane prepared from lecithin/n-decane free from cholesterol. The capacitance of a black film also depends on the proportion of hydrocarbon solvent remaining in the film.

TABLE 2. A comparison of some physical properties of biological membranes and lipid bilayers. (After Henn and Thompson, 1969)

Physical property	Biological membranes (20–25°C)	Bilayers (36°C)
Electron microscope image	trilaminar	trilaminar
Thickness, $\mathring{A}$	$60 - 100$	$60 - 75$
Capacitance, $\mu F/cm^2$	$0 \cdot 5 - 1 \cdot 3$	$0 \cdot 38 - 1 \cdot 0$
Resistance, Ωcm^2	$10^2 - 10^5$	$10^6 - 10^9$
Dielectric breakdown, mV	100	150–200
Surface tension, dynes/cm	$0 \cdot 03 - 1$	$0 \cdot 5 - 2$
Water permeability, μ/sec	$0 \cdot 37 - 400$	$31 \cdot 7*$
Activation energy for water permeation, kcal/mole	$9 \cdot 6*$	$12 \cdot 7*$
Urea permeability, μ/sec $\times$ 10^2	$0 \cdot 015 - 280$	$4 \cdot 2\dagger$
Glycerol permeability, μ/sec $\times$ 10^2	$0 \cdot 003 - 27$	$4 \cdot 6\dagger$
Erythritol permeability, μ/sec $\times$ 10^2	$0 \cdot 007 - 5$	$0 \cdot 75\dagger$

* 25°C.
† 20°C.

E. X-RAY DIFFRACTION STUDIES

The technique of X-ray diffraction has provided much valuable information on thickness and spacing in membrane systems. It has the advantage, unlike conventional electron microscopy, of being applicable to biological membranes in a hydrated condition. Possible changes in membrane structure due to dehydration, i.e. artefacts of drying, may thus be avoided.

For many years attention has been concentrated on biological specimens in which membranes are regularly stacked in parallel and which are thus most suitable for study by this technique. The majority of X-ray work has been done on one tissue, the myelin sheath of various types of nerve fibre which provides a natural, repeating stack of membranes. The electron–density profile through myelin layers obtained from low-angle X-ray diffraction studies may be accounted for by a sandwich type of structure of alternating lipid bilayer and non-lipid layers, with the separation of the lipid phosphate groups being about 5 nm across the bilayer. On this model myelin from rat optic nerve (periodicity 16 nm) would be distinguished from rat sciatic nerve (periodicity 18 nm) in terms of the thickness of the non-lipid layers in the region where outer surfaces of glial or Schwann cell membranes come together during myelin formation. There appears to be no evidence for repeating sub-units within the plane of the lipid bilayers of nerve myelin. By contrast, low-angle X-ray diffraction studies on another naturally–occurring stack of membranes, those of the outer segment of the frog retina, have provided results consistent with a square array of spherical particles, 4 nm in diameter, within the plane of the membrane. The presence of this array of particles has been confirmed by parallel studies made with the electron microscope and it appears that the particles may be the photopigment molecules of the retinal receptor system. The X-ray diffraction technique has recently been applied to artificially–stacked, "condensed" preparations of biological membranes. Studies have been made, for example, on centrifuged preparations of erythrocyte membranes and of the membranes of microsomal preparations from muscle, which have been partially dried before structural study to achieve close apposition of the membranes—after experiments to determine how far such preparations can be dried before degradative changes occur.

Finean (1969) has remarked that the profiles deduced from analyses of low-angle X-ray diffraction impose some limitations on the type of organization that is likely to apply to the lipids of these membranes. The most accurate values obtained from the thickness of the hydrocarbon layers are in the range of 2-3 nm. These values indicate that the lipid layer is in a highly disordered state : i.e. the hydrocarbon chains have a liquid-like conformation similar to that in liquid paraffin. Nevertheless overall considerations of diffraction and spectroscopic data suggest that small regions of limited order, with a more "crystalline" character, might occur.

Recently Wilkins has drawn attention to the fact that the continuous diffraction obtained from suspensions of disorientated, randomly arranged membrane fragments can be used to give structural data. Application of the X-ray method to suspensions of membranes would bring a wide variety of different types of hydrated membranes within the scope of the technique. However, although X-ray diffraction is very useful in studies on membranes, the derived electron-density profiles relate to the average structure present in the membrane systems studied. As a result, small but significant regions of a non-bimolecular leaflet structure could occur in a membrane without being detected by X-ray diffraction. Furthermore, the technique provides data on a time–average basis because current equipment requires several hours of exposure to the X-ray beam to obtain a satisfactory diffraction pattern from a membrane preparation. Rapid changes in structure or organization occurring in seconds or minutes in response to particular conditions or treatment will therefore pass undetected.

X-ray diffraction has been used to study the organization of isolated membrane lipids. Both dry phospholipids and phospholipids dispersed in water to form the so-called myelin figures have been investigated. Early work indicated that the lipids are in the form of sheets in these preparations: each sheet being a biomolecular leaflet with a thickness of about twice the length of the constituent molecules. When water is present, it is located between the polar groups of adjacent lipid layers. More recent studies by Luzzati and other workers have shown that aqueous preparations of phospholipids, like soaps, can adopt a variety of structural organizations not all of which are lamellar, depending on the concentrations of lipid and of water and on the temperature. Small changes in temperature or chemical structure can readily produce changes in the structures of these lipid–water systems, demonstrating that lipid systems are capable of the behaviour expected of the constituents of biological membranes, and suggesting that phase changes within membranes may be of physiological and biochemical importance. This work has been extended to phospholipid–protein–water preparations which also show a variety of phase structures.

F. ELECTRON MICROSCOPY STUDIES

The advent in 1954 of a commercial electron microscope with a resolution better than 1 nm and the development of microtomes capable of cutting ultra thin sections provided experimental tools that have been used extensively for work on biological membranes. Despite the utility of electron microscopy and the aesthetically attractive appearance of the micrographs, the images of membranes obtained in this way unfortunately do not provide an unequivocal indication of the underlying structure of membranes. Each of the several processes involved in the preparation of tissues for electron microscopy may cause changes in composition or structural organization of the various components of membranes and thus cause difficulties in the interpretation of ultrastructure. Studies on isolated membrane components are beset by comparable difficulties.

The trilaminar structure observed with the electron microscope in cross-sections of biological membranes appeared initially to be consistent with the Danielli model for membrane structure, and the experiments of Stoeckenius on thin sections of isolated lipids, with and without added protein, are of particular interest in this connection. He found that membraneous structures prepared from brain phospholipids exhibited a trilaminar appearance, in thin sections after fixation with osmium tetroxide, which was made up of two electron dense layers each being $0 \cdot 8$ nm wide, separated by a less dense layer of $2 \cdot 5$ nm thickness. When protein was present, the two dense layers each increased in thickness to $2 \cdot 5$ nm, the resulting image being closely similar to that found with preparations of natural membranes (Fig. 1). It was therefore concluded that the phospholipid molecules were arranged in the form of a bilayer which could be coated on both surfaces by protein, under appropriate conditions,

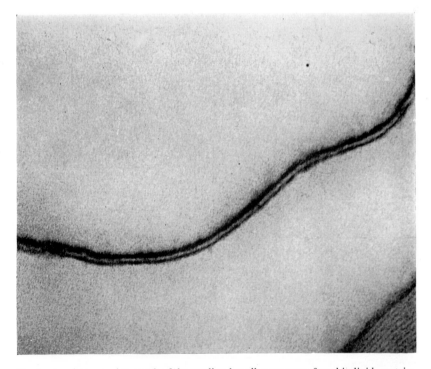

Fig. 1. An electron micrograph of the smallest lamellar structure found in lipid–protein–water preparations. This structure is practically identical in appearance with the "unit membrane" as seen in electron micrographs. Reproduced with permission from Stoeckenius, W. (1962). The molecular structure of lipid–water systems and cell membrane models studied with the electron microscope. *In* "The Interpretation of Ultrastructure" (R. J. C. Harris, ed.), pp. 349–367. Academic Press, New York and London.

thus yielding a structure corresponding to that of the Danielli model. Using potassium permanganate as a fixative, Robertson observed the trilaminar structure of two dense lines about 2 nm wide, separated by a lighter space of about $3 \cdot 5$ nm wide, in thin sections of many biological membranes including those of sub-cellular organelles. He postulated that this uniform appearance confirmed that the basic Danielli model probably applies universally to all animal membranes and that biological membranes could all be described in terms of a "unit membrane". By contrast, Korn (cf. Korn, 1969), in a critical review, has concluded that the dense lines seen in membranes fixed with osium tetroxide reveal very little about the molecular orientation of the phospholipid in the original membrane. It has also been questioned whether the manganese atom is responsible for any of the observed electron density in tissues fixed with $KMnO_4$. This problem has recently been underlined by the results of optical studies on the fixation of membranes by potassium permanganate, which suggest that treatment of intact red cell membranes with this fixative leads to an extensive loss of helical character in the membrane protein.

That biological membranes do not all have the same dimensions has been shown, for example, by Sjöstrand who classified membranes into three kinds on the basis of their width in osmium-fixed tissue. (1) Mitochondria and rough endoplasmic reticulum membranes: 5 nm wide. (The inner and outer membranes of the mitochondria have different permeability properties, are associated with widely differing functions and, quite probably, have different structures. In negatively-stained preparations of mitochondria the inner membranes are seen to be studded with large numbers of particulate components, about 9 nm in diameter (see Chapter 6)). (2) Smooth-surfaced endoplasmic reticulum, including Golgi membranes: 6-7 nm wide. (3) Plasma membranes and zymogen granule membranes: 9-10 nm wide. Sjöstrand also distinguishes membranes which function primarily as barriers (e.g. plasma membrane, myelin sheath) and exhibit the characteristic trilaminar image in thin section from the membranes of mitochondria and endoplasmic reticulum ("primarily metabolically active membranes") which exhibit a globular sub-structure in cross section. He has proposed that the latter membranes consist of globular particles of lipid in the form of micelles, and that the septa between the globular units in mitochondrial membranes may represent, at least in part, the enzymes of the electron transfer chain. In recent years the appearance of sub-units in membranes has been noted by many investigators using a variety of different electron microscope techniques, including freeze-etching (c.f. Branton, 1969). The appearance of sub-units in some membranes and their absence

from others provides some indication that the sub-units are not artefacts. It remains possible, nevertheless, that the structures of some membranes are more labile than others and hence relatively readily subject
to artefact formation. In freeze-etching, frozen tissue is fractured
with a chilled knife. The fractured specimen is then "etched" by
subliming away volatile materials, usually water, from between the
non-volatile components of the sample. The exposed surfaces are
shadowed with a heavy metal and a carbon replica made for electron
microscope study. Branton's experiments with red blood cell ghosts
labelled on both surfaces with covalently-bound ferritin strongly
suggest that cleavage of membranes by freeze-etching occurs along a
plane in the interior of the membrane and not at either surface, since
ferritin molecules were not observed on fracture faces. The particles
observed by this technique, which may be globular micelles of lipid
but are more probably part of the membrane protein, would thus be
located in the interior of the membrane. Such particles have been
observed with erythrocyte membranes: myelinated membranes are
relatively free from particles.

The procedure of freeze-etching is currently the subject of scrutiny
to determine whether or not it also introduces artefacts. Possible
sources of artefact being considered are local heating at the tip of the
fracturing knife, and the possibility that glycerol—used to reduce
ice crystallization in the specimen during freezing—may affect
membrane structure.

The negative-staining technique of electron microscopy, in which
the specimen is not fixed or sectioned but visualized by negative-
contrast with an electron-dense salt of amorphous structure, has been
applied to preparations of isolated membrane lipids. Although the
use of this technique in studies on lipids has been criticized on the
grounds that dehydration of lipoproteins usually lead to a structural
reorganization, experiments on the effects of pH and fixatives on
negatively stained preparations of lipid indicate that drying artefacts
are not a major hazard. This result may depend on rapid drying
effectively preserving the structure of the specimen in a "glass" of
negative-stain, since there is evidence that the inorganic stain loses
water more rapidly than biological materials. Interesting tubular
structures have been observed by Lucy and Glauert in aqueous
dispersions of lecithin mixed with cholesterol and negatively-stained
with potassium phosphotungstate. These tubes (external diameter
11 nm) are apparently made up of five parallel rows of globular lipid
micelles: each micelle having a hydrophobic, lipid core of about
40 nm diameter. The stability of the globular sub-units making up
the tubular assemblies may depend on specific interactions between

C

TABLE 3. Some models for the structure of biological membranes

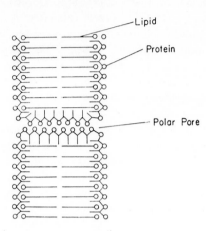

(1)   The original Danielli model for biological membranes was proposed in 1934. This figure illustrates a more recent version which incorporates a trans–membrane pore formed by two polypeptide lamellae. The basic structure of the membrane is a bimolecular lipid leaflet stabilized by adsorbed protein monolayers (see section on origin of the bilayer concept, page 43). An important feature of the Danielli model is the presence of electrostatic interactions between the polar head groups of the phospholipid molecules and polar amino acid residues of membrane proteins. Reproduced with permission from Goldup *et al.* (1970).

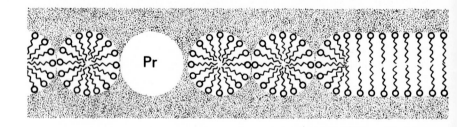

(2)   A cross–sectional view of a membrane model in which globular micelles of lipid are in dynamic equilibrium with a bimolecular leaflet, see section on electron microscopy studies, page 48 (Lucy, 1967). Interactions between the lipid molecules and the layers of protein or glycoprotein (shaded) are polar in character; aqueous pores, about 0·4 nm in radius, traverse the membrane between the globular lipid micelles. One globular micelle of lipid has been replaced by a globular protein molecule which may be a functional enzyme. Reproduced with permission from Northcote (1968).

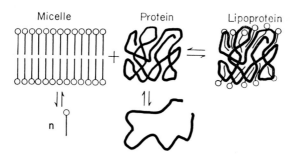

(3)   This figure depicts equilibria involved in the hydrophobic association of surfactant lipids and membrane proteins as presented by Benson. The horizontal equilibrium describes an association of a phospholipid bilayer with a globular protein that gives rise to a lipoprotein unit in which the hydrocarbon chains of the lipid associate hydrophobically with protein (c.f. section on origin of the bilayer concept, page 43). It is proposed that, in chloroplast and cell membranes, these globular lipoprotein subunits associate to form a two-dimensional membrane with a hydrophobic interior. Similar models have been proposed for mitochondrial membranes, see Chapter 6. Reproduced with permission from Benson, A. A. (1966), *J. Amer. Oil Chem. Soc.* **43**, 265–270.

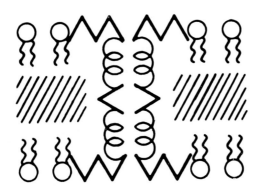

(4)   A generalized model, based on circular dichroism and optical rotatory dispersion spectra of membranes, proposed by Lenard and Singer (c.f. Wallach, 1969), see the section on studies with optical, thermal, nuclear magnetic resonance and spin-labelling techniques, page. 55. In this model the polar head groups of the phospholipids, together with all of the ionic side chains of the structural protein are on the exterior surfaces of the membrane. Sequences of the structural protein consisting predominantly of non-polar side chains are in the interior of the membrane, together with the hydrocarbon chains of the phospholipids and cholesterol. In particular, the helical portions of the protein are in the interior where they are stabilized by hydrophobic interactions. The cross-hatched areas are assumed to be occupied by relatively non-polar constituents (hydrophobic amino acid residues or lipids). Reproduced with permission from Lenard, J., and Singer, S. J. (1966), *Proc. Natn. Acad. Sci. U.S.A.* **56**, 1828–1835.

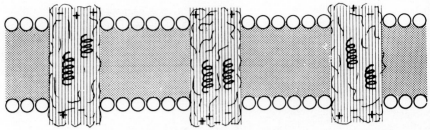

(5)   An example of several recent models which, like that proposed in 1952 by Parpart and Ballentine, is based on a mosaic pattern for the organisation of the lipid and protein moieties of membranes (see section on studies with optical etc. techniques, page 55). The protein is represented by the lined structure, largely in the random coil conformation with helical portions of the protein internal to the membrane and the positive and negative ionic residues at the external faces. The balls represent the polar head groups of the phospholipids; the stippled region represents hydrophobic lipid. In this model a substantial fraction of the phospholipids and proteins can change structure independently of one another. Reproduced with permission from Glaser, M., Simpkins, H., Singer, S. J., Sheetz, M. and Chan, S. I. (1970), *Proc. Natn. Acad. Sci. U.S.A.* **65**, 721–728.

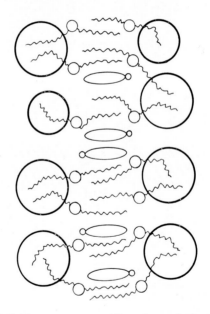

(6)   In this model, lipids are present as a bilayer but each phospholipid has one chain directed into the non-polar membrane interior while the other chain interacts with a hydrophobic bonding site on a membrane protein. Cholesterol is distributed among the inner hydrocarbon chains. On this hypothesis, membrane proteins would have strong hydrophobic associations with membrane lipids, but the proteins would not necessarily be in the membrane interior. In addition, membrane proteins would only become "insoluble" after their hydrophobic binding sites associate with membrane lipids, c.f. section on chemical composition, page 41). Reproduced with permission from Deamer, D. W. (1970), *Bioenergetics*, **1**, 237–246.

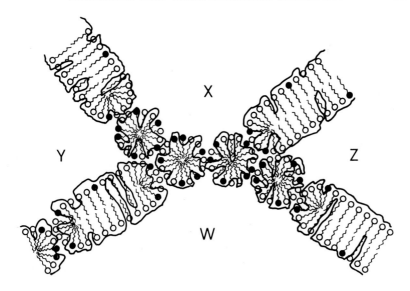

(7)   A model designed to illustrate a possible mechanism by which biological membranes may fuse. Areas X and W, which were previously both membrane-bound, are in the process of being united by membrane fusion. The model shows one way in which lysolecithin (● with tail) may interact with lipoprotein membranes, and cause membrane fusion by inducing changes in the organisation of both the lipid and protein moieties of the two membranes. The presence of the lysolecithin may lead to the formation of globular micelles of lipoprotein which facilitate fusion. Reproduced with permission from Lucy, J. A. (1970), *Nature, Lond.*, **227**, 814–817.

the lipid and the negative stain. It remains possible, however, that conditions leading to the formation and stabilization of globular micelles of phospholipid may occur within natural membranes *in vivo*: the micelles then being held in the plane of the membrane instead of associating to form tubular structures. A micellar model for the lipids of biological membranes has therefore been proposed in which globular micelles of lipid are in dynamic equilibrium with the bimolecular leaflet configuration (see Table 3).

G. STUDIES WITH OPTICAL, THERMAL, NUCLEAR MAGNETIC RESONANCE AND SPIN-LABELLING TECHNIQUES

Early studies with polarized light on the myelin sheaths of nerve fibres indicated that the hydrocarbon chains of the lipid molecules are arranged in parallel rows, perpendicular to the layers, thus providing support for the Danielli model.

As pointed out in the section on the origin of the bilayer concept (page 43) it was proposed in the Danielli model that the first layer of protein adsorbed to the lipid bilayer may be either extended or denatured. Recent studies on membranes made by infrared spectroscopy do not support the presence of extended conformations, however, since the infrared spectra of erythrocyte and Ehrlich ascites carcinoma cell membranes do not show the strong band at 1630 cm^{-1} which is correlated with the β-conformation for protein structures.

Lately, measurements of the circular dichroism (CD: the difference in molecular extinction coefficient between left and right circularly polarized light) and optical rotatory dispersion of membranes (ORD: variation of optical rotation with wavelength) has been employed in investigations on the configurations of membrane proteins and on the interactions between lipid and protein moieties of membranes. The utility of these techniques depends on the fact that the ORD and CD spectra of different structural configurations in proteins (e.g. α-helical, β-conformation, disordered) have specific characteristics. When a number of conformations are present simultaneously, the CD and ORD spectra have intermediate band positions, shapes and amplitudes so that, in principle, the conformational structure of a protein can be determined by comparing its ORD and CD spectra with those of synthetic polypeptides of known configuration. The ORD and CD spectra of membranes from diverse sources are generally similar to one another but, unfortunately, they exhibit a number of anomalous and apparently contradictory features which makes their interpretation difficult. Schneider and co-workers have recently concluded that the characteristic CD spectrum of membranes represents a distortion of a normal α-helical spectrum due to light scattering. With red blood cell membranes they obtained a normal CD spectrum on reducing the particle size of their preparation by sonication: conversely a characteristically distorted spectrum was obtained for globular, non-membraneous helical proteins by introducing scattering. Using the CD spectrum of sonicated red cell ghosts, they obtained the following approximate estimates for the average secondary structure of the membrane protein: 50% α-helix, 45% random coil, 10% β-conformation.

Both Lenard and Singer, and Wallach and Zahler have proposed models based on optical studies which differ significantly from the Danielli model. Thus helical sequences of protein, consisting predominantly of non-polar residues, are located in the interior of the membrane where they interact hydrophobically with the hydrocarbon chains of the phospholipid molecules and relatively non-polar lipids like cholesterol (see Table 3). Interestingly it has recently been found that treatment of human blood cell membranes with purified phospholipase C has no detectable effect on the CD spectrum of the membrane, indicating that no major change in protein conformation accompanies the degradation of 60-70% of the phopholipids. On the basis of this work it has been proposed by Glaser *et al.* that globular proteins, perhaps non-covalently bound to about a quarter of the phospholipids, are interspersed in the membrane in a matrix consisting of the remaining lipids that are organized in a discontinuous bilayer (Table 3).

Differential scanning calorimetry has been used, for example, to compare the properties of the membranes of *Mycoplasma laidlawii* with those of aqueous dispersions of protein-free membrane lipids. In this technique, a calorimeter records the differential power input necessary to heat both the sample and a thermally inert reference at the same rate. If an endothermic transition occurs in the sample, more heat must be applied to the sample than the reference, and a peak is recorded with area proportional to the heat of transition. The membranes of *Mycoplasma laidlawii* and the dispersed lipids exhibited thermal, phase-transitions at the same temperatures. These transitions were reversible and appeared to result from comparable, cooperative changes in the state of the lipids in the two systems. Since the transitions were similar in the two systems, it was concluded that the majority of the lipids in the membranes studied are arranged as in the protein-free, aqueous dispersions of membrane lipids, i.e. in a bilayer, and that the hydrocarbon chains of the lipids associate with one another rather than with membrane proteins. By contrast, investigations on the interactions between lipids and proteins in erythrocyte membranes made by nuclear magnetic resonance spectroscopy (nmr) indicate that the lipid hydrocarbon chains are restricted in their molecular freedom. This restriction could arise from phospholipid–phospholipid, phospholipid–cholesterol, or lipid–protein interactions. The fact that the extracted lipid, including the cholesterol, gives an nmr spectrum in which there is considerable chain movement suggests, however, that the restriction of movement in red cell membranes may be the result of lipid-protein interactions.

Spin–labelling of membranes is a useful technique of recent origin. A spin-label is a molecule containing a chemical group that has an electron spin resonance (ESR) spectrum. Thus the unpaired electron of the nitroxide group ($>N-O$) has a simple ESR spectrum and this group can be introduced into a wide range of molecules including analogues of membrane lipids. Alternatively a spin label such as N-(1-oxyl-2,2,6,6-tetramethylpiperidinyl)-maleimide (Mal-6), which reacts selectively with sulphydryl groups may be used in a study of structural changes in membrane proteins containing sulphydryl groups. When labelled molecules are inserted into natural membranes, they can be employed to provide information on the dielectric nature of the micro-environment of the label, and on the orientation and rotational movement of the label in the membrane. Furthermore perturbations in the structure of membranes caused by exogenous agents, such as anaesthetics, can be investigated by studying the corresponding changes in the environment of the spin label. The technique can also be used to determine whether structural changes in membrane proteins occur when the proteins are solubilized by differing chemical agents.

H. MEMBRANE SYNTHESIS AND DYNAMIC ASPECTS OF
MEMBRANE BEHAVIOUR

Korn (1969), in a review article, has outlined a number of intriguing questions which are currently of major interest in relation to the biosynthesis of membranes. Thus, is new membrane synthesized on a template of existing membrane? Is there a common site of membrane biosynthesis, perhaps the rough endoplasmic reticulum, with subsequent membrane flow to other regions of the cell? Are membranes assembled in one step, or are there intermediate steps in which components are added to an incomplete skeletal structure? Are the protein and lipid components incorporated separately in membranes, and if so in which order, or are they added together as a lipoprotein? Is the order of the macromolecular components random within the membrane or is there a unique arrangement under cellular control? As indicated by Korn in his review, tentative answers to some of these questions are beginning to appear. For example the notable differences in enzyme and chemical composition of membranes within any given cell strongly suggests that some membranes are not interconvertible, although some may be. This in turn implies more than one site of synthesis. It is to be expected, however, that ideas on these complex phenomena may change markedly as more information is obtained.

The complexities of membrane synthesis are illustrated, for example, by studies on the rough- and smooth-surfaced endoplasmic reticulum membranes. Thus, Siekevitz and Palade and their colleagues obtained evidence in 1967 for transfer of newly synthesized membrane components from the rough to the smooth membranes (Chapter 31). They also found that the lipids of these membranes turn over more rapidly than, and independently of, membrane proteins and that the half-life of fatty acids in membranes appeared to be much longer than that of the glycerol component of membrane phospholipids. Work on the problems of membrane synthesis has recently been reviewed by Cook (1971) who comments that experiments on the biogenesis of membranes emphasize the dynamic aspect of membrane chemistry, which is too often overlooked in studies of membrane structure made with physical techniques.

Another topic of present interest is the extent to which the structural organization of the components of any given membrane may vary with time, since it seems unlikely, for example, that the many facets of the behaviour of the plasma membrane of a cell can all be accounted for in terms of an unvarying arrangement of molecules in the membrane. It has been suggested that globular micelles of lipid may be in a dynamic equilibrium with the bimolecular leaflet configuration (model 2 of Table 3), and that membrane–active substances, having small variations in molecular size, shape, hydrophilic–hydrophobic balance, lipid–solubility and electrostatic charge, might be expected to produce relatively large effects on the organization of a membrane that possesses the ability to change from a lamellar to a micellar structure. The fusion of biological membranes in such differing processes as cell fusion, cell division, fertilization, secretion and endocytosis are examples of highly specialized, dynamic behaviour in membranes. These processes would appear to require the lipids of the fusing membranes to be appropriately organized. It has therefore been proposed that agents capable of causing membrane fusion do so by inducing, directly or indirectly, a phase change in lipids of one or both the membranes concerned. More specifically, for fusion to occur a relatively high proportion of the lipids of both membranes may need to be organized in micelles of lipid or lipoprotein (Table 3). Support for this hypothesis seems to be provided by the finding that the micellizing agent, lysolecithin, will induce cell fusion *in vitro* in several different types of cell: both multi-nucleated syncytia and heterokaryons being formed within 30 sec of treating cells with lyso-lecithin (Fig. 2). It is conceivable that lysolecithin is concerned in some of the processes of membrane fusion occurring in living cells.

III. Transport Through Membranes

A. INTRODUCTION

A cell can only function normally if its chemical requirements, e.g. amino acids, vitamins, oxygen, ions, etc. are met and if the waste products that are not needed and cannot be utilized further are removed from the environment of the cell. However, transport to and from the cell is not sufficient in itself. There must also be an appropriate passage of substances in both directions, across the plasma membrane. Similarly, at the sub-cellular level, materials pass in and out of sub-cellular organelles. Transport in this chapter will, however,

Fig. 2. An electron micrograph of part of a heterokaryon containing the nucleus of a hen erythrocyte (E) and the nucleus of a mouse LS fibroblast (F) formed on treating a mixture of the two cell types with lysolecithin. Reproduced with permission from Poole, A. R., Howell, J. I. and Lucy, J. A. (1970), Lysolecithin and cell fusion, *Nature, Lond.* **227**, 810–813.

be discussed principally at the level of the plasma membrane (transport in mitochondria is discussed in Chapter 6). The reader will also need to look elsewhere for an account of those numerous and important processes of "transport" in which substances circumvent membraneous barriers, not by travelling through the membrane but by the occurrence of membrane fusion : such processes, which include phagocytosis, pinocytosis, and secretion, have been reviewed by Bennett (1969). Equally, it has not been possible to include in this chapter an account of communication between cells via junctional membranes, which have been shown to be between 1 000 and 10 000 times more permeable to ions than the adjacent regions of plasma membrane forming a barrier to the exterior milieu (reviewed by Loewenstein, 1970).

Substances pass through biological membranes by a variety of mechanisms depending, among other factors, on the chemical nature of the substances passing through the membrane and on the structure of the membrane concerned. It is therefore possible to classify transport through membranes under several different headings. Recent research on experimental models for biological membranes, and on antibiotic molecules as chemical models for transport "carriers" in membranes, has shed considerable light on the possible mechanisms involved in transport processes *in vivo*. A discussion of some of the contributions which work of this kind has made is included in this chapter, in addition to discussion of studies on biological membranes.

B. DIFFUSION OF LIPID-SOLUBLE SUBSTANCES

We have already seen in the section on the origin of the bilayer concept (page 43) that, unless special mechanisms are involved, water-soluble substances enter cells much less readily than lipid-soluble substances. This was originally thought to indicate that membranes are essentially lipoid barriers and work with lipid bilayer membranes has reinforced this view. However, other models of biological membranes have been proposed which have protein in the interior of the membrane rather than lipid (see Table 3). In such models the protein in the interior of the membrane is nevertheless hydrophobic. Until the structures of differing membranes have been more closely defined than at present it might be better to regard the ease with which lipid-soluble substances cross membranes by diffusion as providing evidence for a hydrophobic, rather than a lipoid, barrier.

C. SMALL, POLAR SUBSTANCES :
DIFFUSION RESTRICTED BY SIZE ALONE

Although a simple bimolecular leaflet model would appear to be too impermeable to water and small water-soluble polar substances to account for the properties of biological membranes, it is sometimes suggested that the bimolecular leaflet of lipid as a theoretical model for the structure of biological membranes does not need to be modified by the introduction of water-filled pores, in view of the fact that water is known to pass through a monolayer of lipid at an air–water interface. This possibility still remains open. It appears, however, that the ratio of penetration of water in monolayers to that in the squid giant axon membrane is approximately 1 000:1, i.e. there is a considerable lack of correspondence between the behaviour of water passing through monolayers and through natural membranes. From a recent extensive series of studies on the rabbit gall bladder, Diamond and his colleagues have concluded that small, polar, non-electrolytes pass through the membrane—without interaction with the hydrocarbon chains of membrane lipids—via a route formed by localized concentrations of membrane polar groups associated with "frozen" water molecules. Permeations by this route depends, to a first approximation, on molecular size.

Solomon and his colleagues have introduced the concept of the equivalent pore radius: i.e. the radius of an idealized water-filled cylindrical pore which is presumed to extend from one side of a membrane to the other. The equivalent pore radius is a measure of the relative restriction offered by the membrane to solvent and solutes, on the assumption that cell membranes discriminate between water-soluble, non-electrolytes by size alone. Solomon has pointed out that it is not necessary to assume that all the pores are uniform and circular, that they all have the same radius, or that they remain fixed in position or in time. Nor need the equivalent pore radius be the actual pore radius. He has also commented that the concept of the equivalent pore radius can be regarded as an attempt to describe in operational terms a physical property of complex biological membranes. The equivalent pore radius may be determined by several different methods, and the values found by these methods are in quite good agreement; a value of $0 \cdot 4$ nm has been obtained for a number of different cells (erythrocytes, nerve cells, intestinal mucosal cells). Interestingly it has been reported that removal of calcium appears slightly to increase the equivalent pore radius, allowing sucrose to enter amphibian kidney cells.

The rate of transport of water across plasma membranes has been extensively investigated and, in general, the spread of values obtained covers a similar range to that more recently observed with lipid bilayers (from 5 to 100×10^{-4} cm/sec) c.f. also Table 2. With bilayers the water permeability depends on the lipid composition of the film.

It also seems that there is no interaction between water molecules in the bilayer membrane. This conclusion is consistent with the bilayer membrane being either an isotropic liquid in which the transferring water molecules dissolve to give an ideal solution, or a structure containing pores of molecular dimensions each of which contains, on average, no more than one water molecule. It is important to note, however, that experimental studies on micro-capillaries and other porous systems have shown that, as the pore diameter decreases, deviations from the properties of bulk water are observable even in pores that are at least an order of magnitude greater than molecular diameters (e.g. an increase in viscosity occurs in 10 nm channels to a value that is about three times the value for water in bulk). Thus the properties of water in any very narrow channels present in membranes may be intermediate between those of bulk water and of ice. In the absence of vasopressin (antidiuretic hormone), the cells of the epithelial wall of the mammalian renal collecting tubule are virtually impermeable to water. Using the toad bladder (from which water can be absorbed in the presence of ADH) as a model system, a value of 4 nm has been found for the equivalent pore radius in the presence of the hormone. It has been suggested that vasopressin may enlarge the effective pores or, in the light of the above considerations, that it alters the properties of the water in the pores.

D. PASSIVE PERMEABILITY TO IONS : LIPOSOMES : IONOPHOROUS ANTIBIOTICS

Different plasma membranes have differing passive permeability properties to ions. In addition, considerable species variation can occur with membranes of the same type : the rate of influx of phosphate into rat erythrocytes being an order of magnitude greater than that for sheep erythrocytes.

With the red cell, electrostatic forces seem to be responsible for the discrimination between anions and cations. There is reason to believe that, without impeding the movement of anions, positive fixed charges impose severe restriction on the movements of diffusible cations between cells and their environment. Membrane-bound calcium ions may partly constitute this fixed charge barrier to the diffusion of cations. In view of the net negative charge of the cell surface, such fixed cations are thought to be located inside the membrane where they line the walls of pores through which the hydrophilic ions are presumed to migrate. However, as Whittam and Wheeler (1970) have remarked to their recent review article, the question arises : is the low permeability towards cations an inherent property of the cell membrane (determined by factors such as fixed positive charges, lipid composition and calcium ions), or does the low permeability depend on metabolism? It seems that while large changes in permeability to cations, resulting for example from variations in pH, can be related to changes in fixed charges, metabolic inhibitors like iodoacetate and fluoride also act synergistically with calcium to cause changes in the passive permeability towards potassium ions.

Interesting observations on the passive permeability of ions in the "liposome" model membrane system have been made by Bangham and other investigators (Bangham, 1968; Sessa and Weissman, 1968). When phospholipids of biological origin are dispersed in an aqueous environment they undergo a sequence of changes. During the initial molecular rearrangements, water and salt have free access into the substance of the liquid crystal: subsequent rearrangements lead to the formation of layered liquid-crystalline structures that apparently consist of numbers of bimolecular lipid leaflets arranged one inside the other. Since these are equilibrium structures, it is thermodynamically probable that each and every bimolecular membrane forms an unbroken sheet—there being no exposed hydrocarbon/water interface. From this it follows that every aqueous compartment is probably discrete and isolated from its neighbour by a closed membrane, and that the outermost aqueous compartment of the whole structure is similarly isolated from the continuous bulk aqueous phase. Thus solutes and water can be sequestered within liposomes and can subsequently diffuse and exchange between compartments, and with the bulk aqueous phase, only by crossing the phospholipid bimolecular membranes. With liposomes prepared from lecithin mixed with dicetyl phosphate, it has been found that (a) anions diffuse out of the liposomes much more rapidly than cations, (b) the exchange diffusion rate of Cl^- and I^- is greater than for F^-, NO_3^-, SO_4^{2-} and HPO_4^{2-}, and is apparently therefore related to ion size; water appears to exchange as fast as, if not faster than, chloride, (c) cations diffuse out of liposomes relatively slowly. The addition of calcium ions to liposomes containing both lecithin and cardiolipin reduces their permeability to potassium ions even further.

At this point attention must be drawn to the value of phospholipid spherules (liposomes) as models for the behaviour of biological membranes. In addition to having the properties already mentioned, liposomes swell osmotically, as do natural membrane-bounded structures. They are susceptible to labilization and stabilization by many agents which act similarly upon a wide variety of natural membranes. These agents include detergents, lysolecithin, steroid hormones, bacterial toxins, antibiotics, narcotics and drugs. As an example, it has been found that the correlation between the capacity of 14 different steroids to disrupt both natural membranes and liposomes is very close. Furthermore cortisone, cortisol, their acetates, and chloroquine all reduce the release of cations from liposomes, an effect in keeping with their action on erythrocytes and lysosomes. It remains difficult, however, to extrapolate from these effects on liposomes to biological systems because the concentrations of steroids required to disrupt liposomes (10^{-3}M) are clearly outside the physiological or even pharmacological range. The value of liposomes as models for membrane studies is enhanced by the investigator being able to vary the composition of the membrane, either by varying the lipid composition or by incorporating trace amounts of test agents. In addition, one may vary the surface potential, and also study the effects of protein on the properties of the spherules.

Some interesting findings relevant to the passive permeability of the red cell membrane have recently been reported by Finkelstein and his co-workers. They have studied the effects of treating lipid bilayers with nystatin, and with amphotericin B : these being polyene antibiotics produced by bacteria of the genus *Streptomycetes*. Nystatin and amphotericin B increase the permeability of bilayer membranes to ions, water and non-electrolytes. It was found that while untreated bilayer membranes had a low permeability to small hydrophilic solutes such an urea, ethylene glycol and glycerol, the nystatin- or amphotericin B-treated membranes showed a graded permeability to these solutes on the basis of molecular size. Since the "cut-off" in permeability occurred with molecules of about the size of glucose, it was concluded that nystatin creates pores having an effective radius of about $0 \cdot 4$ nm. Calculations indicated that less than $0 \cdot 01\%$ of membrane area was occupied by pores in a treated membrane, even when the antibiotic had reduced the membrane resistance from $10^8 \Omega cm^2$ to $10^2 \Omega cm^2$. Furthermore, membranes treated with these antbiotics were more permeable to anions than to cations, although they were not completely selective for anions. These several remarkable similarities between the properties of the red cell membrane and bilayer membranes containing either of these two antibiotics have prompted Finkelstein and his colleagues to comment that it might not be completely fruitless to attempt to extract molecules having structures resembling those of polyene antibiotics from red cell membranes.

While work with nystatin and amphotericin B has shown that these substances can make lipid bilayers relatively permeable to anions, studies with other antibiotics have revealed that some cyclic antibiotic molecules which produce permeability changes in bacterial and erythrocyte membranes have the remarkable property of enabling membranes selectively to distinguish between different cations. The action of these antibiotics on synthetic bilayer membranes has been the subject of recent work, and a macrocyclic antibiotic of particular interest is the cyclic polypeptide, valinomycin. Very low concentrations of valinomycin (e.g. 10^{-11} to 10^{-6} mole/1) have a large effect on the permeability of a lipid bilayer, and in solutions of alkaline metals will produce changes in the film conductance which are proportional to the concentration of antibiotic. (As already indicated above and in the section on electrical properties, page 45, the addition of antibiotics to lipid bilayers reduces their resistance to the level of that of natural membranes.) Even more strikingly, valinomycin confers selectivity properties on bilayer membranes which enables them to exhibit markedly different permeability properties for univalent cations. With valinomycin, the permeabilities are in the order

$$H^+ > Rb^+ > K^+ > Cs^+ > Na^+ \simeq Li^+$$

The precise ratios reported vary, but the selectivity coefficient for potassium ions relative to sodium ions is approximately 400. Valinomycin also influences the permeability to potassium ions of mitochondria and liposomes.

In general a cyclic, ionophorous antibiotic has hydrophobic residues on the outside of its ring which allow the molecule to be soluble in a lipid environment. The inside of the ring is polar : the molecule thus provides a hydrophobic shield which enables highly polar ions to pass readily through the non-polar regions of membranes when they are complexed with the antibiotic. It seems that these substances have selective permeability properties because their three-dimensional geometry allows them to interact specifically with certain ions. The inwardly-directed, polar carbonyl groups apparently interact either with the inner hydration shell of hydrated cations or—in the case of the smaller cyclic antibiotics—they may replace the hydration shell completely. The limited transport by valinomycin of Na^+ compared with K^+ has been ascribed to the smallest space that can be enclosed by a cubic array of eight internal oxygen atoms being too large to permit a close fit with the unhydrated sodium ion. Haydon (1970) summarized in a review article three possible ways by which formation of an ion-antibiotic complex may facilitate ion transfer across a lipid film. Antibiotic molecules may associate in the form of a stack of molecules aligned across the membrane such that the ion may jump from one molecule to the next. Alternatively the complex, which has a diameter less than the thickness of the hydrocarbon part of the membrane may in effect dissolve in the liquid hydrocarbon and diffuse across. Yet another possibility is that the hydrocarbon part of the film may become constricted at the site of the complex, so that a distinction between a "pore" consisting of a single complex on the one hand, and a diffusing complex on the other, may become impossible. In fact, experimental evidence suggest that different antibiotics may function by differing mechanisms. It seems that valinomycin may behave as a mobile carrier while, conversely, a pore may be formed across the membrane, under certain circumstances, in the presence of the antibiotic alamethicin.

Alamethicin is a cyclic peptide composed of 18 residues. The primary structure of the molecule has recently been determined, and a three-dimensional structure has been proposed by Hartley and his colleagues on the basis of their primary sequence determination. Alamethicin also confers selectivity for cations rather than anions, but between cations the selectivity is poor. Interest lies in the fact that, in the presence of alamethicin, the conductance of a lipid bilayer in symmetrical bathing solutions is strongly dependent on the potential across the membrane: a low conductance corresponding to a small or zero potential difference. At low applied voltages alamethicin may exist in monomer form and behave as a cation carrier within the bilayer, while at high voltages some molecules of alamethicin may aggregate to provide a trans-membrane pore. Mueller and Rudin have found that the steady state conductance varies with the sixth power of both antibiotic and ion concentration. This and other electrokinetic and chemical data suggest that six or more alaemethicin molecules form either carriers or tubular channels through which ions flow across the membrane, and that assembly and disassembly of these aggregates by the voltage or chemical factors regulate the conductance. The properties of alamethicin are similar to those of an allosteric protein specifically designed to form ionic coordination complexes within lipid membranes. Alamethicin can induce action potentials in lipid bilayer membranes in the presence of protamine, polylysine or spermine in the aqueous phase. These properties of alamethicin-treated bilayers, and their possible interpretation at the molecular level, are of importance in view of their resemblance to the properties of nerve membranes. As remarked by Haydon (1970), however, it must be emphasized that at present there is little or no evidence that a normal cell membrane embodies any of the macrocyclic or other substances which give such interesting imitative results in lipid films. It is nevertheless likely that whatever substances are employed by cells for the control of transport in membranes (both active and passive), they are likely to be present in extremely small quantities.

b

E. FACILITATED DIFFUSION (PASSIVE MEDIATED TRANSPORT)

Facilitated diffusion, which is sometimes referred to as mediated transport, is a transport process that involves carrier molecules which facilitate movement of material across biological membranes. Although facilitated diffusion is mediated by carrier molecules, it is not uphill transport, i.e. transport only occurs down a concentration gradient in the direction of decreasing concentration, and the process does not consume metabolic energy.

The most studied example of facilitated diffusion is probably the transport of glucose in erythrocytes: simple, passive diffusion of glucose across the red cell membrane playing only a minor role in glucose transport. The transport of glucose across the erythrocyte membrane therefore exhibits characteristic features of facilitated diffusion. For example, the system possesses a high molecular specificity for transported molecules. This in itself provides strong evidence that simple diffusion is not involved. While L-arabinose and D-xylose enter red cells readily, for instance, D-arabinose and L-xylose enter very slowly. That glucose enters liver cells by facilitated diffusion may be indicated by the much more rapid entry of D-glucose compared with L-glucose.

Facilitated diffusion shows saturation kinetics, since a carrier molecule may be saturated with the substance being transported in the same way that an enzyme becomes saturated with its substrate. This behaviour indicates that carrier molecules, like enzymes, possess specific active sites to which the transported substance undergoes reversible binding. Indeed, the kinetics of glucose penetration into red cells are consistent with the assumption that sugar molecules are first bound to a mobile carrier, and then transported across the membrane by diffusion of the sugar-carrier complex. By contrast, the rate of transport by simple diffusion alone is directly proportional to the concentration gradient, and does not exhibit saturation. Very recently it has been reported that the kinetics of Cl^- transport across the red cell membrane shows both saturation and competition phenomena, indicating that transport of Cl^- may also be due to the operation of a carrier.

The phenomenon of "counter-transport" provides good evidence for the existence of a carrier molecule. When a rat heart muscle preparation is perfused with L-arabinose, the intracellular concentration of arabinose drops on the addition of glucose in the system. The arabinose is not metabolized: it is apparently transported out of the cell by the carrier molecules that transported glucose inwards.

In contrast to the active transport of Na^+ and K^+ ions in the erythrocyte, the facilitated diffusion of glucose into this cell is not inhibited by inhibitors of glycolysis. Nevertheless facilitated diffusion can be inhibited either competitively, or non-competitively by causing irreversible chemical changes in the structure of the carrier molecule with, for example, dinitrofluorobenzene which blocks free amino groups. Phloretin is an interesting inhibitor of glucose transport in red cells. A comparative study has been made of the effectiveness of a number of diphenols related to phloretin as reversible inhibitors of mutarotase (from intestine) with their ability to behave as competitive inhibitors of the glucose carrier in human erythrocytes. A marked parallelism was found, and it was concluded that the structures of the active centre of mutarotase (which interconverts α- and β-forms of D-glucose and other sugars) and of the receptor site of the glucose carrier in red cells are probably extremely similar if not identical. The actual identity of the carrier remains unknown, although Stein and his colleagues have prepared a protein fraction from red cells that binds D-glucose better than L-glucose. Calculations based on this latter work indicate that there are 800 000 binding sites on the membrane per cell, and that each site can transport 180 glucose molecules per second.

F. ACTIVE TRANSPORT

Active transport, which may be described as carrier-dependent, metabolically-dependent, uphill transport, resembles facilitated diffusion in that transport depends on the combination of the transported substance with a specific carrier. However, with active transport, metabolic energy is used in the form of ATP to modify the configuration and/or chemical structure of the carrier so that it has a high affinity for the transported material on the accepting side of the membrane and a low affinity on the other side. This enables the process to transport substances against the concentration gradient: i.e. to accumulate material from one side of the membrane and to concentrate it on the other. After releasing the transported material, the carrier effectively moves back either empty, or in combination with a different molecular or ionic species, and the cycle is repeated.

Active transport is of great importance to cellular economy and function. It enables the cell to accumulate materials from an environment where they may be present in short supply. Conversely ions can be secreted by active transport, as in the kidney, while the maintenance of cellular volume depends, in general, on the active transport of sodium ions out of cells. Active transport also enables cells to sustain constant and optimal internal concentrations of inorganic electrolytes, particularly K^+ which is essential for the optimal functioning of several important intracellular activities including the transmission of nerve impulses. The importance of active transport is such that a significant proportion of the metabolism of the body at rest is devoted to active transport processes. It has been reported that when the active transport of Na^+ in kidney preparations is completely inhibited by the cardiac glycoside, ouabain, the rate of respiration is decreased by up to 80%. Since the rate of respiration in intact cells is directly geared to the rate of utilization of ATP, it is apparent that most of the ATP production of kidney cells is used in transporting Na^+ ions.

Erythrocytes, like most cells, maintains a high internal K^+ concentration, and a low internal Na^+ concentration in the face of a low external K^+ and a high external Na^+ by means of active transport. The Na^+/K^+ transport mechanisms of erythrocytes have been the object of much study because of the relative ease of isolating the erythrocyte membrane, and because of the possibility of examining the vectorial properties of the transport system by utilizing the phenomenon of reversible lysis which enables the investigator to reseal erythrocyte ghosts filled with any chosen ion, inhibitor, or metabolite. Following the discovery by Skou in 1957 of a Na^+-, K^+-, Mg^{2+}-dependent ATPase activity in crab nerve, which is directly associated with ion transporting activity, it is now generally considered that the membrane-bound enzyme system which catalyses the Na^+-, K^+-, Mg^{2+}-activated hydrolysis of ATP is the same as, or represents an important component of, the system responsible for the metabolically driven extrusion of Na^+ ions and uptake of K^+ ions through the plasma membrane of many types of cell. ATPase activity associated with active transport is particularly active in excitable and secretory tissues, such as brain, nerve, muscle, kidney, salivary gland, and the electric organ of the electric eel.

The association of Na^+-, K^+, Mg^{2+}-dependent ATPase activity with the active transport, Na^+/K^+, system of the erythrocyte membrane has been demonstrated by the following observations. There is a close correspondence between the conditions necessary for ion transport, and for enzyme activity. For example, the concentration for half maximal activation by K^+ (in the presence of Na^+), is 3mM for ATPase activity in red cell membranes and 2·1mM for transport in intact red cells. Both transport, and ATPase activity, are inhibited by calcium, and also by ouabain. Among the best evidence for the relationship between the two systems was the demonstration by Gallahan and Glynn that inorganic phosphate can be incorporated into ATP by reversal of the sodium pump. Another interesting facet of the Na^+/K^+ transport system in red cells is that ATP reacts with the transport system on the inside of the membrane, while both the ADP and the inorganic phosphate produced are released on the inside of the membrane. For each molecule of ATP consumed, $3Na^+$ ions are transported out, and approximately $2K^+$ ions are transported into the cell.

It appears that the ATPase reaction comprises two essential steps, with reaction (1) occurring at the internal surface of the cell membrane and reaction (2) taking place at the external surface.

$$(1.) \quad ATP + enzmye \xrightleftharpoons{Mg^{++}, Na^+} enzyme{\sim}P + ADP$$

$$(2.) \quad enzyme{\sim}P + H_2O \xrightarrow{K^+} enzyme + Pi$$

There is, however, evidence that there are two forms of the phosphorylated intermediate, enzyme$\sim$P, one form being produced irreversibly from the other in the presence of Mg^{2+}. Reversal of reaction (2) is not found with fragmented membranes.

The first of the two reactions shown above is a Mg^{2+}-, Na^+-dependent phosphorylation, and this is the step which is inhibited by Ca^{2+}. The second reaction is a simple K^+-dependent hydrolysis, and evidence has been obtained that ATPase preparations from a variety of tissues are, in fact, able to catalyse the hydrolysis of a number of different phosphate esters in the presence of K^+ ions: this hydrolytic activity is inhibited by ouabain. Several studies have indicated that an acyl phosphate grouping is involved in the phosphorylated intermediate. The most likely candidates for such an acyl phosphate would seem to be L-glutamyl-γ-phosphate and L-aspartyl-β-phosphate: recent data indicate that it is the γ-carboxyl of a glutamyl residue of a membrane protein which is phosphorylated by ATP. However, whether the phosphorylated protein represents one of the actual ion carriers, or whether it is only an intermediate in the transfer of energy to the pump system, has not yet been elucidated. Owing to the membrane-bound nature of this enzyme system, attempts to obtain purified ATPase preparations are beset by numerous problems. Solubilized preparations have nevertheless been prepared, with the aid of deoxycholate, which retain ATPase activity that remains capable of complete inhibition by ouabain.

A most comprehensive review article on sodium–potassium activated adenosine triphosphatase and cation transport has been produced by Bonting (1970), while an interesting critique of models for sodium/potassium transport has also been published in a recent review article by Caldwell (1970).

It has been demonstrated for a wide variety of tissues that the extrusion of Na^+ from cells is coupled to the uptake of K^+ from the extra-cellular space. The movement of chloride ions has usually been considered to be secondary to the ion fluxes developed by this Na^+-K^+-coupled cation pump. By contrast, Whittembury and Proverbio have found that 1mM ouabain strongly inhibits the Na^+-, K^+-exchange in cortex slices from guinea pig kidney, but has little effect on the extrusion of Na^+ accompanied by Cl^- and water. On the other hand 2mM ethacrynic acid inhibits mainly the extrusion of the Na^+ with Cl^- and water, and only marginally affects the Na^+-K^+ exchange. These investigators have therefore postulated the existence of two separate Na^+ "pumps" which function to transport excess Na^+ from the intracellular space of the kidney cortex cell. Namely, one pump that exchanges intracellular Na^+ for extracellular K^+, and a second that extrudes Na^+ accompanied by Cl^- and water, the latter being an important mechanism for volume regulation. Since both pumps are inhibited by 2,4-dinitrophenol and anoxia, there may nevertheless be a common link between the two systems.

Since calcium plays an important part in several cellular functions, including muscular contraction, the mechanisms that control the levels of calcium ions in cells are of considerable importance and interest. The concentration of calcium ions is extremely low in most living cells, including erythrocytes, nerve cells and muscle cells. Although the permeability of the plasma membrane of the red cell to calcium is very low, the membrane is nevertheless not totally impermeable : active transport of calcium ions outwards is therefore necessary to maintain the internal concentration at a lower value than the external level. Schatzmann and Vincenzi have demonstrated the presence in human erythrocytes of an active ATP-dependent transport process for calcium ions, and they found that the ATPase activity of the membrane was stimulated by internal, but not by external, Ca^{2+}. It appears that the calcium pump is independent of the sodium pump in these cells, and that the ATPase concerned in calcium transport is insensitive to Na^+, K^+ and ouabain. A different situation applies in squid nerve axons where Hodgkin and Baker and their colleagues have discovered that some, or possibly all, of the energy for the extrusion of Ca^{2+} from squid axons may come from the downhill movement of Na^+ into the axon from the more concentrated solution outside. In addition, there is a component of Na^+ efflux which is coupled to the influx of Ca^{2+} ions in the squid axon : this coupled movement is insensitive to ouabain.

With chick intestine, Adams and Norman have obtained evidence indicating that the transport of calcium in normal ileal tissue is also an active process operating against an electropotential gradient. Although they observed that calcium was not actively transported in the vitamin D-deficient chick, a carrier was nevertheless apparently present at the brush border side of the intestinal mucosal cells of both vitamin D-treated and deficient chicks. It seems that vitamin D may function by increasing the efficiency of the transport mechanism for calcium uptake, although this may be achieved indirectly through the synthesis of RNA and protein rather than through the direct participation of the vitamin in the transport process.

We have already seen that the movements of Na^+ and Ca^{2+} are inter-related in squid nerve axons. The involvement of sodium ions in the transport of other materials appears, indeed, to be quite widespread. For example, the active transport of sodium ions out of cells is necessary for the active transport of amino acids and glucose inwards in intestinal cells by processes which probably involve combination with specific carriers. The transport of glucose and of amino acids in the intestine are examples of secondary active transport, in that energy is not used directly in the inwards transport of the (glucose; Na^+) or (amino acid; Na^+) complexes but for the subsequent outward transport of Na^+. According to the mechanism proposed by Crane (1968) the two inward-moving substrates, viz. (glucose and Na^+) use a common, membrane-bound carrier. Inside the cell the Na^+ and the sugar dissociate from the carrier molecule. The sodium ions are then transported out of the cell by the sodium pump which maintains a low intracellular concentration of Na^+ ions. A gradient for sodium ions is thereby maintained across the cell membrane, from outside to inside, and this enables (glucose and Na^+) to continue to enter the cell. By this means, high concentrations of glucose can be absorbed into the cells. A similar mechanism appears to apply to amino acid transport in the intestine (c.f. Ring, 1970). The intestinal absorption of both sugars and amino acids is inhibited by ouabain at the serosal side of the intestine owing to inhibition of the sodium pump by the cardiac glycoside and the subsequent dissipation of the concentration gradient of Na^+ ions between cells and extracellular fluid.

REFERENCES AND RECOMMENDED READING

Bangham, A. D. (1968). Membrane models with phospholipids. *Prog. Biophys. Mol. Biol.* **18**, 29–95.

Bennett, H. S. (1969). The cell surface: movements and recombinations. *In* "Handbook of Molecular Cytology" (A. Lima-de-Faria, ed.) pp. 1294–1319. North-Holland, Amsterdam.

Bonting, S. L. (1970). Sodium–potassium activated adenosine triphosphatase and cation transport. In "Membranes and Ion Transport" (E. E. Bittar, ed.) Vol. 1, pp. 257–364. Wiley, New York.

Branton, D. (1969). Membrane structure. *Annu. Rev, Pl. Physiol.* **20**, 209–238.

Caldwell, P. C. (1970). Models for sodium/potassium transport: a critique. In "Membranes and Ion Transport" (E. E. Bittar, ed.) Vol. 1, pp. 433–461. Wiley, New York.

Cook, G. M. W. (1968). Glycoproteins in membranes. *Biol. Rev.* **43**, 363–391.

Cook, G. M. W. (1971). *Annu. Rev. Pl. Physiol.* **22**. 97–120.

Crane, R. K. (1968). Digestion and absorption of carbohydrates. In "Carbohy. drate Metabolism and its Disorders" (F. Dickens, P. J. Randle and W. J. Whelan, eds.) Vol. 1, pp. 25–51. Academic Press, London and New York.

Finean, J. B. (1969). Biophysical contributions to membrane structure. *Quart. Rev. Biophys.* **2**, 1–23.

Goldup, A., Ohki, S. and Danielli, J. F. (1970). Black lipid films. *Recent Progr. Surface Sci.* **3**, 193–260.

Haydon, D. A. (1970). The organization and permeability of artificial lipid membranes. In "Membranes and Ion Transport" (E. E. Bittar, ed.) Vol. 1, pp. 64–92. Wiley, New York.

Henn, F. A. and Thompson, T. E. (1969). Synthetic lipid bilayer membranes. *Annu. Rev. Biochem.* **38**, 241–262.

Kaplan, D. M. and Criddle, R. S. (1971). Membrane structural proteins. *Physiol. Rev.* **51**, 249–272.

Korn, E. D. (1969). Cell membranes: structure and synthesis. *Annu. Rev. Biochem.* **38**, 263–288.

Loewenstein, W. R. (1970). Intercellular communication. *Sci. Amer.* **222** (5) 79–86.

Lucy, J. A. (1967). *J. Theor. Biol.* **7**, 360–373.

Northcote, D. H. (1968). Structure and function of membranes. *Brit. Med. Bull.* **24**, 99–184.

Ring, K. (1970). Some aspects of the active transport of amino acids. *Angew. Chem. Int. Ed. Engl.* **9**, 345–356.

Sessa, G. and Weissman, G. (1968). Phospholipid spherules (liposomes) as a model for biological membranes. *J. lipid Res.* **9**, 310–318.

Stoeckenius, W. and Engelman, D. M. (1969). Current models for the structure of biological membranes. *J. Cell Biol.* **42**, 613–646.

Van Deenen, L. L. M. (1969). Membrane lipids and lipophilic proteins. In "The Molecular Basis of Membrane Function" (D. C. Tosteson, ed.) pp. 47–78. Prentice-Hall, New Jersey.

Wallach, D. F. H. (1969). Membrane lipids and the conformation of membrane proteins. *J. Gen. Physiol.* **54** (1) Part 2, 3s–26s.

Whittam, R. and Wheeler, K. P. (1970). Transport across cell membranes. *Annu. Rev. Physiol.* **32**, 21–60.

3. The Cell Surface

G. R. MOORES and T. A. PARTRIDGE

Department of Cell Biology, University of Glasgow, Scotland
Experimental Pathology Unit, Charing Cross Hospital Medical
School, London, England

The cell surface occupies a key position in the economy of the cells and can be considered the mediator between the cell and the environment. It enables the cell to maintain its internal environment, the "constancy" of which is as essential to the life of the cell as it is to the life of the whole organism. An important function of the cell surface lies in the part it plays in cell behaviour particularly, in the case of multicellular organisms, in the integrative aspects of cell behaviour. The surface of the cell is generally considered to be important in cell locomotion, adhesion, cellular recognition and specificity, in the regulation of intercellular communication and in the specialized function of nerves and muscle : i.e. resting and action potentials. Finally it seems possible that changes in the cell surface are responsible for some of the changed properties of malignant cells.

The problem of what constitutes the surface or boundary of a cell arose with the cell concept of living organisms and the definition of the cell surface is still problematic. Permeability studies and electron microscopy have established the existence of a lipoprotein barrier, the cell membrane (= plasma membrane = plasmalemma).

The term cell surface is often used synonymously with cell membrane.

As visualized by electron microscopy the cell membrane has a thickness of about 10 nm. However the apparent depth of the cell surface depends on the method of investigation used. Any definition of the cell surface is really an operational definition, dependent on the function or property under review. In discussing the cell surface, we are in fact looking at the surface regions or periphery of the cell. The cell membrane, as the term is normally used (see Chapter 2), is, without doubt, of prime importance in cellular functions but it does seem that some aspects of cellular behaviour are mediated by surface components not normally included in a description of the cell membrane, though the term can be extended to cover all such components.

However, as indicated, "cell surface" is being used in the sense of a functional entity and refers loosely to any part of the cell periphery (Weiss, 1967) that seems to have a role in cell behaviour. The first part of this chapter considers what is known about the constitution of the surface of animal cells and the second with those aspects of cell behaviour in which the surface is thought to play a part. Unfortunately it is, as yet, not easy to correlate the constitution of the cell surface with its functions.

I. CONSTITUTION

A. THE APPEARANCE OF THE CELL SURFACE

Although we are considering the cell surface as a functional entity, one of the classical aims of biology is to relate structure to function. Thus it is important to know what structure can be seen at the surface of cells.

The major feature revealed by electron microscopy is the plasmalemma already discussed (see Chapter 2). The outermost staining layer is often considered to represent the true exterior of the cell. However a number of studies have revealed material external to the classic triple-layer structure. The cellulose wall of plant cells and the lipopolysaccharide coats of bacteria are well established examples of structural material outside the plasma membrane. However the demonstration in animal cells of "surface coats" of mucoplysaccharides which are an integral part of the surface, with regard to both structure and function, is more recent. *Amoeba* has been shown very clearly to have a layer of "fuzz", bearing an array of filaments, extending some 200 nm beyond the plasma membrane. In mammals the brush borders of intestinal epithelia bear a surface coat, the glycocalyx. This appears as a network of delicately branching filaments covering the microvilli and it is thought that it may protect the cells from attack by digestive enzymes. Similar coats have been seen on other absorptive cells and also on some cultured cells. In some cases such as the erythroblast, the "fuzz" is restricted to small patches on the surface. However even with the electron microscope it is not usually possible to see any structure outside the plasma membrane of cells. But other evidence is available which suggests that non-stainable material is present round many cells. Colloidal particles have been shown to be picked up in a region extending 20–30 nm beyond the stained plasma membrane. Studies of the surface charge of cells (see below), combined with enzymic treatment also suggest the presence of mucopolysaccharides on many cells. As will be seen later chemical and immunological studies of the cell surface indicate that there is what has been termed a forest of glycolipids and glycoproteins at the cell surface. These are now largely accepted as normal components of the plasma membranes though their staining and antigenic properties suggest that the glycoproteins are, in part, outside the hydrophobic interior of the membrane (Pardoe, 1970). They constitute a very much sparser layer than the classic surface coat, seen with the electron microscope, which probably represents the extreme case. Finally the possibility that some surface coats consist of material adsorbed from the medium or are a result of cell injury cannot be wholly excluded.

Until recently our view of the cell surface was based upon thin-sectioning, but new techniques have given us a face view of the cell surface structure. One of the most successful techniques has been freeze cleaving and etching. Using this technique the erythrocyte has been found to have 8–13 nm particles, thought to be of glycoprotein, on its interior and exterior surfaces. These are found in clusters, the greater part of the surface being bare. This technique may reveal more and has already demonstrated the complexity of some intercellular junctions (see below).

There is a tendency to assume that not only the shape of the cell, but the shape of the surface is determined by the packing of cells. However a combination of thin-section and surface scanning micros-copy have revealed that the surface may have definite shape. The most notable example is in the presence of microvilli. These have long been known from the brush border of intestinal epithelial cells where they are slender, cylindrical processes of about 100 nm diameter and usually about 1 μm in length. They are also found on other cells with an absorp-tive function. Microvilli of similar dimensions have been found on cultured animal cells, where they are more numerous in rounded cells. It has been suggested that they form a store of plasma membrane.

Transient structures are often observed at the cell surface, pseudo-podia being the most obvious and surface folds have been observed on the ruffled membrane of cultured fibroblasts. Invaginations which give rise to pinocytic vesicles are found in a variety of cells concerned with macromolecule uptake (Chapter 7).

B. SOME ASPECTS OF THE CHEMICAL ANALYSIS OF THE SURFACE

The major components are the lipids and proteins of the cell membrane (Chapter 2), but other minor components may be important. We have seen that the surface of the cell is covered by a layer of polysaccharide. Chemical studies of isolated cell membranes and of enzymatically cleaved fragments have shown that these are present in the form of glycolipids and glycoproteins, and constitute only a minor component of the membrane (about 3% by weight). The position of their protein moiety in the cell surface is a matter of dispute, but it is generally assumed that the lipid moiety of the glycolipid is located within the hydrophobic region of the membrane. Glycolipids consist of a polysaccharide chain linked to a lipid, usually ceramide, and are found as cerebrosides, sulphatides and gangliosides, the chain being inaccessible to glycosidic enzymes. In glycoproteins the polysaccharide chain is linked usually to serine or threonine and is accessible. The composition of the chain has been found to vary. Many of the sugar chains contain sialic acids often as the chain-terminating group. The most common sialic acid is N-acetylneuraminic acid. Agglutination of cells by the specific animal agglutinins and plant lectins has been used to study the composition of the chains.

It now seems that at least in terms of the polysaccharide chains, the glycolipids and glycoproteins of different cell types differ. Further and perhaps more interestingly, those of transformed and other neoplastic cells apparently differ from those of their normal counterparts and are more like the fetal glycoproteins and glycolipids.

RNA has also been reported as a trace component of the cell surface sometimes in the form of ribosomes. Actomyosin-like protein has been found at the surface of some cells.

C. ENZYMES OF THE CELL SURFACE

A number of enzymes have been found to be associated with the plasma membrane. Most common is a Mg^{2+}-dependent ATPase possibly associated with the actomyosin-like protein. Na^{+}-K^{+}-Mg^{2+} dependent ATPases, 5'-mononucleotidases, alkaline phosphatases, phosphodiesterases and nucleotide–sugar transferases have also been reported. The Na^{+}-K^{+}-dependent ATPase is thought to function in active transport (see below).

The discovery of the enzyme adenyl cyclase on cell membranes is of particular interest with regard to hormone action (see below). Membrane associated enzymes are presumably important in surface functions though as yet little is known of the way they act.

D. SURFACE CHARGE

The cell surface bears a net negative charge. This is a result of ionogenic groups at the cell surface which are ionised at physiological pH. The electrostatic field set up by these charges results in a surface potential, ψ_o. The surface charge attracts counter ions (cations) from the bulk solution and an ionic atmosphere is formed round the cell. This is termed the double layer. Further, cations may be specifically adsorbed at the surface giving a potential, $\psi\delta$, at the plane of adsorption. The surface potential falls exponentially across the double layer which, under physiological conditions, has a thickness of about 1 nm (see Fig. 1). The surface potential $\psi\delta$ decreases with an increase in the ionic strength of the medium, the surface charge being balanced by an increased concentration of counterions in the double layer.

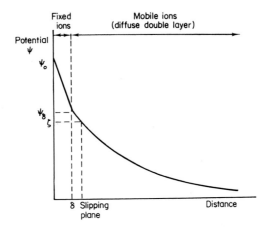

Fig. 1. Schematic representation of the decrease of the potential, ψ, with distance from a charged surface. ψ_o = potential at the surface $\psi\delta$ = potential at boundary of layer of fixed ions, at distance, δ, from surface.
ζ = zeta potential; potential measured by electrophoresis.

The surface potential can be measured by electrophoresis, in which the cell moves under the influence of an applied electrical field and its mobility is measured. In fact the cell carries some counter ions with it and the potential measured is that at the edge of this layer of associated ions (slipping plane), the zeta potential (ζ). A number of assumptions have to be made to get an estimate of the surface potential, including that of assuming the cell to be a sphere. The possible presence of microvilli poses a problem and all measurements involve a fair amount of uncertainty.

The measurements on a large number of cells give values for the surface potential ranging from –8·5 mV to –38 mV, with a modal value of -20 to -25 mV which corresponds to a charge density of about 5 500 e.s.u./cm^2. Measurements of the electrophoretic mobility under different conditions of ionic strength, pH and after enzymic treatment have given a crude picture of the cell surface. This evidence suggests that the charged groups are not located in a single plane at the cell surface but that they are arranged at different levels within the cell surface over a depth of about 1–2 nm. The most important group contributing to the surface charge of cells is the carboxyl groups of sialic acid which are thought to be about 5 nm apart on the erythrocyte surface. Amine groups, phosphate groups, and the carboxyl groups of some amino acids are also thought to contribute to the surface charge. Malignant cells have generally been reported to have a higher negative surface potential than normal cells.

E. CELL SURFACE ANTIGENS

Another line of investigation which has yielded information concerning the range and diversity of molecules at the cell surface is the study of cell surface antigens. It is generally held that antibody molecules, whether humoral or cell-bound, do not penetrate the surfaces of undamaged cells and, therefore, that any manifestation of antibody-antigen interaction initiated on living cells involves anitgens located on or outside the cell membrane. Most of the antigens, thus demonstrated on the surface of cells and which have been chemically characterized, prove to be polysaccharides, proteins, glycoproteins or glycolipids. The common human blood groups antigens are determined by oligosaccharide chains, although the evidence for this in the case of the Rhesus antigens is only tentative. A and B antigens are found on many epithelial cells as well as on erythrocytes and are also produced in a soluble form secreted in mucous secretions e.g. saliva, gastric juice of some individuals; in the cell-bound form the oligosaccharide chains are attached to a lipid backbone and in the secreted form to a polypeptide chain. The specificities are determined by the terminal non-reducing sugars and A and B substances appear to be identical except for this immuno-dominant terminal group which is N-acetyl-D-galactosamine in A substance and D galactose in B substance. Lewis blood group substances which are serum antigens that adsorb onto the surfaces of red blood cells, bear structural resemblances to A and B substances and to H substance (which antigen is found in high concentrations on the red cell surface in group O individuals) and it is suggested that they may all be variations on a common basic synthetic sequence. MN blood group substances are also determined by polysaccharide chains which are attached to a protein backbone and, like A and B substances, closely resemble one another.

A large number of other antigens have been identified on cell surfaces by various means. Perhaps the most important and extensively studied of these are the histocompatibility antigens, which are concerned in the rejection of tissues transplanted between individuals of the same species. The H-2 system on mice comprises a series of histocompatibility antigens whose specificity is genetically determined by a single complex gene locus or a closely linked cluster of loci. They are found on the surfaces of erythrocytes and all nucleated cells so far investigated, and appear to be largely restricted to the surface, for live cells have 80% of the absorptive capacity against anti-H-2 antisera of disrupted cells. More than 30 separate antigenic determinants in various assortments are identifiable in the antigenic phenotypes of inbred strains of laboratory mice, and since most of them seem able to segregate independently of each other there is considerable scope for antigenic uniqueness of mice on the basis of this system alone. In man the HLA antigens, which are also histocompatibility antigens, present a similar picture of variability and distribution as far as they have been studied, except that they are lost from erythroblasts during maturation into erythrocytes.

In addition to the histocompatibility antigens, a number of specific antigens are found associated with tumours, either virally induced where the antigen tends to be common to the particular virus, or chemically induced in which case each tumour appears to display its own unique antigenicity. Evidence has also been found in mice of surface antigenic specificities which may reflect the tissue types of cells : for example antigen of thymus-derived as opposed to bone-marrow-derived lymphocytes, and an antigen which is found in epidermis but not in leucocytes.

Use of fluorescein- or ferritin-conjugated antibodies to localize surface antigens has shown that many of them are not homogeneously distributed on the cells but give a patchy labelling pattern or sometimes a series of stained spots of fairly regular size and spacing which may be characteristic for the particular antigen. This suggests that cells have a degree of mosaic structure in their surfaces, at least with regard to the distribution of their surface antigens.

II. Functions

A. TRANS-MEMBRANE POTENTIAL

It has been found that the interior of many cells is at a potential of about -50 to -100 mV with respect to the extracellular fluid. This is the resting potential. Measurements of squid axon and frog muscle have also shown that there is a disparity in the ionic compositions of the cell interior and the extracellular fluid. In the latter the major ionic constitutents are sodium and chloride, whilst inside the cell sodium is

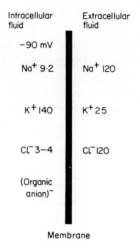

Fig. 2. Electrolyte concentrations (in mM/1.) and potential difference across a cell membrane (frog muscle).

largely replaced by potassium and chloride may be replaced by organic anions. The potassium ion gradient is responsible for the resting potential. This gradient is assumed to result from the properties of the cell membrane. For if we consider a membrane that is permeable to potassium ions but not to sodium ions, with solutions as shown in Fig. 2 on either side, potassium ions will move across the membrane under the concentration gradient. However this movement will result in a potential across the membrane, with more positive ions on one side of the membrane, and further movement of potassium ions is resisted by an electrostatic force. The potential difference caculated* for the cells on the basis of such behaviour is greater than the experimental value and this discrepancy can be explained by the fact that the cell membrane is permeable to sodium ions. But the ionic gradient is maintained by active transport: by the so-called sodium–potassium exchange pump. Allowance for the sodium permeability reduces the calculated potential difference by about 12 mV. Sodium ions are extruded from the cell and it can be shown that the efflux of sodium depends on a parallel influx of potassium. This exchange is often on a molar basis of $3Na^+:2K^+$ but other ratios have been measured. This pump is inhibited by respiratory poisons such as 2, 4-dinitrophenol and depends on a supply of ATP. The Na^+-K^+-Mg^{2+} dependent ATPase found at the surface of many cells has some of the

*From Nernst equation: $E = \frac{RT}{F} \ln \frac{Ko}{Ki}$, where E = potential difference, R = universal gas constant, T = absolute temperature, F the Faraday and Ko and Ki are the potassium concentrations outside and inside respectively.

characteristics of the Na^+/K^+ pump, for example both are inhibited by the cardioactive glycosides such as ouabain. A model has been proposed in which the ATPase is phosphorylated inside the cell in the presence of Mg^{2+} and Na^+ by ATP to give a Na^+-phospho-protein complex. The enzyme complex to the exterior carries the sodium and it is then dephosphorylated in the presence of K^+, releasing Na^+ and giving a K^+-enzyme complex. This complex carries the potassium into the cell where it is rephosphorylated releasing the K^+. The enzyme thus shuttles across the membrane taking K^+ in and Na^+ out (Fig. 3). These pumps are not confined to excitable tissue such as nerves and muscles but have also been found on erythrocytes. Also ouabain-sensitive, Na^+/K^+ dependent ATPases have been found on a number of cells, which suggest that the sodium-potassium exchange pump may be present at the surface of most cells. The specialization of this mechanism with the concomitant resting potential reaches its peak in neurones. The action potential will not be dealt with in this chapter.

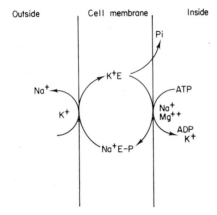

Fig. 3. Postulated model for the transport of sodium and potassium ions across the cell membrane (Pi = phosphate, E = enzyme i.e. ATPase, E-P = Phosphorylated Enzyme).

B. CELL ADHESION

One of the most important properties of cells is that they will stick to one another and to a wide range of substrates. This is an obvious prerequisite of the multicellular existence. It is also necessary for any cell that moves on a solid substrate to be able to form adhesions, albeit transitory, with the substrate. Adhesion is to some extent, a surface activity though the nature of the mechanism is very much a topic of current controversy.

Electron microscope studies on the adhesions of cells have led Curtis to classify cell adhesions into two broad groups: (i) those with a gap of

E*

10–20 nm between plasmalemmae and (ii) those with a gap of 2 nm or less. Although large intercellular spaces, often filled with extracellular material such as collagen, are found, many tissue cells abut onto adjoining cells by one of these two classes of adhesion.

The first type of adhesion is the one most frequently seen and is usually the major adhesive structure between cells. This is a relatively weak adhesion in that the cells can be fairly easily separated. The 10–20 nm gap between cell boundaries often extends over large areas of the cells and shows a quite uniform spacing. It is sometimes termed the zonula or fascia adhaerens depending on the extent of the adhesive structure (zonula = belt, fascia = band). It has long been controversial whether this gap has a real existence and, assuming that it has, whether it is filled with colloidal material or mainly with extracellular fluid with no structural, macromolecular components. The existence of the gap is now fairly firmly established; not so the gap material. The problem is tied up with that of the surface coat. Conventional E.M. stains show up no material in the gap, although the use of Lanthanum stains has indicated the presence of high molecular weight material between some cells. A real problem lies in showing that any intercellular material exists in the living organism and has not leaked out during the preparation of specimens. A significant piece of evidence for the idea that the gap contains no appreciable, structural material is Brightman and Palade's demonstration that haemoglobin or ferritin molecules will permeate the gap within 12 min. when specimens are perfused before fixation, though not if perfused after fixation. It would thus seem that the gap material is of low viscosity.

Specialized structures are found often, associated with the regions of cell adhesion described above. One such is the desmosome or macula adhaerens = adhering spot) (Fig. 4). In sections as seen by electron microscopy this appears as two dense symmetrical plaques about 10 nm thick on the inner layers of the apposing cell surfaces. There is a gap of 10–20 nm between the outer leaves of the plasmalemmae and this gap is often bisected by a thin electron-dense line. Bundles of tonofibrils usually converge on the dense plaques within the cytoplasm. In general they only cover small areas of contact and reconstructions have indicated oval patches of 10 nm by 250 nm (hence the term macula).

The second type of adhesion defined by Curtis appears to be stronger, with the cells being difficult to separate. Two classes of structure have been seen for this type of adhesion. In one, no gap is visible between the cells and the outer layers of the two plasmalemmae seem to be fused. This is termed the "tight junction" (zonula or fascia occludens). It is seen as one of the elements of the junctional complex of epithelial cells, the zonula occludens and, between other cells in particular muscle

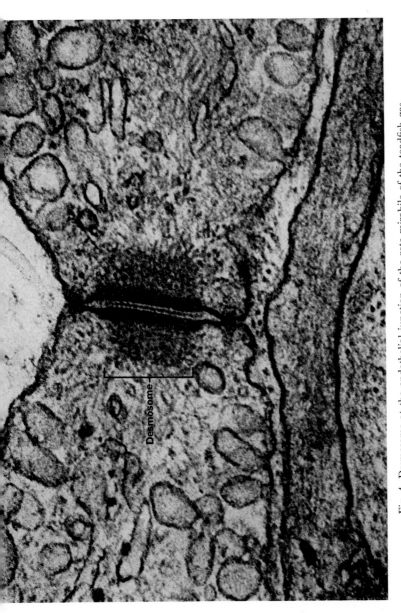

Fig. 4. Desmosome at the endothelial junction of the rete mirabile of the toadfish gas-bladder. Collidine buffered gluraraldehyde and osmium fixation. Uranyl acetate and lead citrate staining. Magnification 95 000 ×. Reproduced with permission from D. W. Fawcett "The Cell", W. B. Saunders Co.

cells, the fascia occludens (Fig. 5). In the second class of structure Lanthanum staining has shown a gap of about 2 nm between the outer leaves of the plasmalemmae. The terms "close junction" or "gap junction" have been used for this type of structure to distinguish it from the true tight junction. An example of a gap junction is the nexus of muscle which has recently been shown to have a complex, hexagonal sub-unit structure.

All the structures described above are usually presumed to have an adhesive function. But we have perhaps been too ready to assume that any surface structure at a junction between cells must be concerned with adhesion. It is known that desmosomes are continually formed and reformed during the migration of cells of the epidermal basal layer. Further the work on communication between cells has indicated that gap junctions are the sites of electrical coupling (see paragraph on Communication). This obviously does not vitiate any adhesive function they may have, and there is evidence that cells stick more strongly at desmosomes. But their role in cell adhesion needs to be investigated more thoroughly.

The experimental evidence on cell adhesion is rather unsatisfactory. Too few truly quantitative measurements have been made. It is not always clear what is actually being measured in any experiment. In particular two types of technique have been used to study adhesion which may not give comparable answers. The first measures the force required to separate cells. This has been severely criticized as it must be shown that the force required to separate the cells is equal to that which holds the cells together and not, for instance, the force required to rupture the cell surface. Also the maximum force measured will depend on whether the cells separate all at once or peel apart (the total work done will be the same however the cells are separated).

The technique more frequently used is concerned with the reaggregation of dissociated cells, usually embryonic cells or sponge cells. The cells are suspended in a suitable medium and shaken so that they collide and adhere. The kinetics of this re-aggregation and the size of the aggregates formed have been used to assess adhesion. By careful control of the conditions it has been possible to use the kinetic method to measure the force or energy* or adhesion. However such measurements have been criticized on the basis that they only measure the initial adhesions when the cells collide and provide no information on the maintained adhesions between cells. It is not known, though, whether there is any difference. But obviously, more quantitative data are needed on adhesion and some agreement is needed on an operational definition of

---

*Force and energy are related by the equation $E = \int F ds$, where E = Energy, F = Force and s is distance.

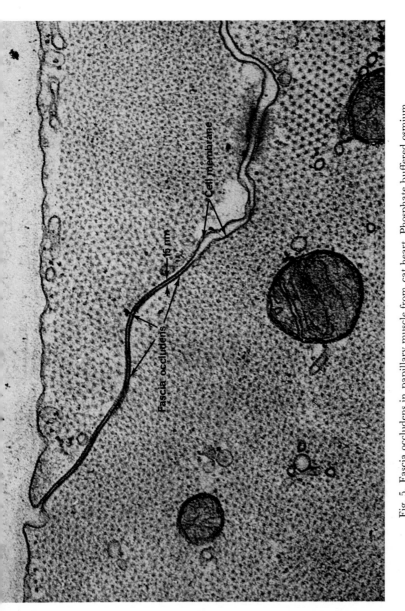

Fig. 5. Fascia occludens in papillary muscle from cat heart. Phosphate buffered osmium fixation. Lead citrate staining. Magnification 62 000×. Reproduced with permission from D. W. Fawcett "The Cell" W. B. Saunders Co. Note: There is a distance of 15 nm between the *inner* aspect of the plasmalemmae of adjacent cells in this area.

intercellular adhesion for the purposes of measurement. The one advanced by Curtis seems suitable: that a measurement of intercellular adhesion is the probability that two cells will form and maintain an adhesion when brought together.

With these caveats in mind, it is possible to summarize certain facts about cell adhesions which emerge from a large number of experiments:

(1)   Cells can be separated mechanically, for example, by mechanical shearing with a pipette.

(2)   Such separations are rendered easier and cleaner by pretreatment of tissues with a wide range of proteolytic enzymes such as trypsin, pronase, pancreatin, papain and collagenase.

(3)   Pre-treatment with the calcium chelating agent ethylene diamine tetra-acetic acid (EDTA) is also effective, in increasing the ease of cell separation. High pH ($> 9 \cdot 5$) has also been used.

(4)   Cells so dissociated will re-adhere if they are brought into contact again.

(5)   Divalent cations, especially calcium, are required for the re-adhesion of some cells, in particular sponge cells and amphibion embryonic cells.

(6)   Under certain conditions some cells require metabolism to re-establish an adhesion: low temperatures and inhibitors of protein synthesis prevent their reaggregation.

(7)   Cells will adhere to a variety of non-living substrates: glass, plastics of various types.

A number of theories have been advanced to account for cell adhesion, using this rather thin slim body of evidence. These are considered below.

Theories of cell adhesion

(a) *Adhesion by molecular groupings.*   It has been proposed that cell adhesion results from either chemical bonds between the cell surfaces or from other molecular interactions between groups on the surfaces. Such groups as carboxyl, amine and phosphoryl ions could be implicated. One of the earliest theories along these lines was proposed by A. Tyler and P. Weiss. They suggested that cell adhesion is similar to an antigen–antibody reaction, whereby molecules on one cell surface possess small areas which are complementary to small chemical groupings in molecules on the apposing cell surface. This leads to a very close approach and binding of the two complementary molecular groupings. M. Spiegel, in fact, was able to prepare antisera that inhibited the reaggregation of cells of the species against which the antisera were prepared but not cells of other species. Strictly, however, this only demonstrates the presence of species-specific antigenic sites, which may not be directly involved in cell adhesion. The antisera may merely mask nearby adhesive sites. S. Roseman has recently shown that the glyco-proteins and glyco-lipids are important for the reaggregation of mouse teratoma cells. He proposes that adhesion is brought about by sugar chains on one cell being bound by the enzyme glycosyl transferase on the other cell.

B. M. Jones has proposed that linking-groups on the cell surface may be under the control of actomyosin-like structures at the cell surface, the adhesion being controlled by the movement of these groups and mediated by the presence or absence of ATP.

(b) *Adhesion by bridging: Cementing.* It has often been proposed that cells are held together by intercellular molecules that bridge the two surfaces. One of the earliest theories postulated that there is an intercellular "cement" binding cells together. A. A. Moscona and T. Humphreys have argued from the action of metabolic inhibitors (see above) that the synthesis of cementing material is required for the re-adhesion of dissociated cells. The disaggregating effect on enzymes is considered to be due to their action on this cement. They have suc-ceeded in isolating a macro-molecular material from suspensions of sponge cells and of chick neural retina cells which promotes the adhesion of those cells. The material seems to be mucoprotein. This theory can account for both types of cell adhesion, as presumably the cement can be laid down in varying thicknesses. However, as detailed above, the true nature of the gap material is still not clear.

A much-favoured bridging theory is that which proposes that calcium ions bridge between apposing cell surfaces. This is based on the finding that sponge cells and amphibian embryonic cells require calcium for their adhesion. The theory proposes that calcium or other divalent cations bridge, probably between carboxyl groups such as that on sialic acid. In the case of the 10–20 nm gap it is proposed that the calcium binds to an intercellular cement. However evidence from the chemistry of metal complexes suggests that for strong binding of calcium, the co-ordination of symmetrically arranged groups, as in a crystal, about the calcium ion would be a more realistic model. A theory has been advan-ced that adhesion is brought about by the co-ordination of cell surface groups about the potassium ions. However it could be argued that the rather specific arrangement of groups that this would require, does not accord with the rather more dynamic picture of the cell surface now emerging.

(c) *Physical forces.* Curtis has proposed that cell adhesion is com-parable to the stability of lyophobic colloidal solutions. The final adhesion of cells is viewed as a result of a balance between opposing forces of attraction and repulsion. The repulsive force is a consequence of the surface charge of cells (above). The charged surfaces of like sign repel each other and thus there is an electrostatic force of repulsion between cells. The attractive force between cells is less familiar. It is the van der Waals–London force (best known from the gas laws), which is a universal force determined by the nature and size of adhering particles. An approximate picture of this force can be gained by imagining a molecule to be frozen instantaneously. The molecule would be found to have dipoles. These dipoles normally fluctuate rapidly and cancel out when averaged over time. But the oscillating dipoles of one molecule polarize those on other molecules and the molecules thus attract one another. This force of attraction is approximately additive and large particles of similar constitution can experience quite a large attractive force. Electrostatic and van der Waals–London forces both decline rapidly with distances, though not in the same way. Electrostatic forces decline exponentially with distance, whereas the van der Waals–London force falls off with the square of the distance. Fig. 6 shows these parameters plotted as the energies of interaction of two charged spheres against the distance between them. Adhesion of the particles will occur whenever the sum of the energies is such that there is an attractive energy minimum (zero force) great enough to resist the disrupting energy of thermal motion. Values in excess of 100 kT* are usually considered adequate. It can be seen from Fig. 6 that adhesion is possible at two separations : (i) at c. 0·5 nm where the van der Waals–London forces greatly predominate (primary minimum) and (ii) at about 10–20 nm where a smaller energy minimum exists (secondary minimum) due to the different relationships of the two parameters with distance. Between these two the electrostatic force of repulsion predominates and constitutes a potential energy barrier for the approach of particles to the primary minimum. The problem in applying such theoretical calculations to cells is that too little is known about the actual magnitude of the van der Waals–London force in this case. However preliminary experiments designed to measure this force have given a not unreasonable value and calculations based on these values have shown that cells should adhere either at a separation of about 0·5 nm or one of 10–20 nm, with an energy in excess of 100 kT, the primary and secondary minima. These obviously correspond to the two types of adhesion discussed. The distance of separation and the energy of adhesion in the secondary minimum depends on the ionic strength and ionic type of the medium, as we have seen that the surface potential and hence the electrostatic force of repulsion depend on these. Increased ionic strength or divalent cations reduce the latter and hence decreases the

---

*kT is a measure of thermal energy, where k is the Boltzmann constant and T is the absolute temperature. $kT = 4 \times 10^{-21}$ joules at 20°C.

separation and increase the non-adhesive energy at the secondary minimum. The role of calcium in this theory is, thus, to reduce the surface charge of the cells. The final separation and strength of adhesion may depend on the surface charge (potential) of the cells.

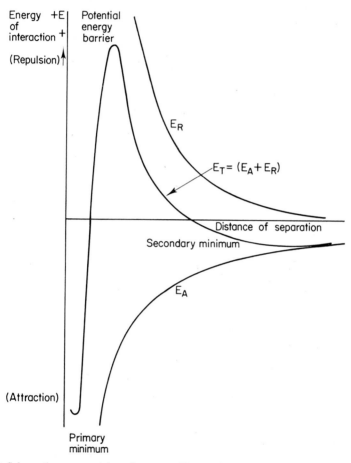

Fig. 6. Schematic representation of energy of interaction of two charged spheres. E_A is attractive energy, E_R is repulsive energy and E_T is total energy of interaction $= E_A + E_T$.

Evaluation of the theories

How far do these theories fit the few facts we know about cell adhesion and our present picture of the cell surface? There is insufficient evidence to distinguish between any of these theories and each is based on facts that are far from conclusive.

The action of proteolytic enzymes may well be on the glyco-proteins of the cell surface, the consequent damage or even change of surface charge being responsible for the separation, rather than the cleavage of intercellular material. Also metabolic inhibitors have only been observed to prevent cell-re-adhesion when proteolytic enzymes have been used and may thus act by preventing repair of the cell surface. This is something of a semantic argument as it is difficult at this level to distinguish between an intercellular molecule and a cell surface molecule, and there may, in fact, be no difference in function. Furthermore, the actual adhesion may depend on the continued metabolism of the cell. The role of calcium does not seem as universal as has been accepted and, as we have seen, it may act other than by bridging. L. Weiss has proposed that by cross-linking groups within the surface, calcium stabilizes the cell surface. Curtis's physical theory can account for most of the facts we know about cell adhesion and has no special requirements of the cell surface but suffers from a lack of adequate evidence on the magnitude of physical forces in living systems.

Of these theories, not all can account for the types of adhesion observed by electron microscopy. This is not necessarily a criticism; there is no reason to believe that only one mechanism is acting. Also it is not known whether the cell to non-living substrate adhesion is the same as the cell to cell adhesion : it may involve the cell laying down a matrix, to which it adheres, on the substrate. The final conclusion must be that we know far too little about cell adhesion to build up a convincing story about cell adhesion and its relation to the cell surface.

Specific cell adhesion

So far we have discussed the general adhesion of cells as a non-specific phenomenon. The question of specificity in cell adhesion : i.e. that cells on one type stick preferentially to cells of the same type and not of another type, arises from studies on the "sorting out" of embryonic cells. Townes and Holtfreter mixed disaggregated cells from different amphibian germ layers and showed that cells sorted out with ectodermal cells at the outside, endodermal cells inside and mesodermal cells in separate sectors in the interior of the aggregate. Many studies on this phenomenon have revealed a number of basic facts. (i) The initial aggregation of cells seems to be non-specific and randomly mixed aggregates are formed; (ii) After about 4 hr the cells change their position in the aggregates and "sort out"; (iii) Cells maintain their identity and there is no transformation of cell types; (iv) Cells are positioned within aggregates according to tissue type, giving a "hierarchy" in which each tissue segregates internally to all those below it and externally to all those above it. Steinberg has given one such hierarchy: germinative epidermal layer > limb bud precartilage > pigmented epithelium of eye > myocardium of heart ventricle >

neural tube > liver, where any member of the series tends to surround any member preceding it; (v) The capacity of the cells for adhesion and movement changes with time.

Specific cell adhesion has been proposed as an explanation of this phenomenon, by Humphreys and Moscona. They propose that cells synthesize specific intercellular cements and have isolated macromolecular components from sponge cells and chick neural retinal cells which enhance the aggregation of the like cells but not of other cells. Roth and Weston (1967) have also shown that cell aggregates will pick up from cell suspensions a far greater proportion of like cells than of unlike cells. This all argues for specific cell adhesion, although obviously as unlike cells do stick to one another adhesion is not totally specific. Roseman's (1970) model for cell adhesion could easily accommodate specificity with different acceptor groups and enzymes on different cell types. Cell adhesions with close contact (Type 1) are often considered to be specific.

However "sorting out" can be explained without recourse to specific cell adhesion, and specific cell adhesion, of itself, does not explain the positioning of cells as exhibited by hierarchical ordering of tissues. Steinberg (1970) has shown that quantitative differences in the adhesive energy of different cell types will result in both sorting out and positioning. He has postulated that the population of cells will tend to assume the state with the minimum free energy. This will be that in which the adhesions are maximized by the cells exchanging weaker for stronger adhesions. This "differential adhesion" theory assumes further that the cells move randomly during sorting out. Thus if two cell types, A and B, are mixed, A cells being more cohesive than B cells, and A to B adhesions being intermediate in strength, A cells will tend to exchange A–B adhesions for A–A cohesions forming an internal mass of A cells and squeezing out B cells to the periphery. Curtis has proposed that sorting out and positioning may result from temporal changes in the adhesive and motile properties of cells during the recovery from the dissociation procedure. Cells are assumed to recover at different rates and to cease migrating on recovery.

Once again we must conclude that there is too little unequivocal evidence on this subject. Specific cell adhesion does occur but how important it is for morphogenesis in the embryo is unknown. Perhaps too much attention has been paid to adhesion as the sole explanation of "sorting out". There is insufficient knowledge of the role of cell movement. In aggregates cells have been shown to move over only short distances, though cases are known where embryonic cells cover quite large distances during morphogenesis (Chapter 22). The sperm egg adhesion of some invertebrates is a case of specific adhesion which

seems to be mediated by chemical binding. In some marine invertebrates the egg produces a specific mucopolysaccharide, "fertilizin"
which has been shown to react with a sperm protein "anti-fertilizin".
This mechanism presumably allows the sperm to "recognize" the egg.
There is no evidence for such a mechanism in vertebrates. Colwin and
Colwin (1964) have shown that this adhesion culminates in cell fusion
by the fusion of the plasma membranes of the egg and sperm.

C. CELL LOCOMOTION

Cells undertaking active locomotion must transmit the forces generated within them to their environment, via the cell surface.

Those cells that swim in a liquid medium, by means of cilia or
flagella appear to propel themselves according to fairly simple principles of hydrodynamics, and the surface is important only as an interface of suitable plastic and elastic qualities to accommodate to the
undulations developed by the underlying motile elements and to maintain the form of the organelles against the resulting shear forces.

On the other hand, in the case of locomotion on solid substrates
performed by many cells, there is no general agreement as to the
mechanisms involved or, specifically, as to the role played by the cell
surface. Three basically different relationships between the cell, the
cell surface and the substrate have been proposed on the assumption
that the locomotive force is generated within the cell and transmitted
directly to the substrate by an adhesive cell surface.

(a) Fig. 7a. The cell surface behaves like tank tracks moving forward on the upper aspect of the cell, rolling over the front edge and
moving backwards on the lower aspect where it adheres to the substrate.

(b) Fig. 7b. New surface is constantly added at the front and
resorbed at the rear, with the result that the cell contents move through
a freshly laid-down tube. According to this model, the surface on all
aspects of the cell remains stationary relative to the substrate and moves
backwards relative to the cell interior.

(c) Fig. 7c. Portions of the cell extend, form attachments and
contract while posterior attachments break : a mode of locomotion resembling that of looper caterpillars. This entails no specific relationship
between the movement of the cell surface and that of the cell except
at the points of attachment to the substrate, but, in the absence of
mechanisms (a) and (b), one would expect that the rest of the surface
would either make no net movement relative to the cell interior or
move randomly.

It should be possible to distinguish experimentally between these
three alternatives according to the fates of marked areas of the surface.

Hypothesis (b) also predicts complete turnover of organized surface with every cell length of advancement, which should be reflected in an equivalent turnover of some types of marker. However, the interpretation of observations of marking experiments is complicated by two factors. Firstly, all methods used almost certainly mark the surface coat only and this may behave differently from the underlying cell membrane. Secondly, it is probable that some, perhaps all, marking techniques affect the behaviour of the surface to which they are attached, for example stabilizing it, increasing its rate of ingestion by pinocytosis or altering its structure in some way. These considerations may help to explain the discrepancies between the results of investigators using different techniques.

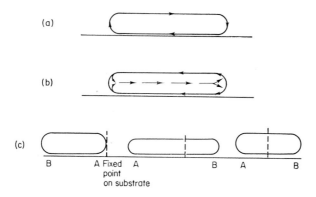

Fig. 7. Relationships between the cell surface, its interior and the substrate (a) Tank-like movement (b) Assembly of new surface at the front: Resorption at the rear (c) Alternate extension and contraction of the cell.

The most extensively investigated type of cell with regard to locomotion is the free living amoeba, in which two of the three above possibilities are supported by evidence from different research groups. Marking experiments, using various sorts of particle, mainly favour the view that new cell surface is assembled quite rapidly at the front ends of pseudopodia, for particles on the dorsal and lateral aspects of these structures are observed to increase their distance from the tips. Some, but not all, of these observations are completely consistent with mechanism (b): the particles are seen to remain stationary relative to the substrate and to accumulate or drop off at the rear end of the animal. The deviations from this behaviour may be explained if it is assumed that mechanism (a) may operate briefly and locally during the formation of new pseudopodia and if the areas of surface not in contact with substrate sometimes move backwards at a slightly different rate from those involved in adhesion to the substrate. Some dyes are found to accumulate at the rear of the cell and to disappear from the majority of the surface in a time consistent with mechanism (b). In contrast, fluorescein-conjugated antibody directed against the surface coat, which is detectable under u.v. excitation with much higher resolution than most dyes and, attached firmly to the surface of amoeba on brief exposure, shows no tendency to move back from the front edge or to accumulate at the rear of actively locomoting amoebae (Wolpert *et al.*, 1964). Furthermore, the even density of staining is maintained for several hours in the absence of any antibody in the medium, and is only eliminated from the surface by pinocytosis with a half time of about 5 hr. Since the amoebae traverse their own length in a minute or two, this is clearly inconsistent with mechanism (b). Fluorescent antibody has also been used to mark the tail regions only and in this case the stain tends to remain at the rear but small areas break off, drift forward and diffuse in a manner that suggests that the surface has the properties of a liquid. In mechanisms (a) and (b) the force for locomotion is transmitted laterally through the surface, and this force could not be transmitted efficiently through a liquid surface, so this evidence is considered to favour mechanism (c), where the locomotor force is transmitted downwards through the surface to the substrate immediately underneath. It is possible to reconcile this with the backward movement of particles by the assumption that they are treated by the cell as part of the substrate on which it is crawling, but the accumulation of some dyes at the rear end is not explicable in this way and the question of the role of the cell surface in locomotion must remain open pending further study.

Few other cells have been studied with regard to this respect of locomotion and, in most cases, even less is known than in amoeba. Individual cells of cellular slime moulds show considerable resemblances to amoebae and observations of the movement of surface irregularities on these cells suggest a type (b) mechanism. The cells actively move within the solid aggregated slugs formed by this group of organisms and, since such cells are in contact on all sides with other cells, a type (b) mechanism offers the most satisfactory explanation of this movement. On the other hand, the mesenchymal cells of the sea urchin illustrate what is probably an extreme case of a type (c) mechanism. Long thin pseudopods are put out and, when they make an attachment, they contract and draw the cell body up to this point of attachment.

A number of tissue cells of various animals have been studied in tissue culture, and subjected to surface marking techniques. Vertebrate fibroblasts crawl on a solid substrate preceded by a flattened extension termed the leading lamella or, because it often exhibits moving ruffle-like thickenings, the ruffled membrane. Particles picked up at the front edge of this lamella are rapidly transported backwards to the region above the nucleus where they accumulate. The speed of this backward transport greatly exceeds the speed of forward movement of the cells, a state of affairs which does not correspond to any of the three proposed mechanisms. However, it has been suggested that a mechanism basically similar to (b) may operate in these cells. Most of the contact between the cells and the substrate occurs in the region of the leading lamella and at the rear end of the cell, most of the middle region being separated from the substrate by a gap of a few hundred nm. It is proposed that new membrane is added to the cell surface from a source at the front edge of the leading lamella and is resorbed in a sink zone somewhere between the source and the nuclear region. The flow between the source and sink on the lower surface results in the forward movement of the cell as in mechanism (b) but forward movement is resisted by the attachments at the rear, so that the cell is held almost stationary relative to the rate of production of surface at the source. Thus the flow of surface on the underside of the cell is impeded by the adhesion of this surface to the substrate and most of the new surface assembled at the source is forced to move over the upper aspect of the lamella, which would account for the particle movement and, perhaps, for the formation of the thickened ridges which constitute ruffles. The outgrowing process of nerve cells is headed by a growth cone which is a variation on the frontal lamella and the amoebocyte from the limpet has a frontal lamella supported by stiff spikes. Both these cells exhibit rapid backward transport of attached particles over the upper surfaces of the lamellae, and probably share the same basic locomotor mechanism with fibroblasts.

As with amoeba, type (c) mechanisms have been proposed for cells with frontal lamellae involving, in this case, the filopodia which can be detected on many cells by electron microscopy, but no direct evidence of such a mechanism in these cells is available and, as with amoeba, further information is required before any decision can be made between these hypotheses.

It has been suggested that the force required for movement may be developed within the surface itself, by surface expansion, or by virtue of the surface tension properties by which the cell surface can compete with the medium for the solid substrate. In amoeba, however, the participation of the internal cell constituents in locomotion has long been recognized, and a number of theories has been proposed concerning the exact site at which contractile proteins expand, relax, gel and solvate. Contractile proteins, resembling the acto-myosin extracts of muscle, have been prepared from several types of cell including amoeba and fibroblasts and, recently, actin-like fibrils have been identified by electron microscopy, particularly in sub-surface positions. In fibroblasts these fibrils are found in a number of compact tracts, situated just inside the membrane, and oriented in the direction of movement, which may be responsible for the tractive property of the surface in these and other cells.

During locomotion on a solid substrate, cells change shape, and therefore area, a great deal. It has been calculated that spread fibroblasts, as seen by light microscopy, may have 4–5 times the area of the same cell in suspension, and it has been suggested that the very numerous microvilli which can be seen on the unspread cells by electron microscopy, are a major store of membrane used for spreading and extension.

D. SURFACE TURNOVER AND BIOSYNTHESIS

The absorbtive activities of pinocytosis and phagocytosis both involve the invagination of cell surface. An actively feeding amoeba engulfs its own surface area in a couple of hours and must renew it at the same rate. In addition, a fast turnover of cell surface is entailed by some mechanisms proposed as models of cell locomotion, calculated as one complete turnover every few minutes in the actively motile amoeba and a complete turnover of the flattened lamella region, comprising about 20% of the area of a fibroblast, every 10 minutes.

The evidence obtained by fluorescent antibody labelling of the surface of amoeba, discussed in the section on locomotion, suggests a much slower rate of surface turnover. Similarly, radioactive labelling of L strain fibroblasts indicates a much slower rate of turnover of the molecular constituents of the surface than is required by the locomotory mechanism. These cells synthesize enough new surface material every 24 hr to double their number, and in actively growing cultures, where the cells mitose approximately once every 24 hr, there is very little turnover: labelled material being lost from the total surface fraction of these cultures with a half time of 7–9 days. In non-growing cultures, on the other hand, the excess production of cell surface is accommodated by release of surface components into the medium. Loss of labelled surface material occurs with a half time of approximately one day, which indicates that turnover of this component exhibits no discrimination as to age. These comparatively slow rates of turnover can be reconciled with a type (b) mechanism of locomotion, only if there is a cycle of efficient re-utilization of membrane materials, probably with little or no degradation of the molecules, and perhaps involving sub-units, or organelles such as vacoules, as carriers between source and sink. There are some observations which suggest that such a cycle of surface turnover might exist. Mouse lymphoma cells *in vitro* show a considerable fluctuation in the concentration of histocompatibility antigens, with a cycle time of about 2 hr, indicating that a significant proportion of the cell surface is absorbed and replaced or, in some way, restructured at a fairly high rate. Marcus (1965) monitored the appearance of new antigen on HeLa cells, following viral infection, by the appearance of virally induced red-blood-cell agglutination sites. These first appear at one point on the periphery of the spread cell, opposite the Golgi region, and spread from there to cover the whole of both surfaces in hours. This is broadly in accord with mechanism (b) of locomotion as modified for fibroblasts, except that the antigen is found to spread over both surfaces at the same rate, whereas the hypothesis of locomotion would predict that the dorsal surface would be covered more quickly.

This system also provides some evidence on the site of assembly of surface material, for fluorescent antibody against the viral antigen reveals that it is present in the Golgi vacuoles prior to the first appearance of haemagglutination sites on the surface. In the course of their function of transporting proteins to the outer surface it is likely that the Golgi vacuole membranes normally fuse with the cell membrane, and it is tempting to think that the secretion of proteins may be a modification of a more primitive function—the production of glycoprotein-coated cell surface. However, there is no direct evidence that the new surface is produced in this way.

E. SPECIFICITY AND RECOGNITION

There are many instances of cells producing a specific response to a specific stimulus in which the cell surface is thought to be directly involved. Specific information can only be carried in molecules which are too large to permeate the cell surface membrane readily, so one suspects that at least the receptor part of the mechanisms resides on the surface, and when the stimulus involves cell–surface–bound molecules the initial interaction is plainly limited to the two cell surfaces.

A number of hormones have been shown to act primarily at the surface of the target cells (Robinson, *et al.*, 1968) to stimulate the membrane-attached enzyme adenyl cyclase to convert ATP to cyclic AMP. This, in turn, initiates the intracellular chain or cascade of enzyme reactions appropriate to that cell-hormone system. Adrenalin, insulin, glucagon, at least some pituitry hormones, and perhaps synaptic transmitter systems work on this principle in which the specificity resides in the hormone, the surface receptor and the intracellular enzymes.

Recognition on contact has long been known in the case of fusion of gametes in lower forms. A similar situation has recently been des-

F*

cribed in yeasts in which only cells of particular mating strains will fuse with one another. Extracts from one strain specifically agglutinate the complementary strain, and an extract of this second strain in turn inhibits the agglutinative activity of the first extract. Both extracts are fairly low molecular weight mucoproteins.

Molecular complementarity, of the type that accounts for the specificity of antibodies to antigens, is obviously an attractive model to apply to specific adhesions between cells apart from gametes and, although it is not widely favoured, it has not yet suffered formal disproof. Most experiments on specificity of adhesion between tissue cells consist of variations on the general technique of dissociating the cells from the tissues and studying the rate and manner of reaggregation and sorting out according to tissues or species-of-origin from mixtures of different cells (see para. on Specific cell adhesion, p. 92). Using this technique, it has been shown that extracts of living cells of one species of sponge will enhance the re-aggregation of that species but not of another species, and it has been proposed that this substance, a mucopolysaccharide, is a species-specific intercellular cement. Similar cements are postulated to explain the segregation of cells according to tissue type but there is no evidence in any of these cases that these substances participate physically in adhesion. Furthermore, it is possible to explain much of this specificity of association without postulating specificity of adhesion (see section on cell adhesion). However it is possible that some sort of surface recognition between cells, using the conformational specificity of the polysaccharides and proteins composing the outermost layers of the surface, plays a part in the attainment and maintenance of particular patterns of association between cells, although it may not be by regulation of adhesiveness.

In the cellular slime mould *Dictyostelium discoideum* there is evidence that aggregation of the amoebae into slugs, which precedes the formation of fruiting bodies, is dependent on the presence of particular surface antigens. Univalent fragments of antibody produced against the amoebae prevent normal contact formation and aggregation, although they do not inhibit their locomotion or chemotactic behaviour. Some mutant strains of this organism do not form aggregates and these apparently lack the antigen for they do not absorb the active antibody fragments. This absorption technique also reveals that the antigen is present in higher concentrations on cells in the aggregation phase than on cells in the feeding phase, and that is probably a protein-carbohydrate complex.

The attachment of some viruses to animal cells has been shown to be quite specific with regard to species, which demonstrates a capacity of the cell surface to be recognized by other surfaces with a high degree of discrimination. It is not known, however, what relationship, if any, this bears to adhesion or recognition between cells. In the case of a bacterial system, *Salmonella* and T4 phage, it is known that viral attachment depends on the presence of O antigen, a polysaccharide of many repeating triose units, each consisting of mannose, galactose and rhamnose. Different strains of phage attach specifically to particular structural variations of O antigen on the surface of the lipoprotein coat, determined by different types of linkage between the sugar units of the chain and the presence or absence of glucosyl or acetyl side groups. In the case of the infecting viral genome becoming incorporated into the bacterial genome, in the process of lysogeny, the structure of the O antigen is altered, under the direction of the viral genetic material, such that other virus particles of the same strain cannot attach and infect.

Some oncogenic viruses alter the surfaces of infected animal cells, and it has been suggested that this results in the breakdown of two important contact dependent characteristics of normal cell populations and thus leads to the assumption of malignant behaviour by these cells. These two characteristics are : contact inhibition of movement, which causes cells to cease locomotion in the direction of contact with other cells, and contact inhibition of mitosis, which prevents growth and mitosis of cells having more than a certain degree of contact with their neighbours. Both of these control mechanisms imply a degree of recognition by normal cells, at least to the extent of distinction between other cells and the inert substrate. Between them they could account for the high level of regulation of growth and movement exhibited by normal cells, as opposed to the uncontrolled growth and invasiveness which characterizes malignant cells, in which they appear to be lacking or greatly reduced in effectiveness. The location of the recognition system for these activities at the cell surface is based largely on circumstantial evidence and is suggested initially by the demonstration in both cases that they are dependent on contact between the cells rather than merely by close proximity. In addition, in the case of contact inhibition of mitosis, there is some evidence linking the behavioural change and the cell surface change resulting from the malignant transformations of cells by Polyoma virus. The surface change consists of the appearance of a site which permits the agglutination of the cells by the plant agglutinin, Concanavalin A, and, since the same agglutination site is exposed on normal cells by trypsin digestion, it is thought to be a subsurface molecular entity which is exposed on virally transformed cells as a result of incomplete synthesis or assembly of the surface structure. Recently it has been shown that contact inhibition of mitosis can be restored to polyoma-virus-transformed cells by treatment with active univalent fragments of Concanavalin A, which cover the exposed agglutination sites. It has been suggested that virally transformed cells lose the ability to recognize one another in the way required for inhibition of mitosis because the polysaccharide structure, which normally covers the agglutination site, is implicated in the recognition mechanism and is missing in the transformed cells. Further, it is suggested that it can be functionally replaced by fragments of the plant polysaccharide, Concanavalin A.

A number of reticulo–endothelial cells have been found to bear on their surfaces immunoglobulin molecules which have particular antigenic specificities. In some cases these molecules are adsorbed onto the cell surfaces from the surrounding medium, but in others it is believed that they are synthesized by the cells bearing them and that the majority of molecules on a particular cell are directed against one antigen. There is some specific evidence that immunoglobin molecules are also carried on lymphocytes which have not been subjected to the antigenic stimulus for which they are specific, and these may act as the receptors for the initiation of the immune response. Although this seems to be a highly specialized function, it is considered by many biologists that the vertebrate immunological system is only a particular development of the requirement of all animal cells to distinguish between self, and the non-self prey, parasite, and predator. Like the other recognition systems discussed, it is probably basically a cell surface phenomenon, and it is notable that among the most potent antigens are various cell surface polysaccharides.

F. COMMUNICATION

In general, stress has been laid on the function of the cell surface as a discriminative barrier through which the cell conducts all its exchanges with the outside, but this view has been modified by the demonstration of intercellular communication (Loewenstein, 1970). Using the conveniently large epithelial cells of the salivary gland of the fruit fly larva, it was found that the resistance to the passage of electric current between two adjacent cells was several orders of magnitude lower than that between the inside of the cells and the bathing medium. This is interpreted as indicating a comparatively free pathway for ionic flow between the cells. Micro-injection of various coloured tracers into one of the cells has revealed the passage of substances of molecular weights of up to 10 000 between neighbouring cells, as if the intervening intercellular membranes constituted no barrier to diffusion. Subsequent studies have shown this to be a widespread phenomenon in normal cells, including those of embryos after the early cleavage stages, but it has been found that a number of malignant cells do not communicate either with each other or with their normal neighbours and it has been proposed that this may be one of the levels at which failure of tissue co-ordination occurs in some types of cancer.

The physical site of the communication pathway between cells is thought to be the gap junctions (see Cell Adhesion section) or, in invertebrates, special structures termed septate desmosomes. But there is no positive evidence to support these proposals. It is known, however, that the communication is broken in conditions which break the adhesion between cells and in a number of cases the low electrical resistance is established within a few minutes of the formation of contact between two cells.

No specific function has yet been demonstrated for these communication pathways, but it has been suggested that the control of mitotic rate and of movement in normal tissues, and some of the embryonic induction processes which require close contact between the participating tissues and the passage of high molecular weight substances between them, may utilize this system.

While the existence of this system does not exclude a role for surface specificity in its functioning, for example, in the determination of whether or not an adhesion is formed, or a pathway is opened, it does not require such specificity. The establishment of adhesion and of communication could be non-specific and, having access to each other's interior, the cells could then use internal recognition systems to co-ordinate their activities.

REFERENCES

Brightman, M. W. (1965). The distribution within the brain of ferritin injected into the cerebrospinal fluid compartments I. *J. Cell Biol.* **26**, 99.

Colwin, A. L. and Colwin, L. H. (1964). Role of the gamete membrane in fertilisation. In "Cellular Membranes in Development". (M. Locke ed.). Academic Press, New York and London.

Curtis, A. S. G. (1966). Cell adhesion. Sci. Prog. Oxford. 54:61.

Humphreys, T. (1970). Biochemical analysis of sponge cell aggregation. In "The Biology of the Porifera". Symp. Zool. Soc. London **25**, 325.

Jones, B. M. (1967). How living cells interact. *Sci. J.* **3**, 73.

Loewenstein, W. R. (1970). Intercellular communication. *Sci. Amer.* **222**, 78.

Marcus, P. I. (1965). Dynamics of surface modification in myxovirus-infected cells. In "Molecular and Cellular aspects of development". (E. Bell ed.). Harper and Row, New York.

Moscona, A. A. (1962). Analysis of cell recombinations in experimental synthesis of tissues *In vitro. J. Cell. Comp. Physiol.* **60**, Suppl. 1, 65.

Robinson, G. A., Butcher, R. W. and Sutherland, E. W. (1968). Cyclic AMP. *Annu. Rev. Biochem.* **37**, 149.

Roseman, S. (1970). The synthesis of complex carbohydrates by multiglycosyltransferase systems and their potential function in intercellular adhesion. *Chem. Phys. Lipids.* **5**, 270.

Roth, S. and Weston, J. A. (1967). The measurement of intercellular adhesion. *Proc. Nat. Acad. Sci. USA.* **58**, 974.

Spiegel, M. (1954). The role of specific antigens in cell adhesion. *Biol. Bull.* **107**, 130.

Steinberg, M. S. (1970). Does differential adhesion govern self-assembly processes in histogenesis? Equilibrium configurations and the emergence of a hierarchy among populations of embryonic cells. *J. Exp. Zool.* **173**, 395.

Townes, P. L. and Holtfreter, J. (1955). Directed movements and selective adhesion of embryonic amphibian cells. *J. exp. Zool.* **128**, 53.

Tyler, A. (1947). An auto-antibody concept of cell structure, growth and differentiation. *Growth* (10th Symp.) **6**, 7.

Weiss, L. (1967). "The Cell Periphery, Metastasis and Other Contact Phenomena". North-Holland Publishing Co., Amsterdam.

Weiss, P. (1947). The problem of specificity in growth and development. *Yale J. Biol. Med.* **19**, 235.

Wolpert, L., Thompson, C. M. and O'Neill, C. H. (1964). Studies in the isolated membrane and cytoplasm of Amoeba proteus in relation to amoeboid movement. *In* "Primitive motile systems in cell biology" (R. D. Allen and N. Kamiya ed.). Princeton University Press, Princeton.

Further Reading

Curtis, A. S. G. (1967). "The Cell Surface". Logos Press and Academic Press, London.

Fawcett, D. W. (1966). "The Cell". W. B. Saunders, Philadelphia.

Pardoe, G. I. (1970). Perspectives in blood group research. *Brit. J. Hosp. Med.* **3**, 393.

Symp. Soc. Exp. Biol. XXII (1968). Aspects of Cell Motility.

Robbins, P. W. (1969). The receptor for a bacterial virus. *Sci. Amer.* **221**, 120.

Trinkaus, J. P. (1969). "Cells into Organs". Prentice-Hall, New Jersey.

4. The Nucleus

A. P. MATHIAS

*Wellcome Laboratories, Department of Biochemistry, University
College, London, Gower Street, London, UK*

I. Structure of the Nucleus

A. INTRODUCTION

The nucleus is the most prominent organelle in a wide variety
of cells. It has a vital role in protein synthesis, through which it
dominates and controls the structure and activity of the cell. It is
essential to the replication of the cell. In executing these functions,
which may be regarded as the genetic functions of the nucleus, it
is responsive to the cytoplasmic environment. Before the advent of
the electron microscope it was already apparent that the nucleus
was complex in its chemical structure and its architecture. The high
resolving power of modern electron microscopes and the application
of refined biochemical techniques, have revealed in much greater
detail the organization of this fascinating organelle. However, much
remains uncertain and ill-defined.

The chemical composition of the nucleus is deceptively simple. It
contains the bulk of the cellular DNA. The content of DNA of the
diploid nucleus is constant for a particular species and is approximately
6 pg for mammals. This represents roughly a tenth of the nuclear
dry weight (of the order of 50 pg). Appreciable quantities of RNA,
in some instances equal to half of the amount of DNA, have been

detected in nuclei from cells active in protein synthesis. The major component of the nucleus is protein. Some of the protein, in the form of histones and non-histone proteins, is associated with the DNA. The metaphase chromosomes contain basic proteins (histones) which are in mass equivalent to the DNA. They also contain rather larger amounts of non-histone proteins. Nearly half the mass of nucleus is accounted for by non-chromosomal proteins, the amounts of which fluctuate over a wide range during the cell cycle. Less than 2% of the nucleus is lipid, and this is confined to the nuclear envelope.

Compounds of low molecular weight such as amino acids, nucleotides, coenzymes, and various metabolites, including intermediates of glycolysis, are found in nuclei. Their amounts vary according to the method of isolation. The concentrations are the same order as those in the cytoplasm. Most biological metal ions have been detected in nuclei. Some may be concentrated in the nucleus.

B. ISOLATION

The ideal method for isolation should provide nuclei in high yield and uncontaminated by other subcellular fractions. The isolated nuclei should retain all the components characteristic of the nucleus in the intact cell. Foreign substances which might have become attached to the nucleus during the isolation procedures should be absent. Finally the isolated nucleus should retain full biological activity. It is virtually impossible to achieve all these ideals in any single method. Methods of disrupting the tissue that are sufficiently severe to give complete breakage of cells usually damage some of the nuclei. In many cells the outer nuclear membrane is continuous with the endoplasmic reticulum, fragments of which often remain attached to the nuclei after homogenization. Although ribosomes and glycogen granules have higher densities than nuclei, they are much smaller and may be removed by taking advantage of the far greater rapidity with which nuclei sediment in a gravitational field. Nuclei may be separated from mitochondria and microsomes by exploiting the differences in their buoyant densities. Thus methods exist to purify nuclei from other subcellular components. However virtually all the media that are used for centrifugation extract nuclear components to varying extents. The redistribution of soluble components of cell homogenates to sites within the nucleus is a major hazard because the nucleus is in effect a polyfunctional ion exchange resin (in the sense that it contains both acidic and basic groups). In assessing methods of purification it must be borne in mind that it is difficult enough to detect gross damage to the biological properties of the nucleus during isolation, let alone to observe subtle changes.

The methods for the isolation of nuclei fall into two major classes divided according to the media that are used.

Isolation in organic solvents

In the first step of this procedure the tissue is frozen rapidly and then lyophilized. The freeze-dried material is ground to break the cells. The nuclei are separated from other components by a series of centrifugations in mixtures of organic solvents of different density such as cyclohexane and carbon tetrachloride or glycerol and α-chlorohydrin. By mixing these solvents in varying proportions, the density of the suspending fluid can be controlled exactly. This method has the advantage that exchange of water soluble components and of macromolecules between nucleus and cytoplasm is virtually prevented. However, during the isolation many proteins are denatured and membranes are destroyed in systems based on cyclohexane.

Isolation in aqueous media

Although solutions of glycerol, citric acid, and various other water-soluble compounds have been employed, solutions of sucrose are the most widely used. The cells of intact tissues, especially the softer organs such as liver, are readily broken by homogenization. Cells in free suspension are usually much more resistant to breakage. Homogenization is carried out in sucrose solutions ranging between $0 \cdot 25$ and $0 \cdot 8$ M. Lower concentrations of sucrose cause substantial losses of protein from the nuclei. A short centrifugation at low speed yields a crude nuclear pellet. This may be purified and the contaminants including whole cells and other subcellular organelles removed, by resuspending the pellet and centrifuging the nuclei through solutions of sucrose exceeding 2M. The high density of the sucrose prevents the sedimentation of light components, such as whole cells and mitochondria. Careful choice of the gravitational field will ensure that nuclei are pelleted without contamination by denser particles, such as the ribosomes. The pH and the ionic composition—especially the magnesium ion concentration—of the sucrose solutions must be carefully controlled to obtain clean active preparations. The purity of the isolated nuclei may be checked by electron microscopy and assays of marker enzymes.

C. COMPONENTS OF THE NUCLEUS

The Nuclear Envelope and Nuclear Pores

The DNA of many bacteria is confined to a region of the cell—often called a nucleoid—but no barrier separates it from the cytoplasm. In contrast, the chromatin of higher organisms is surrounded

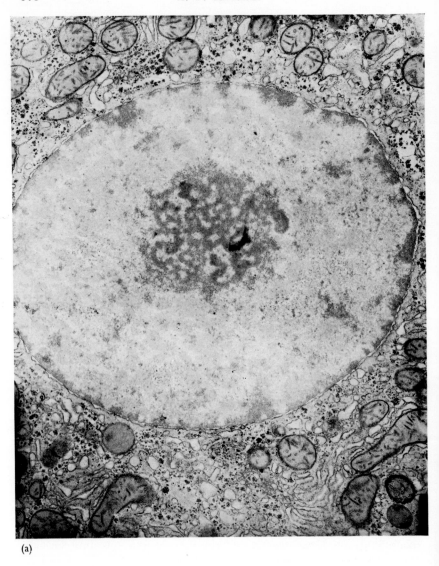

(a)

by two membranes which define the envelope of the nucleus and offer
a potential barrier to the interchange between nucleus and cytoplasm (Fig. 1). These membranes, composed of lipoproteins and
cholesterol, enclose not only the chromatin, but also the other components of the nucleus, such as the non-chromosomal proteins and
the nucleolus. The membranes are, in fact, unit membranes, each of
which is about 8 nm in thickness, separated from one another by a

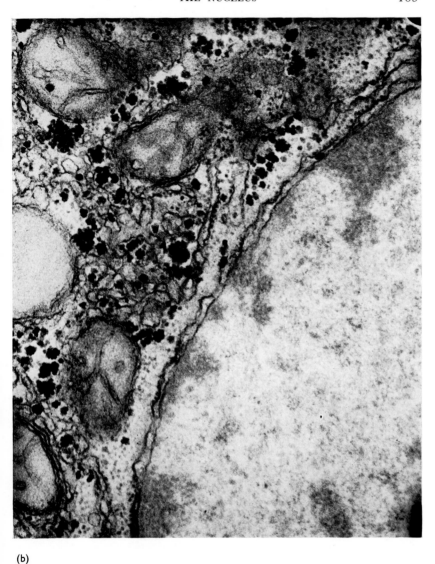

(b)

Fig. 1. (a) Electron micrograph of an intact nucleus of a rat liver parenchymal cell. The nucleolus occupies the centre of the nucleus and is surrounded by euchromatin. The heterochromatin is concentrated on the periphery ($\times 30\ 780$). (b) Higher magnification ($\times 75\ 240$) reveals details of the nuclear membrane, the nuclear pores and the concentration of heterochromatin near the membrane. Micrographs by D. W. James and J. Wynter.

space of variable width ranging from 10–50 nm. The outer membrane is supposedly continuous with the endoplasmic membrane. Usually the outer surface is partially covered by ribosomes. Examination of isolated nuclei in the scanning electron microscope indicates considerable wrinkling or corrugation of the nuclear envelope of some types of cell. The inner membrane appears, in some cells, to be supported and strengthened by a layer of filaments called the fibrous lamina. The outer and inner nuclear membranes can be removed by detergent leaving the perinuclear chromatin and pore apparatus, which is discussed below, in position. The perinuclear chromatin may form a continuous integument which, together with the fibrous lamina, is capable of maintaining the shape of the nucleus. Cylindrical channels lead through this layer from the nuclear pores to the interior of the nucleus.

A prominent feature of the nuclear envelope is the presence of large numbers of conspicuous circular structures or "pores", sometimes seen in hexagonal arrays (Fig. 2). They vary from 60–90 nm in diameter, and may be larger in oocytes and some other cells. The outer and inner membranes are fused together around the margin of the pores. In many instances the edge of the pore is surrounded by eight spherical granules arranged around the margin (Fig. 3). These peripheral granules are sandwiched between eight pairs of granules situated on both the nuclear and the cytoplasmic sides of the pore. Heterochromatin appears to be absent from the immediate vicinity of each pore although it seems that there may be a number of sites of attachment of chromatin fibres at the margin of the pore possibly to each of the spherical granules (Fig. 1). Often the pore appears to be blocked by a central granule which does not fill the entire central space of the pore complex. The remaining space may be filled by an amorphous material. In principle, this could constitute a septum or diaphragm which might prevent free diffusion across the membrane. The swelling of nuclei in hypertonic media has led some biologists to believe that nuclei have osmotic properties. However, the swelling may well be due to changes in the degree of condensation and hydration of the chromatin. There is little convincing evidence that the nuclear membrane is a barrier to the diffusion of compounds of low molecular weight into the interior of the nucleus. Certainly only a small proportion of the total space of isolated nuclei is impenetrable by sucrose solutions. The existence of specific transport mechanisms for some metabolites has not been rigorously excluded. Many large molecules, such as proteins, appear to pass freely through nuclear membranes, possibly via the pores. Some particles may be excluded if they exceed critical dimensions or if their electrical properties are such that they are repelled by the granules of the pore complex. In addition to any barriers posed by any limitations to permeability of the nuclear envelope, the concentration of materials within the nucleus will be influenced by the existence of binding sites for individual molecules in the nucleus. The number of these sites, the specificity and tightness of binding, and the occupancy of these sites are important factors in determining the concentration of materials, other than DNA, within the nucleus. The dimensions of the pores are large enough to permit passage of ribosomes, which, as we shall see,

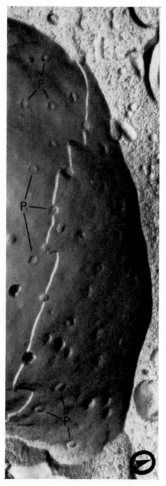

Fig. 2. Freeze-etched replica of a portion of the nucleus of a mesenchymal cell. Numerous nuclear pores P are seen on the two membranes of the nuclear envelope, × 39 000. Courtesy of Dr. E. Katchburian.

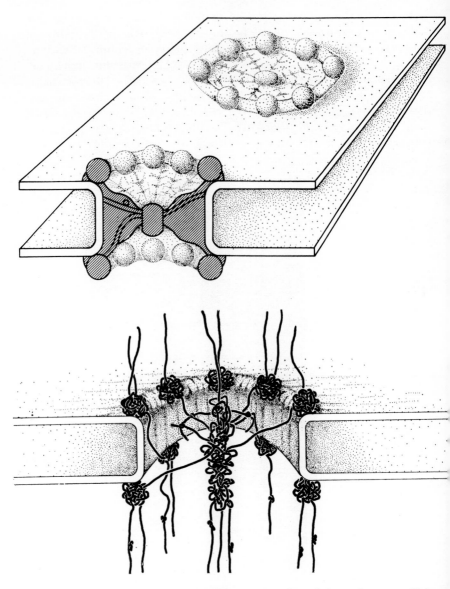

Fig. 3. These diagrams show two possible representations of the nuclear pore. Eight regularly spaced annular granules lie upon either side of the margin of the pore. A granule is located in the centre of the pore. Amorphous material projects from the wall of the pore towards the centre and may form an equatorial layer of diaphragm. The upper drawing emphasises a pore in which the components are compact. In the lower drawing the emphasis has been put on the fibrillar aspects, and the granules are represented as coils of filaments with fibrils projecting in several directions. The central granule may be composed of ribonucleoprotein moving from nucleus to cytoplasm. Reproduced with permission from Franke (1970).

are assembled within the nucleus. However, it seems unlikely that ribosomes pass through in an uncontrolled fashion.

During prophase the chromosomes tend to condense onto the nuclear envelope, which then disappears, to reform in telophase from flattened vesicular structures which are probably related to the endoplasmic reticulum. These align along the arms of the chromosomes and eventually fuse together.

Chromatin

Nuclei readily absorb basic stains, which reveal the presence of both fibrous and granular components. Because they react with Feulgen's stain, both components are known to contain DNA. DNA accounts for 30% of the dry weight of the fibres at interphase. The other major component is protein and there is also a small amount of RNA. This fibrous material, which is the form adopted by the chromosomes during interphase, is called chromatin. It consists of long fibrils 20–25 nm in diameter. In some areas these are closely packed and are in a condensed state referred to as heterochromatin (Fig. 1). The condensed chromatin tends to be concentrated around the periphery of the nucleus. In other regions the chromatin fibres are more loosely arranged and diffuse. This is described as euchromatin. In some nuclei continuity of nucleoprotein fibrils between condensed and dispersed chromatin is observed. The chromosome in metaphase seems to be almost entirely in the condensed form.

The fibres are not of uniform thickness throughout their length. They average approximately 23 nm. Scattered along the fibres are protuberances or bumps, probably caused by twisting of the fibres. It has been claimed that each fibre contains a single double-stranded helical DNA molecule which is coiled upon itself. The double-stranded DNA molecule 2 nm in diameter, which is increased to $3 \cdot 5$ nm by the attached protein, may be wound into a supercoil 8–10 nm in diameter which in its turn is coiled again to give the fibre with a dimension of 23 nm. This twofold supercoiling would have the effect of shortening the DNA molecule to 2% of its length in the fully extended state. Supercoiling and the subsequent folding of the supercoils enable the genetic material to be fitted into a nucleus not much more than a millionth of its total length. A human cell contains DNA that would stretch for 3 metres if fully extended. This is accommodated in a nucleus roughly 5 μm in diameter. Some of the fibres have points of attachment to the nuclear membrane near the edges of the pores. The chromatin fibres are closely packed or condensed in the metaphase chromosome in which they wind back and forth following tortuous paths. After telophase certain regions of individual chromosomes open up to yield diffuse euchromatin.

The chromosomal proteins may be defined as those present in the metaphase chromosome. Apparently at this stage the chromosomes contain some acid-insoluble (i.e. non-histone) proteins that are not found in interphase chromatin. Extraction of interphase or metaphase chromosomes with acid yields the acid-soluble histones, contaminated with a small amount of non-histone chromosomal protein. The histones are basic proteins which lack tryptophan and are associated with both eu- and hetero-chromatin. They may be fractionated by various methods. Their molecular weights lie in the range 15 000–20 000. A DNA molecule having a molecular weight of 1×10^6 would have approximately 50 histone molecules attached to it. The histones form five distinct groups (Table 1). They are present in roughly equivalent proportions.

TABLE 1. Nomenclature of histones

			Mole %	
			Arg	Lys
F1	I	Lysine-rich	2·5	26·3
F2b	II	Moderately lysine-rich	6·7	16–17
F2a2	IIb	Moderately lysine-rich	11·5	10·5
F2a1	IV	Arginine-rich	14·1	9·7
F3	III	Arginine-rich	13·4	8·8

The ratio of histone to DNA of the metaphase chromosome is similar to that of interphase chromatin, although there may be some changes in the proportions of the different types of histone. In spermiogenesis the lysine-rich histones are replaced by the arginine-rich histones and then by the protamines which contain up to 80% arginine. During metaphase, chromosomes seem to have more non-histone protein bound to them than during interphase. Corresponding fractions of the histones from different tissues and species closely resemble one another in amino acid composition and sequence. The amino acid sequence of histone F2a1 from calf thymus differs from that of pea seedling in only two amino acids out of 102. Other histones have been sequenced, including an unusual histone, histone F2c (V), so far found only in the avian erythrocyte. In histone F2a1 and F2b the basic amino acids (and glycine) are concentrated in the NH_2-terminal half of the molecule. The COOH-terminal is rich in hydrophobic and aromatic residues. However in histone F1, the reverse distribution is found. The histones are probably attached to DNA predominantly by electrostatic linkages involving the negatively charged phosphate oxygens of DNA and the positively charged side chains of lysine, arginine and histidine. This suggests that part of each histone—the basic part—is bound to DNA, and the remainder is free to interact with other molecules including proteins. The basic section of the histones may be accommodated within the large groove of the DNA double helix. Attachment of histones is supposed to generate supercoils in DNA. It seems likely that these condensed structures are maintained by both the histones and the non-histone proteins. Although the ratio of positively charged amino acid residues in chromatin protein to negatively charged phosphates in DNA is roughly 1 to 1, it appears that not all the DNA is masked by protein. Extensive stretches of DNA, amounting to half the total, may be exposed.

In the intact cell the histones are modified by the attachment of acetyl and phosphoryl groups, some of which are in a state of rapid turnover. These reactions are also observed in isolated nuclei. Acetyl residues are found on α–NH_2 groups and on the ε–NH_2 of lysines in certain specific positions in the chains. Different specific enzymes are involved in the acetylation of the individual acceptor residues which are concentrated in the arginine-rich hisones. It has been shown that phosphorylation occurs of particular serine-hydroxyls. Another modification is the methlyation of ε–NH_2 groups of some lysines. Histone F3 contains –SH groups which are in the reduced form in the interphase chromatin, but are oxidized in the metaphase chromosomes.

The chromatin contains another group of heterogeneous and poorly characterized proteins. In their overall composition, this group of proteins contains nearly twice as many acidic as basic amino acid residues. These non-histone proteins can be resolved into at least thirty components, some of which are phosphoproteins. There is some similarity between the non-histone proteins of different organs and creatures. It is possible that a degree of organ and species specificity is concealed behind these apparent resemblances. A number of the non-histone proteins are enzymes. For example, DNA polymerase activity is found in this fraction.

The nucleus contains a considerable amount of protein which does not appear to be associated with chromatin. Some of this may be accounted for as ribosomal proteins. Some of these non-chromosomal proteins are nuclear enzymes, and some appear to be phosphoproteins.

The nature of the RNA that is found in chromatin, which is, or will become, ribosomal RNA, will be discussed later. There is another type of RNA called chromosomal RNA which is found in association with chromosomal proteins. It may be bound to non-histone protein. It contains 40–60 nucleotides, 8–10% of which are dihydrouridylic acid. There is evidence for heterogeneity in sequence. This type of

E

RNA, which has also been detected in the cytoplasm, will hybridize to about 5% of the DNA. Some of this chromosomal RNA may be degraded transfer RNA. An additional type of RNA has been described recently. Eleven or more species of RNA ranging from 4–8 S, some of which are rich in uridine, have been found in nuclei of mammalian cells.

Chromatin is soluble in distilled water and also 1 M sodium chloride in which it dissociates. It is insoluble in 0.14 M NaCl which will extract ribonucleoprotein particles and soluble proteins from the nucleus.

The Nucleolus

The presence of dense granules within the nucleus was first detected by light microscopy. These granules, named nucleoli, which contain protein, give a positive reaction with stains specific for RNA, but only a very weak indication of the occurrence of DNA. They disappear during prophase when they become visibly associated with the "nucleolar" or satellited chromosomes, probably at the nucleolar organiser regions. The latter are identifiable as secondary constrictions on certain chromosomes and they contain the genes for ribosomal RNA.

Nucleoli may be prepared by disrupting isolated nuclei. This can be achieved by controlled sonication at low temperature. The nucleoli are freed of chromatin by repeated sedimentation or by digestion of the latter with deoxyribonuclease. About 80% of the mass of the nucleolus is protein, the remainder being RNA (11%) and DNA (8%).

The nucleoli of most cells are built up from trabecular (beam-like) structures, called nucleonemas, which contain dense particles about 15 nm in diameter embedded in a less dense matrix. These particulate or granular elements, which contain RNA, resemble cytoplasmic ribosomes. There are also fibrillar structures about 10 nm wide. A light space, or *pars amorpha*, occurs between adjacent nucleonemas and at the periphery of the nucleolus which is not bounded by a membrane. Some nucleoli are in a more compressed state and lack nucleonemas. In these nucleoli there is a uniform distribution of dense ribonucloprotein particles. Chromatin fibres apparently penetrate the nucleolus and occasionally may be seen within the structure as clumps of fibrils.

The number of nucleoli per cell varies from one cell type to another within the same organism. In rodent liver the number of nucleoli increases but may not double at each twofold increment of ploidy. The mean nucleolar size also increases when comparing diploid with tetraploid, suggesting a degree of collaboration in the formation of nucleoli at the higher level of ploidy.

Hybridization experiments show that the nucleolar DNA includes DNA coding for ribosomes (rDNA) as well as other cistrons. Several hundreds of these rDNA (45S) cistrons are found in each nucleolus in mammalian cells. The high GC content of nucleolar DNA is compatible with its relationship to ribosomal RNA. The rDNA consists of tandem arrays of sequences coding for 28S and 18S rRNA with some spacers between the cistrons. It is transcribed as a 45S rRNA precursor. The nucleolar genes are among the last to stop functioning in mitosis, showing inactivity only during anaphase and metaphase in the majority of cells. They resume activity early in G_1, the phase of the cell cycle that follows mitois (see p. 125). A great increase in the amount, or amplification, of nucleolar DNA has been demonstrated in many oocytes. In amphibia this is manifested by proliferation of the nucleoli, and up to 2 000 nucleoli have been seen in a single nucleus.

Nucleoli contain RNA polymerase and a ribonuclease which, because of the presence of an inhibitor, is in a latent form. NAD pyrophosphorylase, an ATPase, and various phosphatases are concentrated in the nucleolus. The occurrence of phosphatases is interesting in view of the high concentrations of inorganic phosphate in some nucleoli. Cells that are inactive in RNA synthesis, for example the avian erythrocyte, usually lack any nucleoli.

Satellite DNA

The haploid content of DNA of *E. coli* has a molecular weight of $2 \cdot 5 \times 10^9$ daltons which is equivalent to approximately 4 million base pairs. It has been estimated that this is sufficient for approximately two thousand genes. For mammals the molecular weight of the haploid DNA is about 2×10^{12} daltons or 3×10^9 base pairs, an increase of the order of 10^3. This would allow for well over 2×10^5 genes. The enzymology of *E. coli* is as diverse as that of the mammalian cell. It is true that a far greater complexity is encountered in higher organisms including control systems that permit differentiation. Nevertheless the massive increase in DNA when contrasting a higher organism with a prokaryote is surprising.

Some of this "excess" DNA may be accounted for in a fraction possessing unusual properties. If all its base sequences are unique the kinetics of renaturation of denatured DNA from a cell should follow a simple law. The initial rate of renaturation should be proportional to the reciprocal of the size of the genome. However the initial rate of renaturation or reassociation of mammalian DNA is faster than that of bacteria. Part of the explanation of this paradox is the presence in the DNA of higher organisms of groups or families of DNA molecules that are repeated many times in the complete genome. The single-stranded DNA, produced by the heat denaturation, from these regions will have a much greater chance of finding the complementary strand during renaturation. Consequently the duplex will reform rapidly. These repetitive or reiterated sequences are found in all higher organisms ranging from protozoa to primates, and they may amount to 30% or more of the total DNA. No two species have the same kinds of these multiple repetitions of similar nucleotide sequences. Part of the repetitive DNA is called satellite DNA because in many species it may be separated from the bulk of the single copy DNA by buoyant density centrifugation, in the form of one or more satellites. One of the satellite DNAs of guinea pig, which accounts for 5% of the total DNA, has been shown to consist of tandem arrays of a sequence which is probably

$$5'\text{—CCCTAA—}3'$$
$$3'\text{—GGGATT—}5'$$

This sequence is repeated roughly 300 times with some heterogeneity within the blocks making up the repeat. The heterogeneity may have arisen by multiple doublings of the basic sequence, during which a number of point mutations have been accumulated. This gives rise to a repeat length of about 2 000 base pairs. The genome contains many thousands of copies of this repeat length. The satellite DNAs are distributed in roughly constant proportion to total DNA throughout all the chromosomes. In metaphase chromosomes the bulk of the satellite is in or near the centromeres. During interphase it is concentrated in heterochromatin and probably also in the vicinity of the nucleoli. The nucleolar organizer regions and the centromeres are the principal sites of heterochromatin. Therefore it appears that the latter may contain a high proportion of repetitive DNA. However it is premature to equate satellite DNA and heterochromatin.

Ribosomal DNA can be regarded as a class of repetitive DNA, consisting as it does of numerous copies of the gene producing the 45S ribosomal RNA precursor. It is possible that the spacer or redundant regions that are trimmed out of this precursor are made on reiterated sequences within the *r*DNA which in some instances can be isolated as a minor satellite. (See p. 130.)

There is evidence that multiple copies of the genes for the various types of histone occur in sea urchins and probably in other organisms. However probably most genes are represented by single copies. In summary the types of DNA in the nuclei of higher organisms may be classified under three headings :—

(i) Satellite, or highly reiterated
(ii) Repeated sequence DNA. This group contains DNA with sequences much longer than those of satellite DNA, but with many fewer copies. Some of it may perform a role in the control of transcription
(iii) Unique sequence DNA, responsible for the synthesis of messenger RNA

II. Enzymology of the Nucleus

A. SYNTHESIS OF DNA

Replication of DNA in bacteria

Before considering eukaryotic organisms, it will be helpful to summarize what is known of DNA synthesis in microorganisms. The present understanding of the mechanisms in these simple systems, although confused, is more advanced than that of complex systems.

In some bacteria such as *E. coli*, and perhaps in all, the genome consists of a one double stranded circular molecule of DNA. This means that it is without either 3′ or 5′ ends. The replication of this structure is semi-conservative. Each of the strands of the parent molecule serves as a template for the synthesis of a complementary strand. The products of a complete round of replication are two molecules, each of which contains one strand of the parent and one newly synthesized polynucleotide chain. The immediate precursors of the nucleotide residues in the nascent chain are the deoxyribonucleoside triphosphates. These are incorporated into the new strand by attack by the 3′–OH at the growth point on the α-phosphate (the one adjacent to the deoxyribose moiety) of the incoming triphosphate, which is selected so that it is complementary to the base opposite in the template strand. The latter is one of the polynucleotide chains of the original molecule. The strand in the process of synthesis is called the primer strand and it grows by stepwise addition of nucleotides with the formation of 3′ to 5′ phosphodiester bonds. Each addition of a nucleotide residue leaves a new 3′–OH at the terminus of the primer which is then available for the next step. The template strand, which is copied in a complementary fashion (i.e. A in place of T and vice versa, and G in place of C and vice versa), will have its phosphodiester backbone running 5′ $\longrightarrow$ 3′ in the direction of growth, whilst the nascent chain will run 3′ $\longrightarrow$ 5′ in this direction. It is important to distinguish between the direction of the phosphodiester backbone and the numbering of the ends of the molecule. In the diagram the 3′ end of the primer molecule is on the right and the 5′ end on the left, whilst the phosphodiester backbone runs from left to right in the 3′ to the 5′ direction.

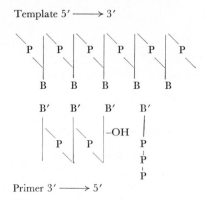

This mechanism offers a ready explanation of the growth of one of the new strands in the daughter molecules. It cannot apply to the other newly synthesized strand. Because of the antiparallel character of the two strands of the DNA duplex, this has to grow along a template whose phosphodiester backbone runs 3′ ⟶ 5′. Suggestions have been made about how this difficulty is overcome, and some of these will be discussed later. At this stage it should be emphasized that there is no evidence for a mechanism in which a chain is built up with a triphosphate end group and where growth occurs by the attack of the 3′–OH of the incoming triphosphate on the α–phosphate of the end group.

An essential for the initiation of the synthesis of DNA is a primer terminus with a free 3′–OH. How then is synthesis begun with a circular double stranded molecule? One possibility is that a small oligonucleotide having the appropriate base sequence and with a free 3′–OH displaces part of the 3′ ⟶ 5′ strand. The oligonucleotide could be composed either of ribo- or α-deoxyribonucleotide residues. The 3′–OH end of this oligonucleotide could then act as the primer terminus for synthesis along the 5′ ⟶ 3′ template. If an oligonucleotide containing ribonucleotides were required to function as a primer in the synthesis of DNA, it would mean that a specific DNA-directed synthesis of RNA, was an essential preliminary to the initiation of the replication of DNA.

An alternative hypothesis for which there is some experimental support, is that before DNA synthesis can start, a specific cleavage of the phosphodiester links of the two chains of the circular molecule occurs in such a way that a 3′–OH and a 5′–phosphate group are generated at each position of hydrolysis. The points of scission within the two chains need not be opposite one another and could be separated by a number of nucleotide residues. The attack would occur at points dictated by a specific base sequence. There is a precedent for this type of site-specific staggered cleavage in the conversion of the circular form of the double-stranded DNA phage, phage λ to the linear structure. An endonuclease attacks one chain of λ DNA at a distance of twelve nucleotide residues from the position of cleavage in the other. The attack is highly specific, only one phosphodiester bond being broken in each chain, to produce an open double-stranded molecule with protruding ends. A 3′–OH end whether protruding or not could constitute a primer terminus. Initiation by this type of primer end, produced by nuclease attack, may be described as template-primed.

Replication of the *E. coli* chromosome starts at a definite and unique position in the genome called the "replication origin". There is evidence that the growth point, or replication fork, moves away from this point in a single direction which is clockwise in the usual representation of the genetic map of this organism. The chromosome is replicated in an orderly sequence, with the site of synthesis moving steadily round the chromosome. Such a unit of replication is called a replicon. There are indications that there is a stable attachment of the replication origin of *E. coli* to the cell membrane from which it can be released by detergents. Recently it has been suggested that growth from the origin may proceed in both directions rather than one. The arrangement of the strands of DNA at the site of replication, in particular the necessity or otherwise for unwinding the double helix, is obscure. The synthesis of phage T_4 DNA involves a protein which appears not to act catalytically but which, by binding of several molecules to adjacent sites on the DNA, may cause a local unwinding.

The problem of the reconciliation of, on one hand, the mechanism of synthesis of DNA, and the fixed direction of synthesis at the growth point, with, on the other hand, the opposite polarity of the two parent strands, led to an investigation of the nature of the newly synthesized DNA. Much of this appears to be in the form of short segments, which are subsequently connected to older parts of the growing chain. It is uncertain whether one or both of the nascent DNA strands are synthesized in this discontinuous fashion. One model suggests that one strand grows continuously along the 5′ $\longrightarrow$ 3′ template, and the other is made in short segments backwards along the 3′ $\longrightarrow$ 5′ template. Alternatively the polymerase could copy one strand for a certain distance and then jump to the other strand and back-track. Cleavage by an endonuclease at the point of transfer to the second strand would provide a site for the next bout of replication. This and various other hypothetical schemes would lead to discontinuity in one or both strands.

It is apparent that breaks in the polynucleotide chains may occur as a result of the action of endonucleases, which nick one or both strands, or by synthesis of DNA in a discontinuous manner. These gaps will have a 3′–OH on one side and a 5′–phosphate on the other. The nucleotide residues to which these groups are attached will be aligned by involvement of their bases in hydrogen bonding to bases forming part of the intact strand. There will be no missing nucleotides. Enzymes that can catalyse the sealing of these gaps by forming a phosphodiester bond have been found and are given the trivial name of DNA–ligases. The enzyme for *E. coli* is dependent on NAD. The joining of the single-stranded break involves the following sequence : —

$$\text{Enz} + \text{NAD} \rightleftharpoons \text{E–AMP} + \text{NMN}$$

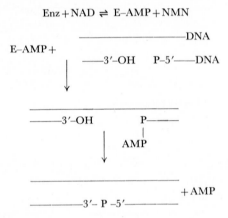

The T$_4$ phage ligase and the enzymes of some higher organisms require ATP in place of NAD.

It remains to discuss the nature of the polymerase involved in the synthesis of DNA. An enzyme, DNA Polymerase I, with many of the anticipated properties has been obtained in highly purified form from *E. coli*. It is a single polypeptide chain with a molecular weight of 109 000 and N-terminal methionine. There are about four hundred molecules in each bacterium. It has a single binding site for the deoxyribonucleoside triphosphates whose affinity for the enzyme depends mainly on their triphosphate group. There is also a single very tight binding site for DNA. In double-stranded DNA this binding occurs only at the ends or at nicks.

In the polymerization reaction the enzyme will add about 10^3 nucleotides per minute per site of synthesis. This should be compared with a maximum rate *in vivo* which is approximately 150 times as fast. The enzyme has other catalytic properties including both and endo- and exo-nuclease activity. The exonuclease activity can be separated from the polymerase activity by splitting the molecule with the proteolytic enzyme subtilisin into two fragments, one of 70 000 containing the polymerase, and the other of 40 000 which is the exo-nuclease. The endonuclease activity may be of value in the repair synthesis of DNA damaged by u.v. irradiation, which causes the formation of covalent links between two pyrimidine nucleotides, and in particular two thymidine residues at position where they are adjacent to one another in the same DNA strand. Repair of the DNA involves excision of an oligonucleotide containing the thymine dimer. The excision step and the synthesis of the replacement oligonucleotide are probably undertaken by this enzyme.

DNA Polymerase I *in vitro* is capable of highly accurate synthesis of long DNA molecules, although with some templates it gives rise to branched products. Nevertheless doubts have accumulated concerning its role in the synthesis of DNA *in vivo*. The doubts were reinforced when mutants of *E. coli* were isolated which possessed less than 1% of the DNA Polymerase I activity of wild strains, yet were capable of replication of their DNA at a normal rate. These mutants are sensitive to u.v. light. This indicates that recovery from the effects of u.v. is in part at least mediated by Polymerase I.

The reservations felt by many workers concerning the function of Polymerase I in DNA synthesis stimulated a search for other systems. Several groups discovered activity for DNA synthesis in a cell-membrane complex from *E. coli*. This activity was distinguished from Polymerase I by its sensitivity to N-ethylmaleimide and its failure to react with anti-serum to Polymerase I. The enzyme which has been named Polymerase II may be solubilized to yield a protein with a molecular weight of approximately 75 000 which is capable of fast prolonged synthesis of DNA in the presence of a primer-template and the four deoxyribonucleoside triphosphates. About 5% of the total DNA polymerase activity of wild type strains of *E. coli* is in form II. As yet it is premature to conclude that Polymerase II is the true enzyme for DNA synthesis or to relegate Polymerase I to a repair function. Indeed there are indications that the latter may be required for normal DNA synthesis to fill in the fairly long single-stranded gaps that may be left after discontinuous synthesis. A third system capable of the synthesis of DNA has been isolated. This also is bound to membranes. Studies of this subcellular aggregate, which have included genetic investigations, lend support to the view that this system may well play the major role in the replication of DNA *in vivo*.

Shortly after synthesis of the DNA chain a few specific bases are modified by methylation. Site-specific methylases are responsible for this.

A number of conclusions concerning the replication of the genome may be drawn from this brief survey of DNA synthesis in bacteria. Firstly, the mechanism of replication of DNA is poorly understood. Secondly, a multiplicity of enzymes is required, including the polymerizing system itself, methylases, endo-nucleases, ligases and possibly proteins that promote unwinding. The last two may be important in recombination. Thirdly the cell may contain several enzymic systems capable of catalysing the polymerization of precursors of DNA. During attempts to isolate these from the cell, some may be lost because of their greater susceptibility to denaturation. The assignation of function to individual activities is fraught with hazard when the overall molecular processes have not been clarified. Also it should be emphasized that the synthesis of DNA is a highly ordered and carefully controlled process, dependent on a multiplicity of biochemical steps which are carefully dovetailed. In the intact cell, once the process is started, it continues to completion. It is possible that the proteins responsible for the various steps in the replication of DNA are integrated into a multienzyme complex. The catalytic or other biological properties of individual proteins removed from this complex could be altered drastically. It may transpire that all three of the polymerases, so far isolated are required for normal synthesis of DNA, each with its specific function.

Replication of DNA in higher organisms and its relation to the cell cycle

Perhaps the most obvious contrast between the genome of a mammalian species and that of a bacterium is the division of the former amongst a number of separate chromosomes. The nucleoprotein fibrils in the chromosomes are very long and lengths up to 2 cm have been reported in human lymphocyte nuclei. Even this is considerably shorter than the length of the total DNA in a single chromosome in man. It is not known if the DNA of the chromosome is a single molecule or if it exists as several discrete molecules, or as a number of molecules joined by linkers that are not polynucleotides. It has been shown that replication is semi-conservative.

Investigations of mammalian chromosomes, using autoradiography and other techniques, have revealed the presence of replicating sections joined in tandem within the DNA fibres. These regions are not equidistant along the DNA molecule. Fork-like growth points are seen which appear on either side of the origin of replication and move away from it in opposite directions. This indicates that replication occurs in both directions. The rate of movement of the forks can be measured and can be used to calculate the rate of replication. This works out as approximately 1μ/min, which is roughly one tenth that of bacteria. From a knowledge of the time taken for complete replication of the genome it is obvious that there must be many sites of initiation of replication within a single chromosome. We can consider each of these to be the initiation site of separate replicons. The number of replicons per nucleus varies from species to species and may approach 10^4 in human cells. The replicons are variable in length, with adjacent replicons sharing a common terminus. The existence of a multiplicity of replicons is confirmed by autoradiography.

A consideration of DNA synthesis in higher organisms leads inevitably to discussion of the cell cycle. This may be defined as the period between the end of mitosis in a cell and the completion of the next mitosis in one or both of the daughter cells. Is it divided into four phases :—

(1) G_1, the gap between the end of mitosis and the start of DNA synthesis.
(2) S, the period of DNA synthesis.
(3) G_2, the gap that separates the completion of DNA synthesis and the start of prophase.
(4) M, mitosis itself.

Certain types of cells in culture and some cells of intact animals, such as those in the lower parts of the crypts of the intestine, divide continuously. Others, for example the adult liver parenchymal cell, rarely divide. They may be regarded as having an exceptionally long G_1 period or they may have been side-tracked from the cell cycle at some point in G_1. These cells can be summoned to return to the cycle by various stimuli such as partial hepatectomy. In many animal cells S phase lasts for 6–9 h and G_2 and M together for 3 h each. G_1 can vary over a wide range and is sensitive to the environment.

The majority of the replicons and especially those in the euchromatin may begin replication together at the start of S phase. The replicons complete the synthesis of their DNA at different times dictated by their lengths. There is contradictory evidence concerning the sites of initiation of DNA synthesis. It has been claimed that incorporation of precursors at the beginning of S phase is predominantly in peripheral regions of the nucleus and the sites of synthesis subsequently migrate into the interior. This suggests that DNA synthesis is initiated at the points of attachment of the chromatin to the nuclear membrane, in a manner analogous to the situation that may prevail at initiation in *E. coli*. Another view is that early DNA synthesis occurs in replicons evenly distributed throughout the nucleus, and late synthesis is concentrated in replicons on the periphery mainly in condensed, or heterochromatin. It seems likely that some at least of the satellite or repetitive DNA is made in the late S phase.

DNA polymerases have been found in cytoplasm of mammalian cells and more or less firmly attached to the nuclei, probably in the non-histone chromosomal protein. Nuclei isolated in non-aqueous media contain half the total cellular polymerase activity. The polymerases have not been fully characterized. Some tissues may contain two forms, one showing a preference for heat-denatured DNA as a template, and the other for native double stranded DNA. The activity with a preference for denatured DNA may increase during times of rapid DNA synthesis. The degree of selectivity towards different templates may be dictated by the presence of nucleases which to differing extents influence the priming activity of native and denatured DNA and by ionic strength and concentration of Mg^{2+}. Some of the cytoplasmic enzyme may enter the nucleus during S phase.

Isolated nuclei are capable of DNA synthesis. The nuclear enzyme may be extracted with 1 M NaCl and further purified. The presence of the nuclear proteins does not prevent replication of the DNA. For example it has been found that nucleohistones will serve as templates for *E. coli* polymerase, although the presence of the histones slows down the rate of replication. In some systems the activity of the nuclear polymerase increases immediately prior to S phase and declines after S phase is complete. However in other cells either there is no change or a fall in activity. Nuclei may be fractionated by zonal centrifugation. Lower DNA polymerase activity is detected in those nuclei active *in vivo* in DNA synthesis, that is traversing S phase, in comparison with the other nuclei. This indicates that assays *in vitro* may not provide a true measure of biological activity.

Part at least of the newly synthesized DNA seems to be made discontinuously in intact cells and probably also in isolated nuclei. Ligases have been found in higher organisms. The joining reaction requires ATP rather than NAD. Nucleases are found in nuclei and may be involved in the initiation of DNA synthesis. The explanation of the requirement for calcium ions for DNA synthesis in cultured cells may lie in the presence in these cells of an endonuclease, activated by Ca^{2+}, which is required to create single-stranded breaks. Finally it should be pointed out that mitochondria which contain one or more circular double stranded DNA molecules have a distinctive polymerase which shows a marked preference for mitochondrial DNA.

DNA polymerases directed by RNA

Several oncogenic viruses have been shown to contain single-stranded RNA. The nucleic acid of these viruses is replicated by an RNA-dependent DNA polymerase. This produces first an RNA–DNA hybrid, then a single-stranded DNA, and finally double-stranded DNA, one strand of which will have an identical sequence (with T replacing U) of the original viral RNA, and the other complementary to it. This type of polymerase has been called "reverse transcriptase". The DNA synthesized by this enzyme either is integrated into the host genome thus causing transformation, or it may serve as a template for the synthesis of viral RNA without integration.

The reverse transcriptase has been detected in extracts in lymphoblasts from leukaemic patients. At one time it was thought to be absent from normal subjects. However recently such activity has been found in cells that show no sign of infection with tumour viruses.

B. SYNTHESIS OF RNA

Synthesis of RNA in bacteria

In both prokaryotic and eukaryotic organisms, all types of RNA are transcribed from DNA with the ribonucleoside triphosphates serving as precursors. The enzymology of transcription is better understood in microorganisms, and may provide a model for the process in higher organisms.

The synthesis of RNA in *E. coli* is catalysed by RNA polymerase, of which there are roughly 2000 molecules per cell. It has been obtained in pure form and shown to consist of several non-identical subunits. Its structure may be represented as $\alpha_2\beta\beta'\sigma$ (ω). This structure is called the holoenzyme. The α subunits have a molecular weight of 41 000 and their precise role in the complex process of RNA synthesis is unknown. The β subunit (molecular weight 155 000) is probably involved in the initiation of RNA synthesis. It reacts with the drug rifampicin—a potent inhibitor of RNA initiation. The β' subunit is 165 000 and is required for binding to DNA. The σ component (86 000) is needed for correct initiation and may be dissociated readily from the remainder which has been called core enzyme ($\alpha_2\beta\beta'$). The latter will bind rapidly and reversibly but in non-specific fashion to DNA. In the presence of σ factor the enzyme binds to the true promoter site and a stable complex is formed. Once bound to the promoter, transcription of RNA can occur provided that the operator is not blocked by the binding of a repressor. (For a description of the operon see Chapter 34). σ acts catalytically in the initiation of RNA chains and is not involved in elongation of the RNA. Possibly it is released after formation of the initiation complex. There may be a number of σ factors of different specificity which appear to recognize different classes of promoter. The formation of the first phosphodiester bond in the nascent RNA chain leaves a triphosphate group at the 5′–end and a 3′–OH as the growth point. In experiments *in vitro* there is a marked predominance of purines at the 5′ terminus of the RNA. The sequence of bases in the RNA is determined by the DNA, one strand of which serves as a template and is copied in an anti-parallel complementary fashion with A opposite T, U opposite A, etc. This form of transcription is described as asymmetric because only one of the two strands of the DNA is transcribed. However it is not necessary that the same DNA strand be copied at all loci. The rate of synthesis, expressed as the number of nucleotides added per second, seems to be of the same order for all types of RNA. The marked variations in the rate of production of different types of RNA must be attributed to different rates of initiation.

The termination of RNA synthesis may involve release (ρ) factors in addition to a chain termination sequence in the DNA template. ρ is a tetramer with a sub-unit molecular weight of 50 000, which may bring about termination by binding to the DNA and blocking chain elongation. The nascent RNA is then released from the DNA whilst the polymerase remains bound. The status of ω factor is obscure. It may be a contaminant.

Recently evidence has been forthcoming of additional factors involved in transcription including ψ (psi) factors. These may be involved as well as σ factors in the specificity of initiation. There could be several types in the bacterial cell, one of which, ψr, may control transcription of the cistrons for ribosomal RNA. ψr is inhibited by guanosine tetraphosphate (ppGpp). This nucleotide occurs naturally and its concentration is dictated by the rate of growth of the organism. The concentration of this inhibitor may regulate the amount of synthesis of ribosomal RNA. It seems that σ factors may be primary determinants of promoter recognition and the ψ factors have a secondary effect. ψ factors may exert their influence by changing the holoenzyme from a form that prefers initiation at one group of promoter sites to a different form with a changed set of preferences. It has been demonstrated that cyclic AMP is required for the synthesis of inducible enzymes in *E. coli* and apparently promotes synthesis of the corresponding *m*RNAs. This enhancement of the synthesis of the *m*RNA for inducible systems involves a protein that binds cAMP with high affinity. The complex may bind to the promoter region of the DNA and permit the subsequent binding of RNA polymerase.

Once the polynucleotide chain has been completed, it may undergo modification at definite sites. These modifications, catalysed by specific enzymes, include methylation of the purines and pyrimidines and of the 2'–OH of the ribose, and conversion of uridine to pseudouridine or thiouridine.

Synthesis of RNA in higher organisms

RNA polymerase. With the exception of a small amount in mitochondria, synthesis of RNA is confined to the nucleus. Isolated nuclei are capable of synthesis in vitro, if provided with ribonucleoside triphosphates. This reaction is dependent on DNA of the nucleus with which the enzyme activity is closely associated. Virtually all the RNA polymerase activity of homogenates made in aqueous media is found in the nuclear fraction. It may be solubilized and has been shown to exist in a number of different forms. Some of these are strongly inhibited by α-amanitin—a bicyclic polypeptide which is the toxic component of the toadstool Amanita phalloides. It inhibits only mammalian RNA polymerase, the various forms of which also differ in their relative activities in the presence of Mg^{2+} and Mn^{2+} ions, and in their response to ionic strength. They are summarized in Table 2.

TABLE 2. Multiple Forms of Mammalian (rat liver) RNA Polymerase

Nomenclature		Nuclear location	Sensitivity to α-amanitin	Preference
AI	⎱ I	No	—	Native
AII	⎰	No	—	Native
*AIII	III	Np	—	
BI	⎱ II	Np	+	Denatured
BII	⎰	Np	+	Denatured
*BIII	—	Np	+	

No = Nucleolar   Np = Nucleoplasmic.
* The characterization of these forms is incomplete.

Preference indicates the form of the DNA template giving greater extent of synthesis in the assay system. The mammalian RNA polymerases are of similar molecular weight to the bacterial enzyme and are also made up of non-identical subunits.

The nature of nuclear RNA. The nucleus contains many types of RNA including

(1) Low molecular weight RNA
  (a) Chromosomal RNA
  (b) 4–8S RNA some of which is rich in uridine
  (c) tRNA and its precursors

(2)   Ribosomal RNA and its precursors 5S, 7S, 18S, 28S, 32S, and 45S. These are usually rich in GC and are synthesized in the nucleolus.

(3)   High molecular weight RNA, so-called "DNA-like" RNA, hsRNA. which is heterogeneous and polydisperse. The sedimentation coefficients range from 200–50 S and probably throughout the whole region down to 10 S in which rRNA is also found. In many species this RNA is rich in AU. Little of this RNA appears as such in the cytoplasm. Much of it is rapidly degraded within the nucleus, and some may be stored in nucleus. It has been suggested that this RNA, in part at least, could be precursors of cytoplasmic mRNAs which range from 8 to about 30 S.

Recently it has been shown that some cytoplasmic messengers contain segments of polyadenylic acid in which more than a hundred residues of adenylic acid are linked to one another. Such sequences rich in adenylic acid are also found in the nuclear heterogeneous DNA-like RNA. Apparently they are synthesized independently of the formation of hsRNA by a nuclear ATP polymerase and subsequently linked to hsRNA or fragments derived from it. The poly-A sequences may have a function in the transport of the mRNA from nucleus to cytoplasm or in the attachment of the messenger to ribosomes or to membranes.

If cells are exposed for a short time to a radioactive precursor of RNA, most of the label is found in the nucleus. The labelled RNA can be extracted and examined by sedimentation in sucrose gradients or by gel electrophoresis. Apart from label in the 45 and 32 S species and a little in rRNA itself, much of the radioactivity is found as a broad zone in regions of greater S values than 45S. The RNA referred to in category 3 is therefore rapidly synthesized and it also has a fast turnover in the nucleus. The functions of this heterogeneous rapidly labelled RNA are obscure.

Biosynthesis of ribosomes (see also Chapter 5) There is firm evidence, based on the technique of DNA–RNA hybridization, that the cistrons for rRNA are located in the nucleolus. These experiments have been confirmed by the discovery of mutants of *Xenopus laevis,* whose cells do not have nucleoli and cannot synthesize rRNA. The organization of the ribosomal cistrons in tandem arrays has been discussed in Section I C3. The ribosomal RNA is synthesized, using these cistrons as template, in molecules of 45 S (M. W. $4\cdot2 \times 10^6$). Methylation takes place very soon after synthesis. It is believed that the RNA is cleaved into smaller pieces at specific points in the chain in accordance with the following scheme : —

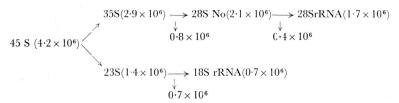

45 S RNA is made in the fibrillar core of the nucleolus and the early cleavages occur here. The later ones may occur in the nucleoplasm. Eukaryotic ribosomes contain between 80 and 100 different proteins, some of which may become associated with the ribosomes before completion of the changes in the RNA. They are probably added in a defined sequence from a preformed pool of ribosomal proteins within the nucleus. The proteins themselves are synthesized largely, and probably exclusively, within the cytoplasm. The half life of a ribosome in adult rat liver is 5 days. Evidently continuous synthesis is required. The concentration of ribosomes in many tissues is responsive to physiological state of the cells.

Mitochondria contain ribosomes, the RNA of which is coded for by the mitochondrial DNA. Mitochondrial RNA synthesis involves a mitochondrial polymerase which is of comparatively low molecular weight. Some, at least, of the mitochondrial ribosomal proteins appear to be made in the cytoplasm, but there is no evidence that mitochondrial ribosomes and cytoplasmic ribosomes have proteins in common.

C. ENZYMES OF NAD METABOLISM

Synthesis of NAD

The final reaction in the synthesis of NAD:—

Nicotinamide mononucleotide $+$ ATP $\rightleftharpoons$ NAD $+$ pyrophosphate

is catalysed by an enzyme—ATP–NMN adenylyltransferase or NAD pyrophosphorylase—which is exclusively located in the nucleolus to which it is tightly bound. The enzyme will also catalyse the condensation of ATP with nicotinic acid mononucleotide. (The formation of the amide group in the conversion of deamido NAD to NAD occurs in the cytoplasm.) Thus the cell derives its entire supply of NAD from the nucleus. Although the level of ATP–NMN adenylyltransferase in many types of nuclei seems to be much higher than that required to maintain the cellular concentrations of NAD, there is a positive correlation between the activity of the enzyme in a tissue and its content of NAD. It is interesting to note that low concentrations of nicotinamide coenzymes are characteristic of rapidly dividing tissues.

Synthesis of Poly(Adenosine Disphosphate Ribose) (Poly ADPR)

There is an enzyme present in the nucleoplasm that catalyses

F

the formation of a polymer of ADPR by transferring this residue from the nicotinamide of NAD to 2′-OH of the adenine ribose of an adjacent residue. The structure of the polymer may be represented as : —

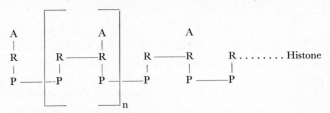

The terminal residue is probably covalently lined to a histone, and n may equal 8–10. The polymerization may involve the formation of an enzyme–ADPribose complex as an intermediate. If such an intermediate is formed it could be attacked not only by the adenine ribose 2′-OH (or the histone acceptor) with the formation of Poly ADPR, but also by water or a nicotinamide analogue. In other words the nuclear NAD glycohydrolase, which also acts as a transglycosylase, may be identical with the enzyme carrying out the polymerization. The product of the reaction would depend on the availability of acceptor groups for the ADPR moiety.

D. OTHER NUCLEAR ENZYMES

Nuclei from many tissues prepared in non-aqueous media contain nearly all the glycolytic enzymes in concentrations in terms of activity/ mg protein that are similar to those of the cytoplasm. Isolated calf thymus nuclei may contain all the enzymes of the tricarboxylic acid cycle and a complete repiratory chain. They also appear to have an active hexosemonophosphate shunt. These properties of lymphocytes are probably not true of most other types of nuclei. To talk of a typical cell may be misleading, but the lymphocyte is undoubtedly very unusual. The nucleus occupies 60% of the cell mass compared to the usual figure of $5–10\%$. The nuclei of most tissues seem to be freely permeable to phosphorylated nucleosides, and probably do not depend on nuclear synthesis of these compounds. Many other enzymes including various nucleases and other hydrolases have been found in nuclei.

III. Nuclear-Cytoplasmic Interactions

A. REGULATION OF DNA SYNTHESIS

Regulation in bacteria

Bacteria suddenly deprived of an essential amino acid complete a round of DNA synthesis that is in progress, but a new one is not started. Addition of the missing amino acid allows initiation of another round. This strongly suggests that DNA synthesis in bacteria requires protein synthesis. There is evidence from the effect of inhibitors that the products of at least two specific types of protein synthesis are required before initiation of DNA synthesis is possible.

It has been suggested that the bacterial replicon contains a structural gene which controls the synthesis of a specific initiator, presumably a protein. This acts by activating a replicator gene or site. Once this is activated, the replication of the DNA attached to the replicator becomes possible and the replicon is copied in its entirety. These two elements, the initiator gene and the replicator, are supposed to be specific for their own DNA. Presumably numbered among the proteins whose synthesis is necessary as a preliminary to that of DNA, is DNA polymerase itself and its associated proteins. The synthesis of these obligatory proteins may regulate the timing of the initiation.

In *E. coli*, a round of DNA synthesis takes roughly 40 minutes at 37°. This period is not greatly influenced by the generation time. In an adequate medium there is usually a gap of approximately 20 minutes between the end of replication of the chromosome and division. When transferred to an enriched medium, the rate of protein synthesis increases more than that of DNA synthesis. The proteins required for DNA synthesis are made available before the cycles of DNA replication can be completed. This leads to the reinitiation of DNA synthesis before a round is complete. Obviously there is an an element of cytoplasmic control which may involve compounds of low molecular weight as well as specific proteins.

Regulation in eukaryotic organisms

In eukaryotic organisms, the onset of DNA synthesis (S phase) is usually preceded by a gap (G₁) which may be regarded as a time of preparation for S phase. During G₁ there appears to be a requirement for RNA and protein synthesis. It is tempting to postulate that enzymes required for the synthesis of DNA itself and for the synthesis of the deoxyribonucleoside triphosphates are made during this interval. However, many of these enzymes appear to be present in adequate amounts throughout most, if not the whole, of the cell cycle. It is feasible that there might be one or a small number of proteins absent in G₂ and M, which are critical for DNA synthesis and have to be made in G₁. As in bacteria, cytoplasmic control is undoubtedly important. However it is not proved that in continuously dividing cells the decision to replicate DNA and proceed to mitosis takes place in G₁. It might have occurred in the preceding cell cycle. A cell that

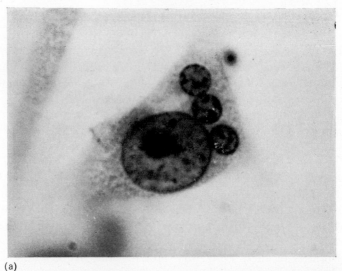

(a)

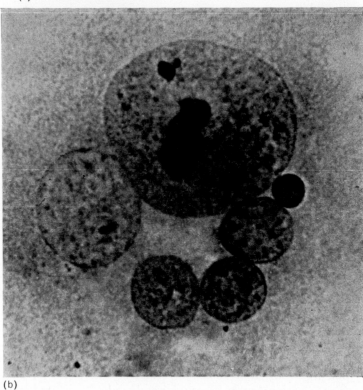

(b)

Fig. 4. (a) A heterokaryon containing three hen erythrocyte nuclei in the cytoplasm of a HeLa cell. (b) A heterokaryon stained in a different way containing five erythrocyte nuclei at different stages of reactivation. The small condensed nucleus is still inactive and the others show various degrees of enlargement. These pictures were taken by Professor H. Harris, a pioneer in the study of the formation and biochemistry of heterokaryons.

has reached the end of a process of development and ceased to divide, such as the avian erythrocyte, has not become inactive because of irreversible damage. If it is provided with a suitable stimulus it may be reactivated. For example, under the influence of inactivated Sendai virus, which is a member of the para-influenza group of myxoviruses, the nucleus of a hen erythrocyte can be introduced into the cytoplasm of a human or a mouse tissue culture cell (Fig. 4). This introduction occurs with negligible transfer of hen cytoplasm. A hybrid cell of this sort is called a heterokaryon. In its new environment in the heterokaryon, namely the cytoplasm of human or mouse cell, the hitherto inert erythrocyte nucleus resumes the synthesis of DNA. This new lease of life is not merely a response to the foreign nature of the new cytoplasm. The resumption of DNA synthesis occurs because the nucleus is now in a cytoplasm that normally supports this process. A marked swelling of the erythrocyte nucleus occurs before the start of DNA synthesis. The increase in volume is due to a massive flow of protein into the nucleus, and is accompanied by dispersion of the highly condensed chromatin of the erythrocyte nucleus.

The healing of wounds is often accomplished by hyperplasia. A system that has been studied extensively is the regeneration of rat liver after partial hepatecomy. Within 24 h after the operation, a high proportion of parenchymal cells begin DNA synthesis in synchrony. Prior to injury these cells were removed from the cell cycle (see p. 125) and can be designated as G_0 cells. These quiescent cells are stimulated in some unknown way by hepatectomy and return to the early G_1 state. After a delay, during which synthesis of RNA and protein occurs, replication of the DNA commences. The trigger mechanism in this and other systems, both normal and stimulated, is unknown. Recently it has been shown that isolated nuclei can be activated for the synthesis of DNA by cytoplasmic protein factors. However the nature and mode of action of these factors is obscure. There is evidence that modifying the nuclear proteins by attachment of chains of Poly ADPR (see p. 131) suppresses DNA synthesis.

Protein Synthesis and the Nucleus

The continuation of DNA synthesis during S phase requires uninterrupted protein synthesis. At least part of this requirement is explained by the observation that the synthesis of histones occurs only in S phase and is tightly coupled to that of DNA. Blocking the latter interrupts the synthesis of histones. These are synthesized on cytoplasmic polysomes. If DNA synthesis is inhibited, the polysomes responsible for histone synthesis rapidly disappear, presumably because of breakdown of their mRNA. Resumption of DNA synthesis leads to reppearance of the histone messengers in the cytoplasm and histone synthesis recommences. The synthesis of non-histone nuclear proteins is not confined to the S phase. This is not surprising because many of them appear to move into the nucleus during G_2.

Mention has already been made of the synthesis in the cytoplasm of the ribosomal proteins and their migration to the nucleus, where the assembly of the ribosome takes place (see p. 130). The histones —characteristic nuclear proteins—are also made in the cytoplasm and indeed it is probable that the non-histone nuclear proteins too are synthesized outside the nucleus. Does any protein synthesis occur within the nucleus? It is true that the nucleus contains ribosomes, mRNA and tRNA. However, the availability in the nucleus of many of the components of the machinery for protein synthesis, albeit in widely separated sites, does not prove that protein synthesis takes place. The balance of evidence is strongly against it, with the possible exception of thymocyte nuclei, which are atypical. The nucleus evidently possesses the ability to concentrate specific proteins from the cytoplasm by raid migration through the nuclear envelope.

B. REGULATION OF RNA SYNTHESIS

Major differences are found in the protein composition of the various types of cell in a multicellular organism. Leaving aside the amplification of genes in differentiated cells, which is possible if unlikely, all these cells have the same genetic information. Yet substantial changes in the pattern of gene activity are evident in comparing one type of cell with another. The differences that have been detected by competitive hybridization in the RNA in the various tissues indicate that at least part of this control of the gene occurs at the level of transcription.

In prokaryotic organisms, transcriptional controls can be explained in terms of the operon theory (Chapter 34). Control may be negative as in the lactose operon, or positive as for example in the arabinose system. Repressor proteins could, in principle, function at the level of translation rather than in transcription. An example of this seems to be the control of the arginine system in $E.$ $coli$ which involves a repressor gene, the product of which acts to restrict the synthesis of the arginine enzymes after the mRNA has been synthesized. Some degree of transcriptional control in microorganisms may be achieved by determining the specificity of RNA polymerase through σ and other factors.

It is dangerous to extend the operon theory—especially if it is considered solely as a negative control system—to multicellular organisms. These differ in many important respects from prokaryotes. In higher organisms there is continuous turnover of macromolecular constituents including proteins, ribosomes and membranes but not, of course, to any large extent of nuclear DNA. Normally this turnover does not occur in bacteria. Less than 1% of bacterial cell protein turns over during exponential growth. The control of the amounts of many enzymes in bacteria is dominated by the need to respond rapidly to changes in the supply of nutrients. In mammalian systems, mechanisms of homeostasis shield the cells of the tissues from abrupt changes. Therefore control systems with a slower response may be acceptable. The disengagement of translation and transcription in eukaryotes (see Chapter 34) offers scope for methods of control not available to prokaryotes. The clustering of genes for enzymes of a metabolic pathway that is sometimes, but not always, seen in microorganisms probably does not occur in higher organisms. Indeed genes for closely related proteins have been found in different chromosomes.

A consideration of the regulation of RNA synthesis is hampered by the difficulties of unequivocal identification of specific mRNAs. In principle this can be done in several ways. One is to demonstrate the complete de novo synthesis of the specific protein either in vitro by ribosomes programmed by the RNA under test or in vivo after microinjection of the RNA into oocytes of Xenophus laevis. The second is to determine the base sequence of the polynucleotide and show that this corresponds to the amino acid sequence of the protein for which it is supposed to code. The third is to test for efficient hybridization of the RNA to a sample of DNA known to contain the appropriate gene. The first two methods are attended by great experimental difficulties. The last, although a comparatively simple technique, is reliable only when the DNA contains the gene in question and not much else. The consequence of obstacles to identification of individual mRNAs has been that many investigators follow changes in enzyme concentration and assume that this reflects parallel changes in the production of its messenger. For certain inducible enzymes in bacteria this may be the case. It is not necessarily true in higher organisms where turnover of protein is normal. An additional complication in attempts to relate measurements of enzyme activity to concentration of mRNA is the existence of preformed pools of precursors to the enzyme, for example apo-enzyme molecules.

The control of protein concentration in eukaryotes may be represented by a simplified model

$$\xleftarrow{\hspace{3cm} k_s \hspace{3cm}}\rightarrow$$

DNA $\xrightarrow{k_1}$ | mRNA [M] | $\xrightarrow{k_2}$ | Protein [P] | $\xrightarrow{k_d}$ | Degradation Products

$\downarrow k_3 M$
degradation

where k_1 is the rate constant for the synthesis of a particular *m*RNA

[M] is the concentration of the *m*–RNA

k_2 is the rate constant for the synthesis of the corresponding protein

[P] is the concentration of the protein coded by the mRNA

k_3 is the rate constant for functional degradation of the mRNA

k_s is the overall rate constant for synthesis and is a zero order constant with respect to [P]. Its value will be determined by k_1, k_2 and k_3

k_d is the first order rate constant for degradation

The amount of protein formed in unit time is $k_2 M$. Under steady state conditions [P] is k_s/k_d. This means that the concentration of protein is dictated by both the rate of synthesis and the rate of degradation. Both these constants are characteristic of individual proteins and of their cellular environment. They can be varied independently of one another by various stimuli such as drugs, hormones and diet. Changes in protein concentration can be achieved by altering either k_s or k_d or both. The protein concentration is not therefore a measure of the concentration of its *m*RNA. Nor can the time of appearance of a protein be correlated with the time of synthesis of its *m*RNA. Experiments concerned with the enucleation of various species of the giant unicellular alga, *Acetabularia,* point to the conclusion that the mRNAs for the syntheses of specific proteins can be stored in the cytoplasm in a dormant form. The translation of the *m*RNA is inhibited. However this blockage does not involve destruction of the messenger, which may be stored on a membrane-bound form. The signal for activation of these inert messengers comes from within the cytoplasm. There is evidence from animal cells which suggests that some mRNAs made at one phase of the cell cycle may be utilized in another although this is not always the case. For example translation of histone messengers seems to follow rapidly on their transcription.

The technical obstacles and the complexities referred to above have prevented the development of a coherent explanation of regulation of RNA synthesis in eukaryotes. Nevertheless certain features of the regulatory systems are beginning to emerge. First, it seems highly probable that different regions of the nuclear DNA are transcribed in different cell types. In other words once a cell has become differentiated only certain genes have the potential for transcription. Second, the rate at which these potentially active genes are transcribed is controlled, and may vary at different times in the cell cycle from zero to a maximum dictated by the activity of RNA polymerase and the availability of precursors. Third, only a fraction of the RNA synthesized in the nucleus enters the cytoplasm. The mechanism by which selection for this transfer is accomplished and the manner of passage through the nuclear membrane or nuclear pores are unknown. Messenger RNA may enter the cytoplasm as complex with protein, and ribosomal sub-units may be involved in the transport. Evidently control of specificity of protein synthesis in the cytoplasm could be achieved by exercising selection of the messengers that are permitted to leave the nucleus. As we have seen, once within the cytoplasm the mRNA may be translated, be degraded or lie dormant. Some of the RNA which turns over in the nucleus might be involved in the control of RNA synthesis. Fourth, much more synthesis of RNA takes place in euchromatin than in heterochromatin. The condensed DNA of the metaphase chromosome or of the heteropycnotic regions of the interphase nucleus is genetically repressed. In line with this, if, during development, a cell line becomes progressively less active in RNA synthesis, it is generally observed that the proportion of heterochromatin increases. Apart from X chromosomes (Chapter 12), heterochromatization is not an invariant property of specific sections of chromosomes. Rather is it a fluctuating state controlled by cytoplasmic signals which cause the condensed DNA to open up and thus become available for transcription. It seems likely that condensation is important both in complete repression and in the control of the rate of RNA synthesis from potentially active genes which are reversibly repressible.

Many attempts have been made to explain these two types of control of gene activity, one of which confers long term inactivity on many of the genes of differentiated cells, and the other which allows fluctuations in the activity of those genes which determine the phenotype of specialized cells. Understandably attention has been focussed on the proteins found in such close association with nuclear DNA. Initially the histones were proposed for the role of regulators of gene activity. This proposition received some support when it was shown that histones inhibit the synthesis *in vitro* of RNA and that chromatin both in the isolated form and in the intact nucleus can be activated as a template for RNA polymerase by removal of its histones. These proteins are important in the condensation of chromatin. However the findings that the histones are relatively few in number, and show remarkable similarities in amino acid sequence in different cell types, and that the proportions of histones in hetero- and eu-chromatin are approximately the same, make it improbable that they control transcription in a selective fashion. The affinity of the histones for DNA may be influenced by the extent of acetylation and other modifications. Acetylation and phosphorylation appear to be more active in euchromatin. The proportion of the sulphydryl groups of histone F3 is high in diffuse and low in the condensed form. Although there are many forms of modification of the histones, the resulting diversity is still inadequate to justify the proposition that histones are responsible for the fine control of transcription.

It has been suggested that attachment of other molecules, such as chromosomal RNA (see p. 115), to the histones could provide a means by which histones might recognize specific base sequences in DNA. Alternatively the histones may have a non-specific structural role with a regular distribution along the DNA, and control of access to and movement along the template of RNA polymerase may reside in the non-histone proteins. In principle these proteins may act either to fortify or to antagonize the actions of the histones or may themselves interact directly with DNA. Until more is known of the structures of the non-histone proteins and the manner in which they and the histones interact with one another and DNA, only speculation is possible. Apart from the chromatin proteins the nucleus contains many other ill-defined proteins which move in and out during the cell cycle and which may also be involved in regulation of gene activity.

The importance of cytoplasmic signals in the control of DNA synthesis has been emphasized on p. 133. They are also important in RNA synthesis. If a heterokaryon is formed from two cells, one of which does and the other does not synthesize RNA, both types of nuclei will make RNA. The reactivation of a hitherto inert nucleus, such as that of hen erythrocyte is accompanied by a substantial increase in volume due to the transfer of proteins from the cytoplasm and the transformation of increasing amounts of condensed chromatin to the diffuse state. This permits transcription from the DNA in the hen nucleus. However, the hen RNA is not transferred to the heterokaryon cytoplasm until a nucleolus develops within the erythrocyte nucleus. Nor are chick specific proteins made in the cytoplasm until the nucleolus is present and actively synthesizing rRNA. This indicates that the flow of mRNA to the cytoplasm is coupled to the movement of newly synthesized ribosomes out of the nucleus.

The interaction between nucleus and cytoplasm is nowhere better exemplified than in experiments in which the nucleus of a differentiated cell is transplanted to an enucleated egg cell. A drastic change occurs in the activity of the transplanted nucleus which is rapidly reprogrammed. This modification of the nuclear activity appears to be determined by the passage of specific proteins into the nucleus.

C. CONCLUSION

Figure 5 summarizes the function of the nucleus and its relation to the cell. It underlines the degree of dependence of the nucleus on the cytoplasm. The role of the nucleus should be seen as one of

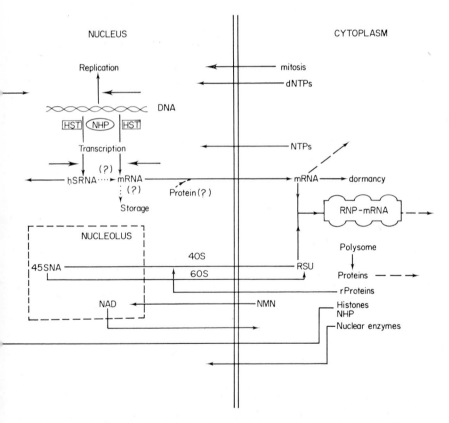

Fig. 5. The relationship between nucleus and cytoplasm. The broad arrows (→) indicate processes in which cytoplasmic signals or controls occur and the broken arrows (———→) degradation. Transfer across the nuclear membrane may involve penetration of both inner and outer membranes, movement through the nuclear pores, or passage into the perinuclear space and hence into the cisternae of the endoplasmic reticulum. Abbreviations: RSU, ribosomal sub-units; RNP, ribosome; HST, histones; NHP, non-histone proteins; hsRNA, heterogeneous RNA; NTP, nucleoside triphosphates.

partnership with the cytoplasm, which enables the genetic material, contained within the nucleus, to find expression and to be reproduced in an orderly fashion.

RECOMMENDED READING

Busch, H. and Smetana, K. (eds) (1970). "The Nucleolus". Academic Press, New York and London.

Du Praw (ed.) (1968). "Cell and Molecular Biology". Academic Press, New York and London.

Harris, H. (1970). "Nucleus and Cytoplasm". 2nd edtn. Oxford University Press.

Lima-da-Faria (ed.) (1969). "Handbook of Molecular Cytology". North Holland Publishing Co., Amsterdam, London.

Pasternak, C. A. (1971). "Biochemistry of Differentiation". Wiley-Interscience, London, New York, Sydney and Toronto.

Siebert, G. (1968). Nucleus in "Comprehensive Biochemistry" (Florkin, M. and Stotz, E. H., eds), p. 1. Elsevier, Amsterdam.

Wainwright, S. D. (1972). "Control Mechanisms and Protein Synthesis". Columbia University Press, New York and London.

5.   Cytomembranes and Ribosomes

P. N. CAMPBELL and A. VON DER DECKEN

Department of Biochemistry, University of Leeds, England
The Wenner-Gren Institute for Experimental Biology,
Stockholm, Sweden

I. STRUCTURE AND FUNCTIONS OF THE CYTOMEMBRANES

A. INTRODUCTION

The cytoplasm of cells, with the exception of the red blood cell, contains a complex network of cytomembranes. One of the functions of these is to synthesize and package secretory products. Not surprisingly, therefore, such membranes are more abundant in those cells in which the secretory function is prominent. Thus the acinar cells of the pancreas, which are responsible for the synthesis and secretion of some of the hydrolytic enzymes required for the digestion of food in the small intestine, are particularly rich in cytomembranes. In contrast the reticulocyte, an immature red blood cell, is almost devoid of cytomembranes. Although reticulocytes actively synthesize protein of which about 80% is haemoglobin, no protein is exported from the cell.

The term cytomembranes is used to describe membranes found within the cell, as opposed to the plasma membrane which surrounds the cell. On electron micrographs all types of membranes sectioned normal to their plane appear as linear profiles, about 5–12 nm thick. They consist of two dark leaflets separated by an intermediate light leaflet. Thus, they are all trilaminar structures. The dark lines are usually interpreted as protein-containing layers and the light line as a lipid containing layer (see Chapter 2 for more detailed discussion). This trilaminar structure has been found almost universally in cells and is usually referred to as the "unit membrane" (Robertson, 1969).

In contrast to the plasma membrane, the cytomembranes are less distinct trilaminar structures and they are usually somewhat thinner (about 7 nm). It has been shown by immunological methods that the protein composition of the plasma membrane is antigenically distinct from that of cytomembranes (Goldblatt, 1969). As to the difference in thickness of the membranes, critical studies have revealed that there are difficulties involved in their exact determination. Comparative studies are acceptable if the same preparation and fixation methods have been used for the electron microscopic studies.

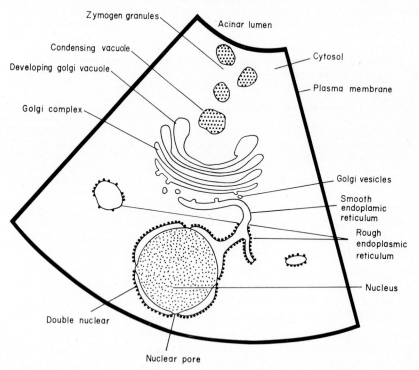

Fig. 1. Diagram of acinar cell of the pancreas, illustrating the organization of the various subcellular constituents involved in the formation of zymogen granules.

The cytomembranes appear to be surrounded by a relatively structureless fluid known as the cell sap or cytosol. This has a high concentration of protein which causes it to be somewhat viscous. The cytosol contains filaments and microtubules which in some cells are very much more abundant than in others.

A schematic view of a pancreatic acinar cell is shown in Fig. 1, indicating the relationships of the various membranes. The acinar cells are grouped around the acinar lumen into which the products of the cells are secreted, the apical region of the cells lining the lumen. In

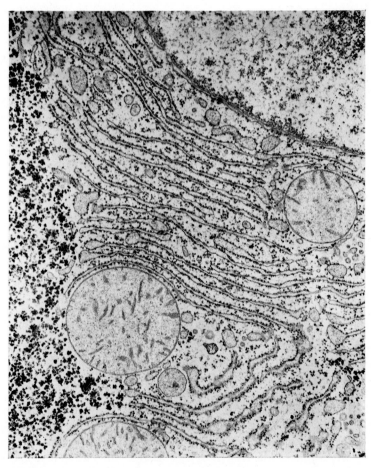

Fig. 2. Electron micrograph of liver from new born rat, × 17 000. (By kind permission of Dr G. E. Palade, Rockefeller Institute, New York). Note the arrangement of the membranes of the rough-surfaced endoplasmic reticulum in tubular cisternal and vesicular forms. The section also shows the ribosomes on the outer membrane of the nuclear envelope.

Fig. 1 we have left out of consideration other organelles that contain membranes, such as the mitochondria and the lysosomes which will be considered separately in chapters 6 and 7 of this book. Electron micrographs occasionally reveal continuity between the outer membrane of the nuclear envelope and the rough surfaced (or granular) endoplasmic reticulum. The adjectives rough or granular denote that ribosomes are attached to the membrane. In places continuity can be demonstrated with smooth surfaced endoplasmic reticulum (which is sometimes called agranular). The adjectives smooth or agranular are used because of the absence of ribosomes. Another form of smooth cytomembrane makes up the Golgi complex, a specialized cellular organelle associated with secretion. In the pancreas secretions are packaged in the Golgi apparatus forming zymogen granules which contain the precursor proteins of the digestive enzymes synthesized by the cells. As precursor proteins chymotrypsinogen and trypsinogen may be mentioned.

In electron micrographs the endoplasmic reticulum often appears pleomorphic, assuming tubular, cisternal and vesicular forms. The precise appearance will vary with the plane of section and in the electron micrograph of the liver cell from a new born rat shown in Fig. 2 this is well illustrated. Secretion is usually associated with dilatation of the endoplasmic reticulum and in some cells (e.g. in the exocrine pancreas of the guinea pig) the secretions are electron-dense and consequently demonstrable electron microscopically (Chapter 31).

Electron micrographs tend to give the impression that the various cytomembranes have a fixed position in the living cell but this is, of course, erroneous for we know from phase contrast microscopy that the mitochondria and secretory granules, at least, are in a state of constant movement, and no doubt this applies to all cell components and is an important factor in their functional interrelationship.

B. NUCLEAR ENVELOPE

The nuclear envelope is the structure that separates the contents of the nucleus from that of the cytoplasm. Not only is the composition of the nucleus different from that of the rest of the cell with respect to macromolecules like nucleic acids and proteins but also with respect to its ionic composition. It is clear, therefore, that the nuclear membranes do not allow the free passage of components to and from the cytoplasm but that such transfers take place as the result of active transport. Electron microscopy reveals that the nuclear envelope consists of two membranes of approximately 7–8 nm thickness each. In Fig. 3(a) is shown an electron micrograph of the nuclear envelope of a cell from the liver of a guinea-pig. The spaces between the two membranes are called the perinuclear cisternae. The nuclear envelope

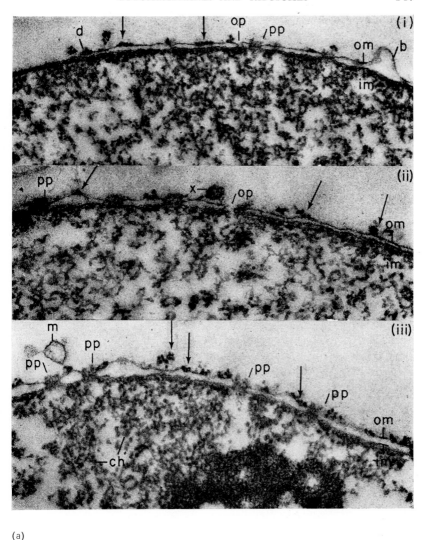

(a)

Fig. 3. (a) Electron micrograph of the nuclear envelope of guinea-pig liver. The outer and inner nuclear membrane are marked om and im, respectively. Ribosomes attached to the outer membrane are indicated by arrows. In (i) a number of nuclear pores are visible at pp and op, some of these appear "plugged" (pp), one is "open" (op) × 52 000. In (ii) "plugged" (pp) and "unplugged" (op) pores are shown at a higher magnification, × 65 000. In (iii) m may be a tenuously attached microsomal vesicle. Part of the nucleolus is visible in the lower right quarter. Chromatin masses are marked ch. × 75 000 (From Maggio et al., 1963), with permission.

E*

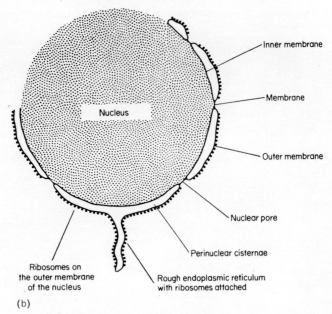

Fig. 3. (b) A scheme to show the names of the various parts of the nuclear envelope.

does not form an uninterrupted continuous sheet around the nucleus. At intervals the two membranes are joined together. As a result discontinuities or openings in the nuclear envelope are observed which are known as nuclear pores. It is probable that macromolecules are transported through the pores but this is not certain for the pores sometimes appear to be closed by a single membrane-like diaphragm see Fig. 3(b). The morphological evidence for the existence of the membrane is not firm but such a membrane would in any case be an essential requirement for active transport. Moreover there is evidence that the nuclear envelope maintains an electrical potential and this again would require a membrane.

While the inner membrane of the envelope has a smooth surface the outer membrane is often studded on its outer surface with ribosomes as shown in Fig. 3. Moreover, as already mentioned in the last section, the outer membrane is at times continuous with the endoplasmic reticulum. The perinuclear space is, therefore, an extension of the space within the endoplasmic reticulum (Stevens and André, 1969). The nuclear envelope is discussed further in Chapter 4.

C. ROUGH-SURFACED ENDOPLASMIC RETICULUM

We have already defined this component and mentioned that the rough-surfaced membranes are abundant in cells that secrete proteins.

As we shall see in a later section, it is possible to study the synthetic ability of the endoplasmic reticulum by radioautography using radio-actively labelled substrates but most experimental procedures involve the disruption of cells and the isolation of fragments of the endoplasmic reticulum. It is necessary to start our description by explaining the rationale for these experiments.

Isolation of the rough-surfaced endoplasmic reticulum

If the plasma membrane of a cell is broken in such a way that the structures within the cell are not completely disrupted it is possible to isolate mitochondria in a form that substantially resembles the state of this organelle in the original cell (Chapter 6). This is because the mitochondrion is delimited by a membrane and is strictly a particulate cell organelle. From our previous description of the endoplasmic reti-culum as a complex three-dimensional network of inter-connecting membranes it will be clear that there can be no question of isolating it in an intact form. In the process of disrupting the cell the membranes of the endoplasmic reticulum are certain to be broken so that the best one can aim for is a process that leads to the isolation of a fraction that is rich in fragments of the endoplasmic reticulum. An electron micrograph of such a fraction isolated from rat liver is shown in Fig. 4. This fraction was obtained from liver by a process that is illustrated in Fig. 5. This shows that the cell has been disrupted, in this case by grinding in a glass mortar with a rotating pestle, and the components separated by differential centrifugation into various fractions. In the case of liver the fraction that is most homogeneous with respect to elements of the endoplasmic reticulum is the so-called microsome fraction. This is not to say that the microsome fraction necessarily contains all or even most of the fragments of the endoplasmic reticulum in the original homogenate. This must depend on the precise conditions for the cell disruption and the size of the resulting fragments but the liver microsome fraction should be virtually free of all other organelles.

If instead of liver some other tissue were to be treated in a similar manner the results might well be very different. Thus the endoplasmic reticulum from guinea pig mammary gland tends to be concentrated in the so-called mitochondrial fraction in the scheme shown in Fig. 5 and there is very little of it in the microsome fraction. In the case of chick oviduct the endoplasmic reticulum is mainly present in the nuclear fraction. For this reason it is quite wrong to use the terms microsomes and endoplasmic reticulum as if they were synonymous. The first refers to a fraction from a disrupted cell that has a certain characteristic with respect to its behaviour on centrifugation and the second is a component of the cell possessing certain structural features.

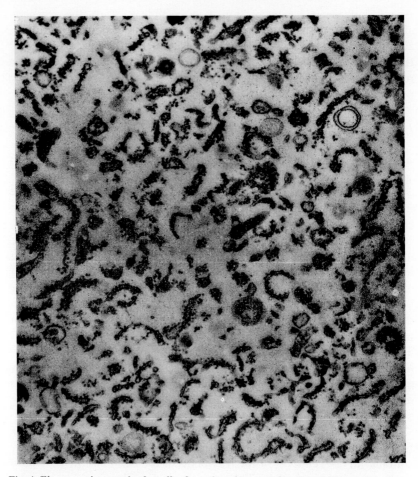

Fig. 4. Electron micrograph of a pellet from the microsome fraction of disrupted rat liver. The most prominent components are pieces of membrane studded with ribosomes. Also present are pieces of membrane without ribosomes (smooth membrane), × 50 000. (Electron micrograph by courtesy of Dr G. E. Palade of the Rockefeller Institute, New York).

The fragments of the rough endoplasmic reticulum that are found in the microsome fraction from rat liver, as seen in Fig. 4, do not usually have the appearance of vesicles although it is true that they do seem to retain soluble material which can be released by further disruption, e.g. by the application of ultrasonic vibrations. In contrast, the microsome fraction from pancreas contains pieces of rough endoplasmic reticulum that are clearly in the form of vesicles as shown in Fig. 6. The appearance of the fragments will clearly depend to some extent on the osmolarity of the medium used for the differential centrifugation.

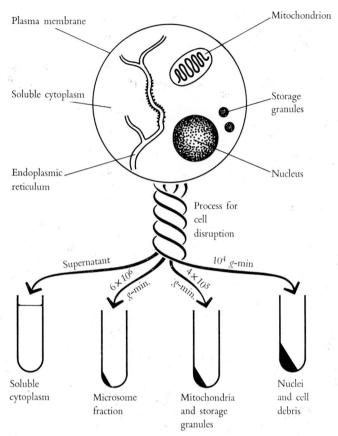

Fig. 5. Fractionation of subcellular particles from rat liver. The liver cell is disrupted by grinding and the particulate material is separated by differential centrifugation. The numbers indicate the centrifugal force g multiplied by the time of centrifuging at that force.

The vesicles are formed by the fusion of pieces of the rough endoplasmic reticulum that are broken during disruption of the cell.

We must now consider in a little more detail the morphological constituents of the microsome fraction from rat liver. In addition to fragments of the rough endoplasmic reticulum this fraction will also contain similar fragments from the smooth endoplasmic reticulum and also some ribosomes and polyribosomes not attached to membrane. We will consider the isolation of and properties of these so-called free polyribosomes later. If the properties of the rough and smooth membranes are to be studied it is of course essential to find ways of separating them. There is a small difference in the density of the two forms of membrane so that it is possible to separate them by centrifugation in

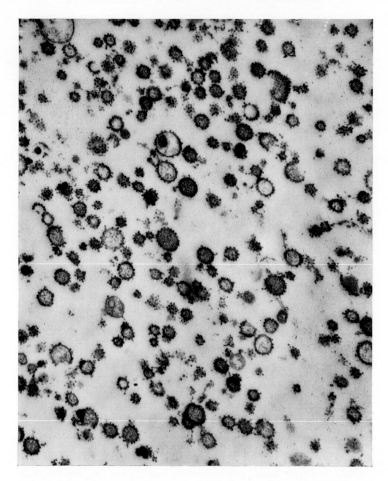

Fig. 6. Electron micrograph of a microsome fraction from Guinea pig pancreas, × 27 000. Note the spherical formation of the vesicles studded with ribosomes. (By kind permission of Dr G. E. Palade, Rockefeller Institute, New York).

a sucrose density gradient until equilibrium is established. This procedure can take about 48 h so that it is not very useful. A more practical way was found by Dallner in Sweden. In this method the rough membranes are made more dense by the inclusion in the medium of caesium in the presence of magnesium. The caesium becomes associated with the ribosomes. The membranes can then be separated fairly quickly on a sucrose density gradient (Dallner and Ernster, 1968).

The ribosomes attached to the rough endoplasmic reticulum may be isolated by treating the microsome fraction with a detergent which breaks up the membrane into very fine particles. Various detergents can be used but the most usual one is sodium deoxycholate. The soluble constituents of the vesicles will be released into the deoxycholate soluble fraction from which they can be recovered. An alternative procedure is to treat the membranes with ultrasonic vibrations. These procedures and the products obtained are shown in Fig. 7. The Fig. also shows that there are certain difficulties in isolating polyribosomes from the membranes of the rough endoplasmic reticulum. If one merely uses deoxycholate then only single ribosomes are obtained. If however the treatment with deoxycholate takes place in the presence of a soluble extract from liver then polyribosomes are obtained. The explanation of this is as follows. In the polyribosomes the ribosomes are linked together by messenger RNA. The latter is particularly susceptible to destruction by ribonuclease so that it may be removed without any apparent damage being done to the ribosomes themselves. The disruption of the membranes by deoxycholate causes the release or activation of a ribonuclease which destroys the messenger RNA thus converting the polyribosomes to ribosomes. The liver cell sap has for long been known to inhibit ribonuclease and this accounts for its role in the successful isolation of polyribosomes.

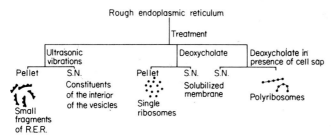

Fig. 7. The effect of various treatments on the fragments of the rough endoplasmic reticulum contained in the microsome fraction from rat liver. (From Campbell, 1970a).

Function of the rough-surfaced endoplasmic reticulum

As we shall see later there is good evidence that both the rough and smooth endoplasmic reticulum are involved in the synthesis of lipids and sterols but we shall deal with this under the smooth endoplasmic reticulum.

The most important function of the rough endoplasmic reticulum is without doubt the synthesis of protein for export from the cell or transport to another site within the cell. Zamecnik and his group in Boston showed in the mid-1950s that the isolated microsome fraction from rat liver was able to effect the incorporation of radioactive amino acids into protein. In order to establish the role of the rough endoplasmic reticulum in the synthesis of proteins destined for export from the cell it was of course necessary to study the synthesis of specific proteins. The liver cell has for long been known to be the site of synthesis of most of the plasma proteins. Of the plasma proteins studied serum albumin has proved to be the most convenient. Not only is it the most abundant of the serum proteins but it is also the easiest to isolate in a pure form. By incubating slices of liver with radioactive amino acid or by injecting an intact animal with such a substrate, Peters in the USA, was able to locate the newly synthesized serum albumin in the cisternae of the rough endoplasmic reticulum. A direct synthesis of serum albumin by the isolated microsome fraction from rat liver was demonstrated by Campbell. It is not possible to effect the synthesis of specific enzymes by the isolated microsome fraction from pancreas but much work by Palade and Siekevitz and their colleagues at the Rockefeller Institute in New York has established that these enzymes too are synthesized by the ribosomes of the rough endoplasmic reticulum (see Chapter 31).

Correlation of structure and protein synthesizing activity of the rough endoplasmic reticulum

So far we have only stated that the ribosomes are attached to the membrane of the reticulum but have not described the mode of attachment. Figure 8 shows a high resolution electron micrograph of the rough endoplasmic reticulum of a mouse liver cell. It is clear from this that the ribosomes are attached to the outside of the membrane and this always appears to be so. As we shall explain in a later section ribosomes are composed of two subunits one of which is about twice the size of the other and so we can call them the large and small ribosomal subunits. The question arises as to which of the two subunits is bound directly to the membrane or whether both are so bound. High resolution electron microscopy excluded the last possibility so that the question is merely whether the large or small ribosomal subunit is attached to the membrane. This was settled by Sabatini, Tashiro and Palade (1966) who treated the bound ribosomes with increasing concentrations of EDTA (ethylene-diamine tetra acetic acid). This substance, which chelates divalent cations, especially Mg and Ca, causes the dissociation of the ribosomes into their subunits (see later section). As the concentration of EDTA was increased the small subunits were first released and then the large subunits. This clearly indicated that the large subunits were attached to the membrane as shown in Fig. 9. This finding was particularly attractive since the growing polypeptide chain is associated with the large subunit. There is indeed plenty of evidence that the newly synthesized peptide chain passes into the cisternae of the endoplasmic reticulum as shown in the

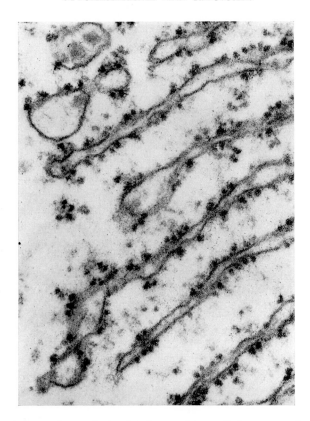

Fig. 8. An electron micrograph showing the attachment of ribosomes to the membrane of the endoplasmic reticulum in a mouse liver cell, × 70 000. (Electron micrograph by courtesy of Florendo, 1969).

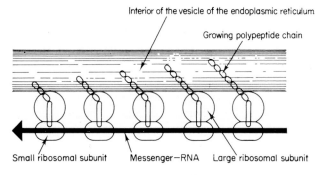

Fig. 9. The structure of the rough-surfaced endoplasmic reticulum. (From Campbell, 1970b).

156   P. N. CAMPBELL AND A. VON DER DECKEN

Fig. 9. Palade has described this undirectional release as 'vectorial' protein synthesis. We shall later describe how this protein which is destined for export from the cell passes through the smooth endoplasmic reticulum to the Golgi complex.

We do not at present fully understand how ribosomes are attached to membranes. It is clear, as shown in Fig. 9, that the nascent growing peptide chains must be concerned in the process but whether they are a sufficient link in themselves is more doubtful. Divalent cations also appear to be involved and messenger RNA itself has also been implicated.

Liver contains a thio-disulphide interchange enzyme which was discovered by Anfinsen and his colleagues. This enzyme hastens the recovery of the enzymic activity of a disulphide containing protein such as ribonuclease after it has been denatured. It appears to play a role in ensuring that the correct disulphide bonds are formed for the biological activity of the protein. Many of the secreted proteins do in fact contain disulphide bonds so that it is interesting that Rabin and his colleagues (Williams, Gurari and Rabin, 1968) have located the disulphide interchange enzyme on the membrane of the endoplasmic reticulum.

The enzymes of the endoplasmic reticulum

Numerous enzymes are associated with the endoplasmic reticulum. Among these are glucose 6-phosphatase, various esterases, some sulphatases and a group of enzymes which make up an electron transport system which is different from and completely independent of that in mitochondria. It includes NAD and NADP-linked systems, cytochrome b_5 and the CO-binding pigment or cytochrome P-450.

Also present are mixed-function oxidases. Fatty acids are oxidized by these enzymes to yield ω-hydroxy fatty acids. For this reaction O_2 is required as well as NADPH. The hydroxylation reaction is followed by desaturation resulting in unsaturated fatty acids which may be utilized for the synthesis of lipids. We shall see in a later section that these enzymes play an important part in the metabolism of carcinogens and drugs.

With respect to the structure of the endoplasmic reticulum the ease with which the various enzymes may be released in a soluble form is interesting. Thus some enzymes, such as the esterases, are released when the membranes, from a preparation obtained by ultracentrifugation, are disrupted by ultrasonic vibrations and we may conclude that such enzymes are held within the vesicles in a soluble form. Other enzymes are released when the membrane is disrupted by a lipase preparation, such as Steapsin which is a crude extract from pancreas. NADPH-cytochrome c reductase and cytochrome b_5 are released by this means. In the case of other enzymes such as NADH-cytochrome c reductase and glucose 6-phosphatase it is not possible to release the enzymes in a form in which they are both truly soluble and also active. This may be because these enzymes require lipid for their activity but is more likely to indicate that they form an integral part of the membrane and are only active against their substrates when orientated in this manner.

D. SMOOTH-SURFACED ENDOPLASMIC RETICULUM

The major difference between the smooth endoplasmic reticulum and the rough is, of course, the lack of ribosomes. The smooth reticulum is found principally in tubular and vesicular forms and is functionally associated with the Golgi complex (see also Chapter 7). The smooth-surfaced membranes appear to be morphologically similar in all types of cell but their biochemical activity varies considerably between different cell types. The most prominent role for the smooth membranes is in the synthesis of lipids and sterols and such membranes contain a variety of phospholipids, cholesterol and neutral fat. These substances tend to be hydrophobic, in some cases markedly so, and yet they are synthesized from small hydrophilic molecules. Thus cholesterol is synthesized from acetate. There are, therefore, great advantages in having the enzymes involved in the synthesis of hydrophobic substances located in lipid-rich membranes since it is then possible for them to interact both with their substrates and their products. The phosphoglycerides are formed by a branching biosynthetic pathway starting from phosphatidic acid and involving cytidine nucleotides. These reactions, too, take place mainly in the membrane of the endoplasmic reticulum. In the intestinal epithelium the smooth endoplasmic reticulum plays a role in lipid transport and this is considered in detail in Chapter 30.

The interstitial cells of the testis, ovary and adrenal cortex are, of course, involved in the synthesis of steroid hormones and this activity is associated with the endoplasmic reticulum. The sarcoplasmic reticulum in skeletal and cardiac muscle is identical with the smooth-endoplasmic reticulum. Each myofibril is surrounded by these membranes, the sarcoplasmic reticulum. In mammalian muscle the diameter of the myofibrils is such that diffusion of substances is possible from the sarcoplasmic reticulum (see Chapter 20).

With respect to lipid synthesis the evidence is not conclusive as to whether the rough or smooth membranes have different roles. If [^{3}H]-glycerol is injected into rats the isotope is incorporated into phospholipids at the same rate in both types of membrane. If the synthesis of cholesterol is followed after the injection of [^{3}H]acetate then ^{3}H is more rapidly incorporated into cholesterol in the rough membranes but the cholesterol content of the smooth membranes is higher. What is certain is that both kinds of membrane are active in this respect.

In liver it is observed that the smooth membranes are associated with regions of glycogen storage. In both starved and fed animals glycogen accumulates near the membranes and both gluconeogenesis and glycogenolysis seems to be related to the membranes. The presence of glucose 6-phosphatase in the membranes must be considered to have some significance but the role of the smooth membranes in carbohydrate metabolism is not believed now to have the importance at one time assigned to them (see e.g. Phillips *et al.*, 1967).

E. GOLGI COMPLEX

This organelle gains its name from Camillo Golgi who, in 1898, revealed its presence in the Purkinje cells of the cerebral cortex of the barn owl. It can be detected in sections of fixed cells impregnated with silver (see Chapter 37, Fig. 7) or in the living cell using phase contrast optics. For many years the very existence of the Golgi complex was disputed and it was not until the advent of the electron microscope that

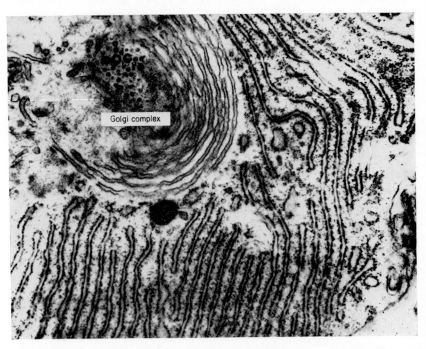

Golgi complex

Fig. 10. Rough surfaced endoplasmic reticulum and Golgi complex in the salivary gland acinar cell of a guinea-pig, ×27 200. (From Freeman, 1964, with permission).

its presence was confirmed. An electron micrograph of the Golgi complex in the guinea pig salivary gland is shown in Fig. 10. The Golgi complex is well-developed in secretory cells and neurones and there is morphological evidence that it is concerned with the elaboration of specific cellular secretions (Chapter 31). Its supranuclear position in secretory epithelial cells is consistent with this interpretation. Furthermore the degree of development of the Golgi complex in a cell appears to vary with the metabolic state of that cell. The complex consists of a stack of disc shaped cisternae one above the other. At one end the cisternae are only slightly curved (convex face) while at the other they are more steeply curved to give a concave face. At their edges the cisternae tend to swell to form vacuoles and this is particularly pronounced at the concave face (Fig. 1).

The composition of the membranes of the Golgi complex is very similar to that of the smooth endoplasmic reticulum. This, together with the functions of the complex to be discussed below, has suggested that the complex is in a continuous state of turnover. Thus new cisternae are probably formed from vesicles derived from the smooth endoplasmic reticulum at the convex face of the cisternae while at the concave face there is a continuous formation of vacuoles.

A major function of the Golgi complex is to serve as an internal transport system for the cell, both with respect to the transport of proteins and also lipids. Thus the work of Palade and his colleagues on the mode of formation of zymogen granules in the acinar cell of the pancreas (described in Chapter 31) has clearly shown that the Golgi complex is a one-way valve within the following membrane systems: endoplasmic reticulum $\longrightarrow$ Golgi vesicles $\longrightarrow$ Golgi vacuoles (secretory vacuoles). Similar systems have been demonstrated in other cells including the β-cells of the Islets of Langerhans; here the cell is concerned with the concentration of insulin and possibly with its conversion from pro-insulin before secretion (Chapter 29). In cells that are specialized for absorption, such as those of the intestinal epithelia, the Golgi complex contains lipid following the synthesis by endoplasmic reticulum of triglycerides from monoglycerides and fatty acids (see Chapter 30).

Not only does the Golgi complex serve a role in transporting and packaging but it also has synthetic activities. Thus it provides the substances required for the formation of the membrane of the vacuoles which are to be secreted from the cell. In the Goblet cells of the colon of rats it has been shown that a protein synthesized at the rough endoplasmic reticulum is passed to the Golgi complex, where it is conjugated with carbohydrate. The resultant mucopolysaccharide is transported as mucigen granules to become the mucus of the colon.

The Golgi is also active in the linking of carbohydrate to protein to form the glycoproteins. Nearly all the proteins which are exported from cells do in fact have a carbohydrate moiety. Some of this carbohydrate is probably linked to the peptide chains while they are still attached to the ribosome but most of the carbohyrdate is added during the passage of the protein through the Golgi complex.

The Golgi complex may also be the site of formation of the primary lysosomes containing hydrolytic enzymes. These enzymes are presumably first made at the rough endoplasmic reticulum (see Chapter 7).

From the above survey of the functions of the Golgi complex various enzymes would be expected to be associated with it. These may be divided into two types. Firstly, those concerned with the synthesis of lipoprotein membranes and polysaccharides and secondly, those which are present because they are being transported through the membrane system. The enzymes which are likely to be unique to the Golgi complex would be expected to be found among those of the first type. It is now possible to isolate from the disrupted liver cell fractions which, according to electron microscopy, are rich in fragments of the Golgi complex. Such fractions possess a high concentration of N-acetyl lactosamine synthetase activity. This enzyme is involved in the attachment of a disaccharide to the polypeptides present in glycoproteins. Some now regard this enzyme as a marker for the Golgi complex. (The role of this organelle in the formation of lactose was recently reviewed by Brew, 1970). Thiamine pyrophosphatase (diphosphatase) is also localized in the Golgi complex and is used as a specific histochemical marker for this organelle. It is probable that not all the cisternae in a single stack of the Golgi complex carry the same enzyme complement. Moreover the enzymic activity of the organelle will change with the metabolism of the cell.

There is no doubt that the Golgi complex is an organelle of considerable importance and much attention is being directed to establishing its many activities. (For reviews see Beams and Kessel, 1968; and Northcote, 1971).

F. ZYMOGEN GRANULES

A great many different products are secreted from various kinds of cells. We are still comparatively ignorant concerning the precise way in which most of these products are transferred through the cytoplasm and find their way into the extracellular space, though a great deal of work has been done particularly on the secretion of the enzyme precursors called zymogens by the acinar cells of the pancreas. In this case the zymogens are concentrated in membrane-bounded vacuoles which can easily be detected even in the light microscope. The zymogens are discharged from the cell into the pancreatic duct. Palade and Siekevitz and their colleagues have established how the zymogens come to be packaged and discharged from the cell and, thanks to their success, we now know much about this phenomenon. The reader is referred to Chapter 31.

II. Ribosomes

A. COMPOSITIONS AND STRUCTURE OF 80S RIBOSOMES AND THEIR RELATIONSHIP TO 70S RIBOSOMES

General properties

Ribosomes are electron dense particles that are found in all living cells and are the site of biosynthesis of protein. We shall be concerned in later sections with the location of ribosomes within the cell and the significance of the attachment of ribosomes to membranes. Here we wish to consider the structure of the ribosomes as such.

Living cells may be divided into two groups, prokaryotes and eukaryotes. The major distinction between these groups is that in the prokaryotes DNA is not confined by a membrane whereas the eukaryotes have a nuclear membrane and chromosomes are confined within this membrane. Another striking difference between the groups is that the eukaryotes possess mitochondria and/or chloroplasts whereas these are absent from the prokaryotes. The prokaryotes are represented by the bacteria and the blue-green algae whereas the cells of plants, fungi and animals are eukaryotes. There is a further difference between the groups in that the ribosomes are much larger in the eukaryotes than in the prokaryotes. It is usual to characterize macromolecules by their rate of sedimentation when submitted to a centrifugal field. The sedimentation coefficient of a particle depends on its size, shape and density (relative to that of the solvent) and is given in Svedberg units. Hence the ribosomes of prokaryotes have an S value of 70 and those of eukaryotes 80. The difference in molecular weight of the particles is much greater than would be indicated merely by reference to the S values. Thus a 70S ribosome has a molecular weight of $2 \cdot 6 \times 10^6$ and an 80S ribosome a molecular weight of $5 \cdot 2 \times 10^6$ under comparable conditions. Thus the one is nearly twice the size of the other.

Chemically ribosomes only contain RNA and protein. The 70S ribosomes are composed of two-thirds RNA and one-third protein whereas the 80S ribosomes contain about equal quantities of RNA and protein.

A ribosome consists of two subunits of unequal size, the large subunit being twice the size of the smaller. The subunit structure of the ribosomes may be observed both by electron microscopy, using a negative staining technique and also by treating the ribosomes in such a way that they dissociate into their subunits (Huxley and Zubay, 1960). The subunits can then be separated by density gradient centrifugation. It used to be thought that the conditions necessary for the dissociation of 70S ribosomes differed from those required to dissociate 80S ribosomes but this is now no longer believed to be the case. The essential condition is to reduce the Mg^{2+} ion concentration to below 1 mM. Table 1 shows the S values and molecular weights for the ribosomes and their subunits. It will be seen that the molecular weight of the large subunit in each case is approximately twice that of the small subunit.

TABLE 1. S values and particle weights of ribosomes

Source	Size of ribosome		Size of subunit	
	S	Mol. wt.	S	Mol. wt.
Prokaryotes	69.1	2.6×10^6	50	1.82×10^6
			30	0.9×10^6
Eukaryotes	81	5.2×10^6	60	3.0×10^6
			40	1.7×10^6

S is as $S°_{20w}$ Sedimentation coefficient corrected for water as solvent at 20°C and extrapolated to zero concentration of protein.

Ribosomal RNA and protein and the structure of the ribosomal subunits

The small subunit from either the 70S or 80S ribosome contains only one molecule of RNA, either 16S or 18S, whereas the large subunits contain two molecules of RNA, 23S or 28S and 5S. The distribution of the RNA is shown in Table 2. (There is evidence that the large subunit from HeLa cells contains also a 7S RNA). Rather little is at present known about the base sequence of the ribsomal RNA except for 5S. Brownlee and Sanger (1967) were responsible for determining the base sequence of the 5S from *E. coli* and the sequence of 5S RNA from other sources has since been elucidated. Apart from saying that it is probable that the structure allows for base pairing by folding the strand so that there are parallel segments, little is known about the structure. The 5S RNA is rather easily removed from the large subunit by treatment with EDTA and cannot be re-associated. There is good reason to believe that it is required for the biological activity of the ribosome but its precise function is not known.

Much work has been done to characterize the proteins of the 70S ribosome and progress has been such that not only have a large number of proteins been isolated but in some cases a function has been assigned to them. Similar work has been performed with 80S ribosomes but progress has been more difficult. All the evidence to hand is that the protein structure of the two kinds of ribosomes will prove to be rather similar and for this reason we will summarize the present state of knowledge with respect to *Escherichia coli* ribosomes.

TABLE 2. RNA content of ribosomal subunits

Ribosome	Small subunit		Large subunit	
	S	Mol. wt.	S	Mol. wt.
70S	16	0.6×10^6	23	1.1×10^6
			5	4×10^4
80S	18	0.7×10^6	28	1.6×10^6
			5	4×10^4

From Spirin and Gavrilova (1969).

Ribosomal proteins may be extracted by a procedure involving treatment with a mixture of urea and LiCl. The urea serves to dissociate the proteins while the LiCl causes precipitation of RNA. When the proteins extracted are examined by disc electrophoresis using polyacrylamide gels many discrete bands are revealed. At first it was doubted whether this really meant that a ribosome contained a mixture of different discrete proteins. This doubt was removed when the proteins were isolated and shown to differ in their amino acid sequence. The mixture of proteins obtained from intact ribosomes is far too complex for analysis so it is usual to extract the protein from the separated subunits. The small subunit contains 20 different proteins, their molecular weight varying from 5 000 to 69 000 but only two have a molecular weight greater than 31 000. The large subunit contains 30 different proteins with a rather similar molecular weight distribution to that of the small subunit. It is possible to calculate from the data the number of moles of each protein per ribosomal subunit and this indicates that some proteins are present in only 0·5 molar quantities while some are present in 1·5 molar amounts. This finding is at present interpreted as indicating that not all ribosomes even in *E. coli* are identical with respect to their protein composition. As might have been expected the ribosomal proteins seem to be species specific. This may be shown by examining the electrophoretic mobility of the proteins extracted from ribosomes of distantly related species or by immunological techniques. Thus rabbits will form antibodies to ribosomal proteins from the brine shrimp *Artemia salina* (class Arthropoda) (Decken and Hultin, 1970).

Ribosomes are revealed as compact dense particles under the electron microscope but unfortunately the resolution is not such as to reveal much about the conformation of ribosomes and their subunits. The

F*

acidic phosphate groups of the ribosomal RNA associate with the proteins which tend to be basic. As already indicated the subunits require Mg^{2+} ions to maintain their association in the ribosome. There are also present polyamines such as spermine which must be regarded as part of the structure and which can to some extent replace Mg^{2+} ions. Monovalent cations especially in combination with Mg also seem to play an important role in preserving the integral association of the RNA and protein.

It is natural to look to another ribonucleoprotein particle, the RNA virus, to try to learn something of the structures of ribosomes. In the virus the RNA is surrounded by a number of protein subunits arranged in a symmetrical manner. It seems certain that this is not so in the ribosome. At one point it appeared that the RNA in the ribosome looped over the outside of a structure composed of a wall of protein subunits. According to this model all the protein subunits would be equally susceptible to attack by proteolytic enzymes. Recent work involving the digestion of the *E. coli* 30S subunit with low concentrations of trypsin or treatment with protein reagents has made this model less likely. It appears that the proteins most closely associated with the RNA are least sensitive to digestion and hence some proteins must be buried within the structure.

B. ASSEMBLY OF RIBOSOMES

Formation of ribosomal subunits in vitro

We have explained in the last section that ribosomal subunits may be obtained by the removal of magnesium from ribosomes. It was shown first with bacterial ribosomes that they would rather easily reassociate to produce ribosomes which in the presence of messenger RNA containing only uridine ribonucleotides were capable of forming polyphenylalanine under suitable conditions of incubation. (UUU is the *m*RNA code for phenylalanine. See Chapter 34). Such experiments with animal ribosomes proved more difficult but it was eventually found that, if the nascent polypeptide chain, attached to the 80S ribosomes on isolation, was first removed by incubation under conditions of protein synthesis in the presence of puromycin (see Chapter 34), the ribosomes could be treated to give subunits which retained activity as with ribosomal subunits from bacteria. In fact it seems that one can prepare hybrid ribosomes made up of subunits derived from different species, so long as the two contributing species are both prokaryotes or both eukaryotes. An experiment that demonstrates the activity of ribosomal subunits obtained from guinea pig mammary gland and rat liver is shown in Table 3.

TABLE 3. Activity of ribosomal subunits from rat liver and lactating mammary gland of guinea pig

Source of subunit	Size	Column							
		1	2	3	4	5	6	7	8
		Incubation medium contained							
Lactating gland	small	×		×				×	
(g-p)	large		×	×					×
Liver (rat)	small				×		×		×
	large					×	×	×	
Radioactivity		254	1068	7524 (1322)	216	764	10052 (980)	8161 (1018)	7383 (1284)

The ribosomal subunits were incubated in the presence of [^{14}C]phenylalanine and and polyuridylic acid so that the results represent the amount of [^{14}C]polyphenylalanine formed (radioactivity as counts per min). The separate subunits have some activity as shown in columns 1, 2, and 4, 5. When the two subunits from the same source are combined there is a marked increase in activity (e.g. column 3 and column 6) which is much greater than that expected from a simple addition of the activity of the two subunits (figures in brackets). A similar result is obtained when subunits from different sources are combined (columns 7 and 8) indicating the formation of active hybrid ribosomes. (From Fairhurst, et al., 1971, with permission).

Nomura and his group are responsible for some dramatic experiments. They have reconstituted the 30S ribosomal subunit of E. coli by mixing together the 16S RNA and the previously isolated ribosomal proteins. The proper assembly of the 20 different proteins with the 16S RNA must involve a very precise series of specific interactions. Nomura has now demonstrated that the proteins link together in a cooperative step-wise manner and that many of the proteins are linked in a precise order. One of the reactions requires a higher temperature. Thus when 16S RNA is incubated with the whole mixture of 20 proteins some of the proteins bind and a so-called RI particle is formed. This particle is then able to receive other proteins after the temperature is elevated to 40° and lowered again to 0°. If the mixture of proteins used in these reconstitution experiments is incomplete the ribosomal subunit is either not re-formed or its physical and/or biological properties are much affected. Experiments aimed at the reconstruction of active 50S particles have not been so successful largely because of the slowness of the reassociation of protein and RNA. Nomura has recently turned to a thermophilic microorganism Bacillus stearothermophilus, and has obtained the reconstitution of the 50S subunit at 60°. Even so the process was very much slower than with the 30S subunit and there is evidence that in the cell the 30S subunit plays a role in assisting in the assembly of the 50S subunit (Nomura, 1970). Similar experiments with ribosomal subunits from 80S ribosomes are under way but have not so far been successful.

Formation of ribosomes in the living cell

In common with all types of RNA, ribosomal RNA is synthesized on a DNA-template using RNA polymerases. While messenger RNA and transfer RNA are synthesized in the nucleoplasm, ribosomal RNA is formed in the nucleolus and this is probably the site where the ribosomal proteins, which are probably synthesized in the cytoplasm, and the RNA associate to give rise to ribosomal subunits. In eukaryotic cells the ribosomal RNA is first formed as a larger molecule with an S value of 45. This is then modified to provide the ribosomal RNA and the process of modification is known as maturation. Much of our knowledge of this subject is due to the work of Penman and his colleagues using HeLa cells and their results are summarized in Fig. 11.

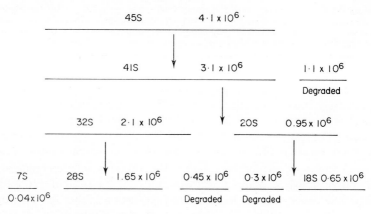

Fig. 11. Maturation of ribosomal RNA in HeLa cells. The numbers to the left represent the S values of the RNA and the numbers to the right the calculated molecular weight. The 7S RNA is hydrogen bonded to 28S RNA in the ribosome. (Modified from Weinberg and Penman, 1970).

Since the molecular weight of the 45S RNA is $4 \cdot 1 \times 10^6$ daltons and that of the 18S plus 28S is $2 \cdot 3 \times 10^6$ daltons nearly half the RNA is discarded during maturation. The 5S RNA is not derived from the 45S RNA but is incorporated into the large ribosomal subunit separately. At present no explanation of this apparent waste of material has been forthcoming. Methylation of certain bases in ribosomal RNA takes place in the precursor molecules but the discarded sections are not so methylated.

C. ROLE OF FREE AND BOUND POLYRIBOSOMES
IN PROTEIN SYNTHESIS

Synthesis of transported proteins

We have already explained how cells that secrete a large amount of protein possess a well developed rough-surfaced endoplasmic reticulum. The liver cell is a typical example of such a cell in that it exports about half the protein it synthesizes in the form of plasma protein. The liver cell is interesting in another respect since experiments have been done which show that other types of protein are also synthesized by the membrane-bound polyribosomes. Thus catalase which is a component of the peroxisomes (see Chapter 6) is synthesized at this site. To the extent that the peroxisomes may be thought of as stay-at-home zymogen granules this finding might have been predicted. More recently it has been shown that the soluble proteins of the mitochondria, e.g. cyto-chrome c, are synthesized in the cytoplasm rather than in the mito-chondria themselves. It is not precisely certain that cytochrome c is synthesized by the bound polysomes but this is very probable. We can conclude therefore that the rough-surfaced endoplasmic reticulum is the site of synthesis not only of protein destined for export from the cell but also of those proteins such as the lysosomal enzymes, that are internally transported to other organelles.

We mention two further examples of proteins that are subjected to transport after their synthesis, and in which it is clear that the mem-brane is of great physiological importance (Campbell, 1970b).

Extracts of guinea pig lactating mammary glands effect the synthesis of the milk protein, α-lactalbumin. This small protein with a molecular weight of about 15 000 is retained after synthesis by the membranous components of the isolated extracts. Ebner was the first to demonstrate that α-lactalbumin was one of the two proteins (A and B) involved in the synthesis of lactose. The reaction is

$$\text{UDP-galactose} + \text{glucose} \xrightarrow{\text{A} + \text{B proteins}} \text{lactose} + \text{UDP}$$

The B protein is α-lactalbumin. Brew showed that the A protein was a transferase enzyme that catalyzes the reaction

$$\text{UDP-galactose} + \text{N-acetyl glucosamine} \longrightarrow \text{N-acetyl lactosamine} + \text{UDP}$$

This enzyme is important in the serial attachment of monosaccharides to form the carbohydrate moiety of many glycoproteins (see above, p. 160). The presence of α-lactalbumin (B protein) changes the specificity of the transferase in the presence of substrate glucose and the net result is the synthesis of lactose.

The transferase, as previously mentioned, is located in the membrane of the smooth endoplasmic reticulum and of the Golgi complex. From this and the fact that α-lactalbumin is synthesized by the bound poly-somes, Brew (1970) has postulated that the lactose synthetase is con-trolled as shown in Fig. 12. As α-lactalbumin passes from the rough to the smooth endoplasmic reticulum its interaction with A protein causes lactose to be synthesized. At the cessation of lactation the syn-thesis of α-lactalbumin ceases and this will at once stop the synthesis of lactose for any residual α-lactalbumin will rapidly pass out of the mammary gland into the milk.

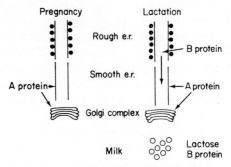

Fig. 12. The role of membrane-bound polysomes in the synthesis of α-lactalbumin and the control of lactose synthesis in the mammary gland. Based on the work of Brew (1970). (From Campbell, 1970b).

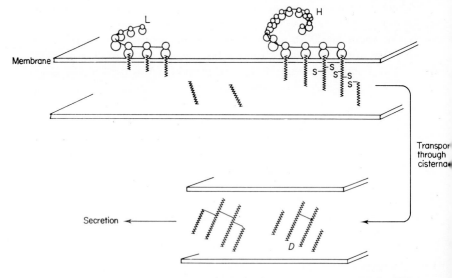

Fig. 13. The role of membrane-bound polysomes in the synthesis of immunoglobulin. From Williamson (1969) with permission.

The second example is taken from the work of Askonas and Williamson (1968) on the biosynthesis of immunoglobulin. This protein is a multichain protein composed of two light chains (L chains) and two heavy chains (H chains). There is a problem of assembling the four chains to a functionally active protein. The chains are linked through disulphide bridges. The present concept is shown in Fig. 13. This shows that the two types of chain are synthesized by distinct polysomes, large and small, attached to membranes. More recently they have shown that an intermediate is formed, this is shown as D in the figure. The formation of the intermediate and of the finished immunoglobulin is obviously greatly facilitated by the containment of the intermediates within the controlled environment of the cisternal space.

Synthesis of protein by membrane-free polyribosomes

In some cells the great majority of polyribosomes are not associated with the endoplasmic reticulum and do not appear to be membrane-bound. Such polyribosomes are said to be free and the reticulocyte and cells of skeletal muscle are typical of cells containing many free polyribosomes. Such cells do not secrete proteins, so that the idea has evolved that free polyribosomes are responsible for the synthesis of retained protein.

Many proteins are synthesized in muscle cells to form the myofibrils. Among these proteins are actin, myosin, tropomyosin and myoglobin of which myosin is the major fibrous protein of muscle. These proteins differ in their molecular weight and have been shown by Heywood (1969) to be formed by polyribosomes of varying size, i.e. containing varying numbers of ribosomes. Myosin has a molecular weight of 200 000 and so the messenger RNA would be predicted to be very long. Provided that the messenger RNA is saturated with ribosomes then one might expect the polyribosomes responsible for the synthesis of myosin to contain 55–65 ribosomes. Heywood working with skeletal muscle from chick embryos did in fact isolate polyribosomes of this size; a record that remains so far unbeaten. Actin has a molecular weight of 60 to 70 000 and is made by polyribosomes of 15–25 ribosomes, while tropomyosin has a molecular weight of 30 to 35 000 and is made on polyribosomes of 5–9 ribosomes. These results suggest that large-size polyribosomes are capable of retaining their structure and activity without the support of membranes.

The question arises as to the origin of retained protein in a cell such as liver which also synthesizes protein for secretion. Is the protein that is synthesized for the internal economy of the cell made on membrane-bound polyribosomes or on polyribosomes that are not associated with the rough endoplasmic reticulum? If all protein is made on the rough endoplasmic reticulum in cells that possess this organelle then some mechanism would have to exist for the separation of the protein destined for export from that to be retained by the cell. The Golgi complex has been suspected of playing a part in this process but on theoretical grounds it would seem simpler for the cell to possess two kinds of polyribosomes.

In electron micrographs of the liver cell, we can see polyribosomes that do not appear to be associated with the endoplasmic reticulum but we cannot be sure of this interpretation. We can also isolate from a cell homogenate two fractions which we term bound and free polyribosomes. Fig. 14 shows a commonly employed method for their preparation taken from the work of Blobel and Potter (1967).

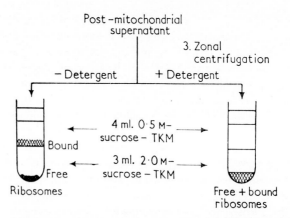

Fig. 14. The preparation from the disrupted liver cell of different polyribosome-containing fractions. From Blobel and Potter (1967). The liver is disrupted and the nuclei and mitochondria removed. To one portion of the post-mitochondrial supernatant is added the detergent deoxycholate to disrupt the membranes of the endoplasmic reticulum. Both suspensions are then centrifuged over two zones of sucrose. After 3 h centrifugation pellets containing polyribosomes are obtained. The fragments of the endoplasmic reticulum with bound polyribosomes are found at the interface of the two zones of sucrose in the preparation lacking deoxycholate.

So far, electron microscopy, analysis of RNA and of protein, and studies on the metabolic turnover of RNA have not revealed any significant difference between the ribosomes in the two fractions. Moreover, there seems little difference in the ability of the polyribosomes in the two fractions to effect the synthesis of total protein either *in vivo* or *in vitro* nor is there a marked difference in the size of the polyribosomes in the two fractions. One might conclude, therefore, that the polyribosomes in the two fractions have a common derivation. According to this the so-called free-polyribosomes would be detached from the endoplasmic reticulum during disruption of the cell and subfractionation. The challenge to the biochemist is to demonstrate that the polyribosomes in the "free-polyribosome" fraction effect the synthesis of a protein that differs from that synthesized by the polyribosomes in the "bound-polyribosome" fraction.

There is now mounting evidence that this objective has been achieved. It is clear that the synthesis of serum albumin is confined to the bound polyribosomes. In contrast the synthesis in the liver of the iron-containing protein ferritin is confined to the membrane-free polyribosomes. In other cells it has been shown that β-lactoglobulin, a protein of milk, is synthesized solely by the bound polyribosomes of the mammary gland.

We can conclude therefore that the segregation of protein for export or retention by the cell is probably effected at the site of synthesis on the ribosome. We do not however know how the mRNA for the two types of proteins comes to be associated either with the free-polyribosomes or with the bound-polyribosomes.

III. Generation of Membranes

A. ENDOPLASMIC RETICULUM

After cell division the endoplasmic reticulum has to increase in amount in order that the daughter cells shall have the same composition as that of their parents. Unfortunately, our knowledge of the behaviour of the endoplasmic reticulum during mitosis is limited and such knowledge as we do have has been obtained from a study of onion root tips by Porter and Machado (1960); though their conclusions may be valid for eukaryotic cells in general. They showed that during the later stage of prophase the endoplasmic reticulum remained outside the spindle and did not appear to participate in the mitotic process. The endoplasmic reticulum is then distributed between the daughter cells.

The biogenesis of membranes is now a very popular field of study for biologists but unfortunately progress has been very slow. This applies equally to our knowledge of the biogenesis of the endoplasmic reticulum and the temporal relationship between the rough surfaced and smooth surfaced reticulum. Since the one is associated with ribosomes, and these are the sites of protein synthesis, it might seem reasonable to conclude that the rough membranes should be made first and that they would give rise to the smooth membranes through the loss of ribosomes. There is sound experimental evidence to support this hypothesis and as an example we consider the work of Dallner et al. (1966).

In the liver of the rat there is a period of rapid cell differentiation between three to eight days after the birth of the animal. Just before birth and immediately after there is an increase in the rough-surfaced membranes. A little later it is the smooth-surfaced membranes that show the rapid increase. By injecting rats with [^{14}C]leucine during these periods and determining the radioactivity of the proteins of the endoplasmic reticulum it can be shown that radioactive protein appears first in rough and then in the smooth reticulum. Another approach is to study the enzyme activity of the endoplasmic reticulum. These authors showed that glucose 6-phosphatase and NADPH-cytochrome c reductase activity which is not present in the liver of new born animals appears first in the rough membranes and subsequently in the smooth membranes. In adult rats these enzymes are equally distributed between the two types of membrane. In order to demonstrate that the increase in enzymic activity is due to formation of new protein and not to the activation of an existing protein the authors used the antibiotics, actinomycin D, and puromycin. The first effectively blocks the synthesis of mRNA which would be required for the synthesis of the mRNA which would need to be translated to form the new enzyme. The second effectively inhibits the translation of mRNA. Both actinomycin D and puromycin prevented the increase in activity of glucose 6-phosphatase and cytochrome c reductase.

While the above experiments provide good evidence for the conversion of rough to smooth reticulum there is some evidence that the two may be formed independently. The fact that the smooth membrane lacks some of the enzymes present in the rough membrane could be accepted on the basis that the enzymes are lost on conversion of rough to smooth membrane, but if it should be shown that the smooth membrane possesses enzyme activity that is lacking in the rough membrane it is harder to see how the former could give rise to the latter. In fact the smooth membranes do seem to have additional enzyme activity. Thus the carbohydrate synthetase that has previously been mentioned as playing an important part in the synthesis of glycoproteins appears to be confined to the smooth membranes, especially Golgi. Another aspect arises from work which has shown that NADPH-cytochrome c reductase is synthesized by both membrane bound and free polyribosomes. The reductase is a constituent enzyme of the membrane of the endoplasmic reticulum. Perhaps the enzyme synthesized by the free polyribosomes is adsorbed on to pre-existing membranes.

These various theories are mentioned so that it may be understood that we at present have much to learn concerning the generation of the endoplasmic reticulum.

B. GOLGI COMPLEX

The Golgi complex is recognized in electron micrographs by the organized arrangement of the membrane-bound cavities. This is in contrast to the endoplasmic reticulum which in structure appears to be either slightly, or very irregular, also described as pleomorphic. The origin and mode of multiplication of the Golgi membrane is not at present clear. The failure lies partly in the technical difficulty of isolating Golgi membranes in a reasonably homogeneous state. Beams and Kessel (1968) have discussed three possibilities for the origin of the Golgi complex: (a) formation *de novo*; (b) division of pre-existing Golgi membranes; (c) conversion of other membrane systems in the cell.

Since the Golgi complex seems to be devoid of ribosomes and thus protein synthetic activity it appears most likely that the organelle is derived from other membranous structures of the cell. Daniels (1964) has studied the origin of the Golgi complex in the amoeba *Pelomyxa illinoisensis*. He came to the conclusion that the membranes are formed by invagination of the plasma membrane which eventually will produce independent entities—the Golgi complex. It remains to be seen whether the results obtained for the origin of the Golgi complex in amoeba are valid also in mammalian cells.

A somewhat different view of the origin of the Golgi complex comes from the electron microscopic data of Claude (1969). His micrographs support the view that pre-existing Golgi membranes of liver may be modified by additions of new membrane components having their origin in the rough or smooth endoplasmic reticulum. In fact, smooth-surfaced extensions of the rough endoplasmic reticulum can often be recognized alongside the Golgi membranes. The secretory pathway in the exocrine pancreatic cell involves the budding of Golgi vesicles from the smooth endoplasmic reticulum and their subsequent fusion with the Golgi, which implies the addition of membrane to the latter.

C. OTHER MEMBRANES

Much attention is being directed to the formation of the mito-chondrial membranes. It is now possible to separate the inner and outer mitochondrial membranes and to show that the composition of the outer membrane is close to that of the endoplasmic reticulum. There is, however, no evidence that the one is derived from the other and much to support the idea that mitochondria increase by budding.

The plasma membrane too may now be isolated and separated from the membranes of the endoplasmic reticulum. It is a membrane of high metabolic activity but no very clear idea as to its origin is yet available.

IV. INFLUENCE OF VARIOUS CONDITIONS ON THE STRUCTURE AND FUNCTION OF THE ENDOPLASMIC RETICULUM

In the previous sections we have considered the structure and func-tion of the membranous constituents of animal cells and may have failed to convey the fact that for the proper functioning of a cell all the various constituents have to act in concert. If for any reason the structure is disturbed then the functioning of the cell will also be disturbed. Perhaps more surprisingly, the reverse is also true. Thus if a drug such as phenobarbitone is administered then there is an increase in the activity of the hydroxylating enzymes that are involved in its detoxication and also a concomitant increase in the amount of membrane associated with the endoplasmic reticulum. The effect of pathological processes and drugs on the endoplasmic reticulum is considered in Chapter 8. In this section we therefore confine our attention to the effect of various physiological changes on the endoplasmic reticulum.

A. STATE OF NUTRITION

Starvation has a marked effect on the structure of the endoplasmic reticulum in the liver cells. Even after a few days the amount of the rough-surfaced membrane is much reduced, the mitochondria are swollen and smooth membrane is found in those regions that previously contained glycogen. Upon refeeding the membranes appear to return to normal within 15 h. Another effect of starvation is to cause a reduction in the size of the polyribosome aggregates so that after about four days the polyribosomes that do exist are small and most ribosomes appear as monomers. Again the polyribosomes will return to their normal size distribution within 8 h of refeeding. The size of polyribosomes must reflect the amount of messenger RNA available for translation into polypeptide chains, and it is apparent that complete starvation reduces the amount of messenger RNA present in the cytoplasm.

On the other hand, if instead of starvation the animals are kept for several days on a protein-deficient but high caloric diet, the effects on the size of polyribosomes are not observed. Under these conditions messenger RNA is available, but the cell has a reduced capacity for protein synthesis due in part to the lack of essential amino acids in the cytosol. In cell cultures within 24 h of partial amino acid deficiency reversible changes in the ultrastructure of mitochondria are observed. However, the endoplasmic reticulum seems unaffected under these conditions although the rate of protein synthesis is decreased considerably. Changes in the ultrastructure of the endoplasmic reticulum are observed only after animals have been deprived of proteins (but not of other food constituents) for a prolonged period of time and, therefore, the changes are considered to be of a secondary nature. Under these conditions, the number of polyribosomes decreases also. The results emphasize that protein synthesis depends on the adequate supply of energy and amino acids and conditions in which one or other or both are reduced will have a marked effect on protein synthesis.

B. HORMONES

A wide range of hormones have been shown to have a marked effect on the rate of protein synthesis in their target organs. Since the rough-surfaced endoplasmic reticulum plays such an important role in the synthesis of protein and the smooth-surfaced reticulum in the transport of protein, it is not surprising that some hormones affect the organization of the cytomembranes and ribosomes in the cell.

The rate of synthesis of a particular protein may be controlled at the level of translation (read-out of the mRNA attached to the ribosomes) or at the level of transcription (formation of the specific mRNA by the DNA-dependent RNA polymerase). Comparatively little is known about the control of protein synthesis at the level of translation. Rather more is known about the mechanism of action of RNA polymerase and so it has been common to determine the effect of the administration of a hormone on the activity of this enzyme. Unfortunately the activity of the polymerase is affected by many circumstances but, even if it is clearly established that a hormone affects its activity, it does not follow that this is the primary site of action of the hormone. A useful technique for differentiating between translational control and transcriptional control is to test the effect of administration of actinomycin D. This drug prevents the formation of RNA, including mRNA, so that if the effect of the hormone is maintained in its presence then it seems more probable that the hormone is acting at the level of translation rather than transcription.

TABLE 4. The effect of some hormones on the induction of rat liver enzymes.

Enzyme	Hormone
Glycerol phosphate dehydrogenase (mitochondrial)	Thyroxine
Phosphoenolpyruvate carboxykinase	Cortisone
Pyruvate Carboxylase	Cortisone
Serine deaminase	Glucagon, hydrocortisone
Tryptophan pyrrolase	Cortisone
Tyrosine transaminase	Cortisone, Insulin, Glucagon

Bacteria can be induced to synthesize a particular enzyme by the addition to the medium of an appropriate substance. Eukaryotic cells can also be induced to synthesize an increased amount of a particular protein but the effects are slower and in general the increases observed are less dramatic than in the case of prokaryotic cells. Enzyme induction and repression in vertebrates occurs mainly in the liver and not in the other tissues such as muscle and brain. This is probably because the liver is the organ that is most rapidly affected by a change in the supply of nutrients. Table 4 shows the effects of some hormones on the induction of liver enzymes. In some cases similar effects can be induced by the administration of amino acids. Thus the activity of threonine dehydrase can be increased 300 fold in this way.

The following is a brief survey of the action of certain hormones on protein synthesis. So far it has not proved possible to demonstrate the effect of a hormone on a subcellular preparation so that all the results described concern the effect of administering the hormone to animals.

Corticosteroid hormones

These hormones, especially cortisone, appear to increase the rate of catabolism of proteins in many extra-hepatic tissues. The effect is to increase the supply of amino acids to the liver. The glycogenic amino acids are utilized for the formation of glycogen so that the proportion of glycogen granules in the liver cells increases. As in Table 4, cortisone causes an induction of the enzymes involved in amino acid metabolism (tryptophan pyrrolase, tryosine transaminase and serine deaminase). In marked contrast to the extra-hepatic tissues the protein synthesizing activity of the liver increases and the endoplasmic reticulum apparently proliferates as can be seen from electron micrographs.

Effects similar to those obtained after the administration of corticosteroid hormones are obtained by exposing the animals to stress. A stress effect is produced by injecting a silicate (Celite) into the intraperitoneal cavity.

Thyroid hormone

The effect of thyroxine is usually studied with thyroidectomized animals. The major response is a rise in the basal metabolic rate with an increase in the rate of synthesis of many enzymes, especially glycerol phosphate dehydrogenase (see Table 4). Not surprisingly, therefore, thyroxine increases the activity of RNA polymerase and there is also an increase in the protein synthesizing activity of the liver microsome fraction under *in vitro* conditions.

Growth hormone

Hypophysectomy causes alterations in rat liver cells and these can be reversed by administration of growth hormone. The cytoplasmic volume decreases and the amount of smooth endoplasmic reticulum is diminished as well as the content of particulate glycogen. A dilatation and disorganization of the rough endoplasmic reticulum is observed as well as a decrease in protein synthesis due to the loss of polyribosomes in the cell. The mitochondria are in a swollen state and when isolated have a reduced P/O ratio. The deficiency in production of ATP would, of course, contribute to the decrease in ability of the cells to incorporate amino acids into protein. Biochemical data indicate a decrease in the rate of synthesis of RNA including both messenger, ribosomal and transfer RNA. In summary, hypophysectomy seems to interfere with cell metabolism as a whole, particularly with protein and RNA synthesis, mitochondrial functions and carbohydrate metabolism. The primary and specific action of growth hormone is, however, not known in detail yet.

Insulin and Glucagon

As shown in Table 4 glucagon induces the activity of some hepatic enzymes. The effect of glucagon is inhibited by insulin (see Kenney, 1970). The induction by glucagon is mediated by cyclic AMP, the concentration of which in the liver is enhanced within a few minutes after administration of the hormone. Certain enzymes can also be induced by the direct administration of cyclic AMP.

The effect of glucagon through the mediation of cyclic AMP on the induction of enzymes is inhibited by actinomycin D. This might be taken as evidence of a direct effect of cyclic AMP on the synthesis of *m*RNA. Against this view is the rapidity of enzyme induction suggesting that there might not be time for new *m*RNA to be synthesized and translated. Some would, therefore, favour the view that cyclic AMP affects the rate of translation of existing messenger RNA. It is not possible at present to choose with certainty from among these alternatives.

Although insulin inhibits the effect of glucagon on the induction of enzyme synthesis in the liver, it has a stimulatory influence on protein synthesis in other tissues, e.g. diaphragm, heart muscle and skeletal muscle. If the ribosomes are isolated by differential centrifugation from the skeletal muscle of a diabetic rat they appear to be less active for protein synthesis than are those from a normal rat. Administration of insulin restores to normal the activity of isolated ribosomes from the skeletal muscle of the diabetic rat. This effect is a rapid one, a mere 5 min. being required between administration of insulin and the isolation of ribosomes. It is not affected by actinomycin, but is prevented by the administration of puromycin which inhibits all protein synthesis. These observations suggest that the action of insulin is to stimulate the synthesis of a protein that is required for the full activity of the ribosomes. So far the identification of the protein has not been established.

C. PARTIAL HEPATECTOMY

If two-thirds of the liver of a rat is removed by surgery the animal lives on and the liver regains its normal size, but not its normal shape, within three weeks of the operation. This phenomenon has fascinated biologists for many years. How is it that the liver just regains its former weight? How does one-third of the liver manage to cope with the synthesis of protein required for a normal size animal? These and many other questions require to be answered. So far as protein synthesis is concerned we appear to have the answers to certain basic questions. If one compares the protein synthesizing activity of an isolated microsome fraction from the liver of an animal 48 h after partial hepatectomy with that of a normal animal, then the so-called regenerating liver microsomes are about twice as active as normal. This increase in activity is not due to some factor in the soluble fraction of the cell, i.e. in the general cytoplasm, but is due to some property of the particles. If one treats the particulate fraction with deoxycholate and isolates the ribosomes then there is no difference in the protein synthesizing activity between the ribosomes from regenerating and normal liver. Thus one can argue that the increase in activity of the microsomal fraction from regenerating liver is due to some factor in the membrane (see Campbell et al., 1967). There have been many attempts to identify such a factor without success. If one examines the size distribution of the polyribosomes from the two types of liver then no clear cut difference can be established so that the difference in activity cannot be attributed to an increase in the size of the polyribosomes in the regenerating liver.

V. CONCLUSIONS

We have been concerned in this chapter with the structure and activity of a fascinatingly complex group of membranous organelles. The fact that we understand as much as we do of their structure is due to the work of the electron microscopists, particularly those who have worked at the Rockefeller Institute in New York. Their work started to yield dividends in the mid-1950s and has continued to do so ever since. The biochemists have worked hard to try and follow up the advances made by the electron microscopists and, led by Zamecnik and his group in Boston, they have done well. They do, however, work under great difficulties in that it is very hard to isolate the membranous organelles in a form in which they retain much of the structure they possess in the intact cell. Each year, however, new structures are isolated and at the time of writing good progress is being made with the isolation of the Golgi complex and the plasma membrane.

Some of our colleagues who work with bacteria, and have made such outstanding progress in our understanding of the fundamental mechanisms in the synthesis of protein and nucleic acid, wonder openly why some of us work with such complex eukaryotic cells. Our answer is that we are intrigued by their complexity, the way each type of differentiated cell is fashioned to serve a particular set of functions, the way their activity may be changed by hormones and perhaps most important of all how eukaryotic cells really do differ from prokaryotic cells. We can only hope that we have managed to infect our readers with some of our enthusiasm.

Acknowledgement

We would like to thank Mr H. Grayshon Lumby for Figs 1, 3(b), 5, 9 and 12.

References

Askonas, B. A. and Williamson, A. R. (1968). Interchain disulphide-bond formation in the assembly of immunoglobulin. Heavy-chain dimer as an intermediate. *Biochem. J.* **109**, 637–643.

Beams, H. W. and Kessel, R. G. (1968). The Golgi apparatus, structure and function. *Int. Rev. Cytol.* **23**, 209–276.

Blobel, G. and Van Potter, R. (1967). Studies on free and membrane-bound ribosomes in rat liver. *J. Mol. Biol.* **26**, 279–292.

Brew, K. (1970). Lactose synthetase: evolutionary origins, structure and control. *Essays Biochem.* **6**, 93–118.

Brownlee, G. G. and Sanger, F. (1967). Nucleotide sequences from the low molecular weight ribosomal RNA of *Escherichia Coli*. *J. Mol. Biol.* **23**, 337–353.

Campbell, P. N. (1970a). Scientific basis of medicine. *Ann. Rev.* 173–188.

Campbell, P. N. (1970b). Functions of polyribosomes, attached to membranes of animal cells. *FEBS Lett.* **7**, 1–7.

Campbell, P. N., Lowe, E. and Serck-Hanssen, G. (1967). Protein synthesis by microsomal particles from regenerating rat liver. *Biochem. J.* **103**, 280–288.

Claude, A. (1969). Microsomes, endoplasmic reticulum and interactions of cytoplasmic membranes. In "Microsomes and drug oxidations", p. 3–39. (J. R. Gillette, A. H. Conney, G. J. Cosmides, R. W. Estabrook, J. R. Fouts, and G. J. Mannering, eds). Academic Press, New York and London.

Dallner, G., Siekevitz, P. and Palade, G. E. (1966). Biogenesis of endoplasmic reticulum membranes. I. Structural and chemical differentiation in developing rat hepatocyte. *J. Cell Biol.* **30**, 73–96. II. Synthesis of constitutive microsomal enzymes in developing rat hepatocyte. *J. Cell Biol.* **30**, 97–117.

Dallner, G. and Ernster, L. (1968). Subfractionation and composition of microsomal membranes: a review. *J. Histochem. Cytochem.* **16**, 611–632.

Daniels, E. W. (1964). Origin of Golgi system in Amoebae. *Z. Mikrosk. Anat. Forsch.* **64**, 38–51.

Decken, von der, A. and Hultin, T. (1970). Immunological properties of isolated protein fractions from Artemia ribosomes. *Exp. Cell Res.* **64**, 179–189.

Fairhurst, E., McIlreavy, D. and Campbell, P. N. (1971). The protein synthesizing activity of ribosomes isolated from the mammary gland of lactating and pregnant guinea pigs. *Biochem. J.* **123**, 865–874.

Florendo, T. (1969). Ribosome structure in intact mouse liver cells. *J. Cell Biol.* **41**, 335.

Freeman, J. A. (1964). "Cellular Fine Structure". McGraw-Hill, New York.

N

Goldblatt, P. J. (1969). The endoplasmic reticulum. *In* "Handbook of molecular cytology". (A. Lima-de-Faria, ed). p. 1101–1129. North Holland Publishing Company, Amsterdam.

Heywood, S. R. (1969). Synthesis of myosin on heterologous ribosomes. *Cold Spring Harbor Symp. Quant. Biol.* **34**, 799–803.

Huxley, H. E. and Zubay, G. (1960). Electron microscope observations on the structure of microsomal particles from *Escherichia coli. J. Mol. Biol.* **2**, 10–18.

Kenney, F. T. (1970). Hormonal regulation of synthesis of liver enzymes. *In* "Mammalian protein metabolism" (H. N. Munro, ed), Vol. 4, p. 131–177. Academic Press, New York and London.

Maggio, R., Siekevitz, P. and Palade, G. E. (1963). Studies on isolated nuclei. 1. Isolation and chemical characterization of a nuclear fraction from guinea-pig liver. *J. Cell Biol.* **18**, 267–291.

Nomura, M. (1970). Bacterial ribosomes. *Bacteriol. Rev.* **34**, 228–277.

Northcote, D. H. (1971). The Golgi apparatus. *Endeavour* **30**, 26–33.

Phillips, M. J., Unakar, N. J., Doornewaard, G. and Steiner, J. W. (1967). Glycogen depletion in the newborn rat liver. An electron microscopic and electron histochemical study. *J. Ultrastruc. Res.* **18**, 142–165.

Porter, K. R. and Machado, R. D. (1960). Studies on the endoplasmic reticulum. IV. Its form and distribution during mitosis in cells of onion root tip. *J. Biophys. Biochem. Cytol.* **7**, 167–180.

Robertson, J. D. (1969). Molecular structure of biological membranes. *In* "Handbook of Molecular Cytology". (A. Lima-de-Faria, ed.), p. 1403–1443. North Holland Publishing Company, Amsterdam.

Sabatini, D. D., Tashiro, Y. and Palade, G. E. (1966). On the attachment of ribosomes to microsomal membranes. *J. Mol. Biol.* **19**, 503–524.

Spirin, A. S. and Gavrilova, L. P. (1969). *In* "The Ribosome". Springer-Verlag, Berlin.

Stevens, B. J. and Andre, J. (1969). The nuclear envelope. *In* "Handbook of Molecular Cytology". (A. Lima-de-Faria, ed.), p. 837–874. North Holland Publishing Company, Amsterdam.

Weinberg, R. A. and Penman, S. (1970). Processing of 45S nucleolar RNA. *J. Mol. Biol.* **47**, 169–178.

Williams, D. J., Gurari, D. and Rabin, B. R. (1968). The effects of ribosomes on the activity of a membrane bound enzyme catalysing thiol-disulphide interchange. *FEBS Lett.* **2**, 133–135.

Williamson, A. R. (1969). The Biosynthesis of multichain proteins. *Essays Biochem.* **5**, 140–175.

RECOMMENDED READING

Campbell, P. N. and Sargent, J. R. (1967). "Techniques in Protein Synthesis," Vol. 1, p. 2–58. Academic Press, London and New York. Introduction by P. N. Campbell and J. R. Sargent, 2–58. Methodological Aspects of Protein Synthesis in mammalian systems by A. von der Decken, 65–123, Biosynthesis of specific proteins in mammalian systems. J. R. Sargent, 133–211.

Fleischer, B., Fleischer, S. and Ozawa, H. (1969). Isolation and characterization of Golgi membranes from bovine liver. *J. Cell Biol.* **43**, 59–79.

Glaumann, H. and Ericsson, J. L. E. (1970). Evidence for the participation of the Golgi apparatus in the intracellular transport of nascent albumin in the liver cell. *J. Cell Biol.* **47**, 555–567.

Grant, J. K. (1969). Actions of steroid hormones at cellular and molecular levels. *Essays Biochem.* **5**, 1–58.

Korner, A. (1967). Ribonucleic acid and hormonal control of protein synthesis. *Progr. Biophys. Mol. Biol.* **17**, 63–98.

Manchester, K. L. (1970). Insulin and protein synthesis. *In* "Biochemical Actions of Hormones". (G. Litwack, ed). 267–320. Academic Press, New York and London.

Munro, H. N. and Allison, J. B. (1964, 1969, 1970). "Mammalian protein metabolism". Vol. 1–4. Academic Press, New York and London.

Petermann, M. L. (1964). "The Physical and Chemical Properties of Ribosomes." Elsevier, Amsterdam.

Tata, J. R. (1970). Regulation of protein synthesis by growth and developmental hormones. *In* "Biochemical Actions of Hormones". (G. Litwack, ed). Academic Press, New York and London.

Young, V. R. (1970). The role of skeletal and cardiac muscle in the regulation of protein metabolism. *In* "Mammalian protein metabolism". (H. N. Munro, ed). Vol. 4, 486–674. Academic Press. New York and London.

6. Mitochondria and Peroxisomes

H. BAUM

Department of Biochemistry
Chelsea College, London, England

I. INTRODUCTION

Virtually all animal or plant cells contain mitochondria. The number per cell may vary from one to many thousands. From one cell type to another, mitochondria exhibit variety of size, form, orientation, mobility and plasticity. In some cells mitochondria may be seen to be in constant movement, streaming with the cytoplasm, changing shape and budding; in other cells the mitochondria appear to be localized and immobile. Mitochondria from different sources also vary in respect of certain functional parameters and in certain details of enzymic complement. Nevertheless all mitochondria, from whatever source, conform to the same basic structural and functional design. It is therefore usually legitimate as well as convenient to treat mitochondria in general terms, and, except where I specify otherwise, this is what I shall be doing in the present chapter.

The name "mitochondria" (*mitos* (Greek) = a thread) was first suggested in 1898 to describe the small, plastic bodies, which, under a variety of other names had already become the object of serious scientific observations. During the next 40 or so years these studies proceeded with increasing momentum and the idea that mitochondria were concerned with the oxidative processes of the cell gradually gained general acceptance. However, it was not until the late 1940s, with the advent of techniques for the isolation of relatively large quantities of the organelle in a relatively undamaged state, that it became possible to undertake systematic studies of the structure and properties of mitochondria.

The functions of mitochondria were found to be unique and central to the process of life and these functions were found to be intimately linked with and dependent upon the membranous structure of the organelle. It soon became clear that, below the level of organization of the intact cell, the mitochondrion was one of the most fascinating and complex of integrated biochemical systems.

Indeed, so complex were the mitochondrial systems and so subtle the interdependence of ultrastructure and function that before progress could be made a new technology of sophisticated physical methods had to be developed and the concepts of a generation of biochemists revised. The fascination and frustrations of these problems gave rise to a new breed of scientists, the "Mitochondriacs", who embarked with enthusiasm and energy (and a certain degree of naivety) upon a campaign aimed at a complete description of the molecular apparatus whereby the mitochondrion achieved its metabolic functions. In their eagerness, numerous teams of these workers, using different preparations of mitochondrial material from different sources, have filled the literature with an indigestible mass of conflicting data and theory.

What I hope to do is to present a reasonably balanced selection and interpretation of what is known and thought about the mitochondrion. You will appreciate that I have inevitably been both selective and partial. Nonetheless, I have also attempted, in some places, to present a compromise view of certain controversial issues. For this reason, and also because the subject has been bedevilled by dogma and the cult of personalities, I have avoided any apportionment of credit (or blame) for the findings and concepts which are mentioned. The reader who pursues the subject further in the literature will form his own views as to the extent and significance of various contributions. In this chapter however I shall concern myself more with where we find ourselves than how we got here.

II. General Features of Mitochondria

A. STRUCTURE

Morphology

(a) *General morphological characteristics.* Mitochondria as seen in tissue section, are typically, rod-like bodies about 3 μm long, and 0.5–1 μm in diameter (Fig. 1). Examination by electron microscopy shows that each mitochondrion consists of a limiting "outer membrane" enclosing an "inner membrane" which lies parallel to the outer membrane and is separated from it by a small peripheral space.

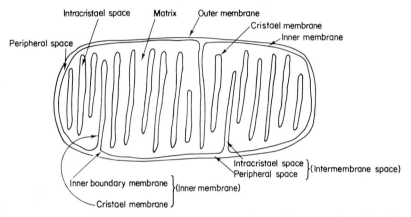

Fig. 1. Diagrammatic representation of a typical mitochondrion, seen in section. The terms in parenthesis are applicable if it is considered that the peripheral and intra-cristael spaces are continuous.

The inner membrane encloses an inner space called the "matrix". Within the matrix lie membranous structures called "cristae", enclosing an "intracristael space". Occasionally, the cristael membrane and the inner membrane appear to be continuous, as therefore do the peripheral and intracristael spaces. The cristae may then be looked upon as invaginations of the inner membrane, in which case the term "inner membrane" may be applied to the whole inner membrane continuum, and the term "inter-membrane space" to the whole of the space separating the inner membrane from the outer membrane. That part of the inner membrane which lies parallel to the outer membrane might then be termed the "inner boundary membrane".

As we shall be mentioning again below, mitochondria from different sources exhibit a variety of morphologies, but in most cases the correspondence between these forms and that illustrated in Fig. 1 is sufficient for the terms defined above to be unambiguous when applied to any mitochondrion in the "typical" or "orthodox" configuration.

We may define the "orthodox" configuration of any mitochondrion as one where the inner boundary membrane and the outer membrane are closely apposed, and where the intra-cristael space is minimal. However, the configuration of the mitochondrial membranes and the relative sizes of the mitochondrial spaces vary both *in vivo* and *in vitro* in response to changes in the metabolic state of the mitochondrion, (see Section on energized shape and volume changes below). In some configurational states there is at first sight a certain ambiguity as to which space is which. This ambiguity can usually be resolved by reference back to the "orthodox" state and by bearing in mind that, so far as is known, the topology of the mitochondrial membranes remains constant in all configurations.

Apart from its use as a point of reference the "orthodox" configuration of a mitochondrion is not necessarily any more "typical" than any other state. What is probably the case is that the conditions usually employed in fixing, embedding and staining sections of material for electron microscopy impose the metabolic state characterized by this particular configuration.

Another morphological problem is the extent to which the peripheral space is continuous with the intra-cristael space. It is quite rare in tissue sections to observe anastamoses between the inner boundary membrane and the cristael membrane, and such junctions where seen are so narrow that it might be argued that they do not necessarily represent a functional continuity of the two compartments (c.f. nuclear pores which appear topologically to connect the cytoplasmic compartment with the nuclear compartment, even though the composition and functions of the two compartments remain distinct). However, electronmicroscopic examination shows that, in some mitochondria under certain *in vitro* conditions, the two spaces become quite clearly continuous. Moreover, in *in vitro* experiments where osmotic effects are demonstrated by virtue of the relative inaccessability of the matrix space to many solutes (see below) the mitochondrion behaves as a two-compartment system, the matrix space representing one compartment and the intra-cristael and peripheral spaces representing the other. It is nonetheless advisable to have reservations on the question of the continuity of the latter two spaces. It might well be that the extent of communication between them is variable with metabolic state and itself constitutes a control feature of the organelle.

Another related question is whether or not the inner boundary membrane and the cristael membrane have identical ultra-structure, composition and function. There is remarkably little unambiguous evidence pertinent to this question, which again is best left open.

A further complication is that there is a growing body of evidence that in certain metabolic states there are areas of close contact between the outer membrane and the inner-boundary membrane. This contact does not appear to be random, but is characterized by a distinct pattern of contact areas over the whole inner surface of the outer membrane.

One final difficulty should be mentioned. The outer membrane of the isolated mitochondrion is permeable to most of the solutes employed in *in vitro* experiments. The intermembrane space however does retain certain soluble protein species, the osmotic pressure of which can only be balanced by impermeant (i.e. macromolecular) solutes added to the suspending medium. Hence there will always be a tendency, under most experimental conditions, for the intermembrane space to undergo osmotic swelling.

(b) *Variations in mitochondrial morphology.* Mitochondria from different sources may frequently be characterized by the morphology

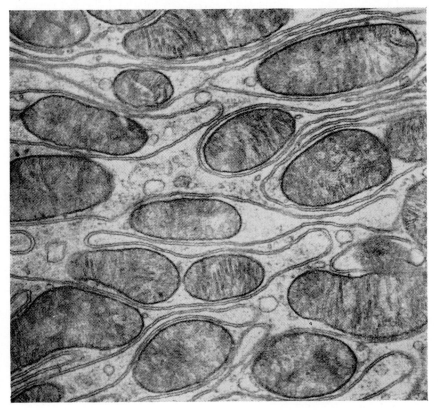

Fig. 2. Kidney mitochondria *in situ* to show arrangement of cristae. Reproduced with permission from D. W. Fawcett (1966).

of their cristae. The "typical" cristael organization as illustrated in Fig. 1 is lamellar with the cristae arranged rather as stacked sheets lying at right-angles to the long axis of the mitochondrion. Mitochondria from mammalian kidney illustrate this regular arrangement very well (Fig. 2). In the case of mitochondria from mammalian liver (Fig. 3), the arrangement, although still lamellar is not as regular, neither as there so many cristae per mitochondrion (see below). Another variant of the lamellar form is seen in mitochondria from the flight muscle of the flesh fly, (Fig. 4), where the cristae are fenestrated. Here the appearance can be close to that of a reticulum of fused tubular cristae.

Tubular cristae of regular, circular cross-section are seen in mitochondria from *Paramecium*, whereas in the case of mitochondria from the adrenal cortex the characteristic organization of the cristae appears to be that of a series of small vesicles joined by short, narrow tubes (Fig. 5).

In considering all of the above arrangements, and the more exotic examples that will be found in the literature, it should always be borne in mind that mitochondrial morphology is never static. Not only is the organelle itself plastic and in many cases constantly changing in shape, but also the inner membrane can exhibit a variety of configurations depending upon its environment and metabolic state.

As we shall see, the inner membrane is the site of oxidative phosphorylation. This can be taken to account for the observations that, whatever the actual arrangement of the cristae, the greater the demand of a tissue for ATP then the greater in general will be the area of cristael membrane relative to the volume of each mitochondrion in the cells of that tissue.

Chemical composition

In this section we shall be concerning ourselves primarily with the lipid and proteins of mitochondria, since these constitute its major structural elements. Mitochondria also contain nucleic acids, whose location and function will be discussed in a later section. In addition mitochondria contain inorganic salts and small substrate molecules, the nature, amount and location of which reflect many factors such as the cytoplasmic solutes of the cell of origin, the composition of the final suspending medium, the conditions of isolation of the mitochondrion and the permeability of its membranes. Finally, mitochondria contain a full complement of all of the cofactors (e.g. NAD^+, ADP, coenzyme A etc.) that they require for their complex integrated functions.

(a) *Total lipid and protein.* Whole mitochondria contain about

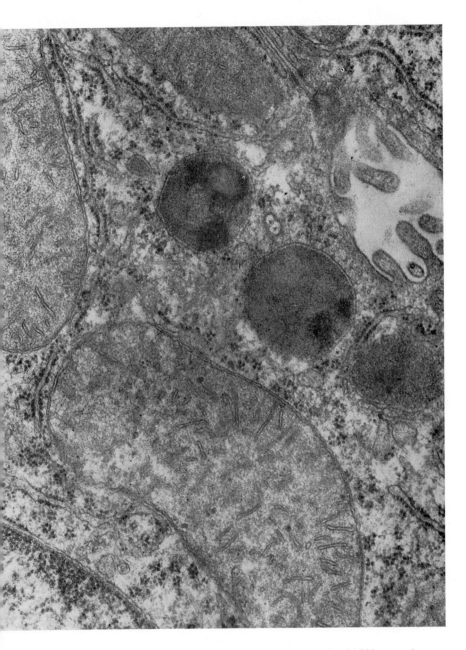

Fig. 3. Liver mitochondria *in situ* to show arrangement of cristae ($\times$ 84 000 approx).

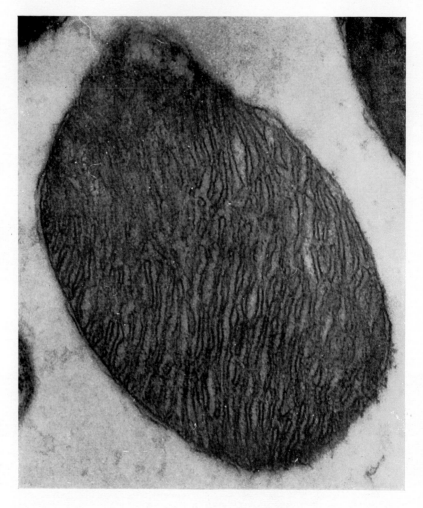

Fig. 4. Electron micrograph of a mitochondrion from the flight muscle of the flesh fly (*Sarcophaga barbata*) × 65 600 reproduced by kind permission of Dr. Ray Pillinger, Shell Research Ltd., Sittingbourne.

200–300 mg of lipid per g of protein. The bulk of this lipid is phospholipid and is associated with the mitochondrial membranes. Depending upon the origin of the mitochondrion, between 15% and 50% of its protein behaves as though it were in solution in the matrix, or the intermembrane space. The remaining protein is more or less firmly associated with the mitochondrial membranes. A proportion of this membrane-bound protein may in fact be detached as water-soluble species without actually disrupting the membranes, and the

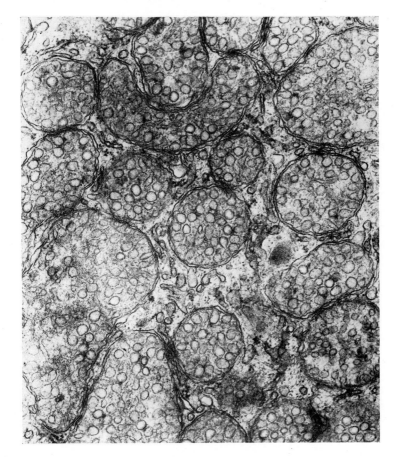

Fig. 5. Adrenal cortex mitochondria *in situ*. Reproduced with permission from Allman *et al.* (1970).

release of this detachable protein is not usually associated with the removal of any membrane lipid. Those proteins that cannot be detached are intimately associated with the membrane lipids and must be regarded as part of the membrane continuum itself, (see below).

There is a characteristic distribution of functions between the various mitochondrial compartments (see below), hence it is not surprising that the chemical compositions of the inner and outer membranes are quite different.

(*b*) *Distribution of lipids.* The outer membrane has a relatively high lipid content (about 1 g lipid per g protein). The major species of phospholipid is phosphatidyl choline, but phosphatidyl ethanolamine and phosphatidyl inositol also constitute significant components. The outer membrane contains about 30 mg cholesterol per g of protein and in general has a lipid composition not unlike that of the endoplasmic reticulum.

The inner membrane contains about 300 mg of phospholipid per g of protein but, unlike most biological membranes, contains virtually no cholesterol. Moreover, the phospholipid composition is quite unusual; there is very little phosphatidyl inositol but a very significant amount (over 20%) of cardiolipin, a very acidic phospholipid.

(c) *Soluble proteins.* Most of the protein species which appear to be in solution in the aqueous compartments of the mitochondrion, as well as those proteins which are membrane bound, but detachable in water-soluble form, have enzymic functions. However, the possibility does still remain that each of these fractions contains a significant amount of a protein, (or family of closely related proteins), which is not catalytic but which serves an organizing or integrative function. It must be remembered that, during mitochondriogenesis (see below) there has to be some mechanism for sequestering into the matrix all of the characteristic soluble enzymes of the mitochondrion which are synthesized on cytoplasmic ribosomes. Moreover, many of the soluble enzymes of the mitochondrion are concerned with catalysing highly integrated metabolic pathways, and it is very tempting to believe that such functional integration is a reflection at least of a loose structural integration.

Several workers have prepared, in large yield from mitochondria of many sources, insoluble protein preparations, of dubious homogeneity, designated "mitochondrial structural protein". If there is any significance at all to this material (and there is a substantial and controversial literature on this) then it is that a portion of it might constitute a denatured preparation of an organizing protein for the soluble components of the mitochondrion; i.e. a non-catalytic protein which might function as the structural template for a macromolecular array of the soluble enzymes of a particular metabolic sequence.

(d) *Membrane proteins.* There has been little characterization of the protein components of the outer membrane itself. Apart from those few species concerned with the known catalytic functions of this membrane, the remaining protein appears to be non-catalytic and may be similar to the "structural protein" of the endoplasmic reticulum. (It is regrettable that the term "mitochondrial structural protein" was assigned to the material that was referred to in the previous paragraph and which probably derives from *soluble* components. This has introduced an element of ambiguity to references to true structural components of biological membranes).

About half of the protein of the inner mitochondrial membrane (following removal of detachable protein) can be accounted for in terms of known catalytic components. We shall be discussing the organization of these components (cytochromes, flavoproteins etc.) in later sections. It is possible that the remaining protein also has a catalytic role. Indeed, it is not very meaningful in such a highly integrated catalytic system as the mitochondrial respiratory chain to distinguish between catalytic and structural or organizational functions. What does seem likely is that a significant proportion of the protein of the inner membrane does not contain characteristic prosthetic groups. Typical of such "non-prosthetic proteins" are the "core proteins" which have been shown to constitute up to 50% of the protein of certain multi-component complexes of the respiratory chain (see below). Chemically, core protein has many similarities to "mitochondrial structural protein", but, unlike the latter, core protein is undoubtedly intimately associated with the structural organization of the inner membrane itself.

Ultrastructure

The ultrastructural organization of biological membranes in general is discussed at length in Chapter 2. It is therefore only necessary here to concentrate on those features that are particular to the mitochondrion. The mitochondrion poses a number of special ultra-structural problems. Firstly, it is a multi-compartment system, the inner membrane of which is impermeable to negative stains so that the fine structure of the inner compartment cannot be visualized without first disrupting the membranes. Hence disruption is frequently a necessary prelude to examination. Secondly, changes in mito-chondrial morphology with metabolic state (see below) may well be a reflection of changes in the ultra-structural organization of the membranes, which may in turn be associated with changes in the conformation of individual membrane components. In other words, the ultrastructure of the membranes will probably depend upon the metabolic state of the mitochondrion. Thirdly, the inner membrane has an exceptionally high content of protein, much of which is organized in functional complexes. These complexes in turn are organized within the membrane so that they interact functionally with one another. Once again it is apparent that, in the mitochondrion, ultra-structure and function are inextricably linked.

(*a*) *Outer membrane.* The problems mentioned above apply primarily to the inner membrane. The outer membrane has only recently been isolated in relatively pure form. As far as is known, this membrane does not seem to differ greatly in ultrastructure from other cytomembranes, such as smooth endoplasmic reticulum. The only feature that may be characteristic of the outer mitochondrial membrane is the presence of areas containing arrays of hollow cylinders 6 nm high, 6 nm in diameter with a central hole about 2 nm in diameter. This feature might relate to the ease with which small molecules penetrate the outer membrane.

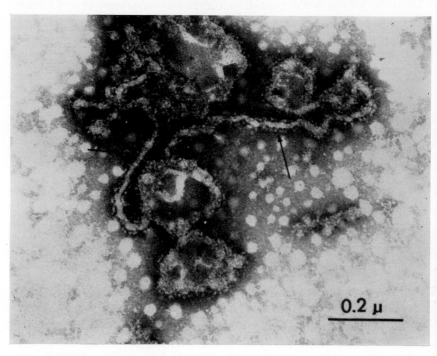

Fig. 6. Negatively-stained preparation of submitochondrial vesicles. Samples were taken from the same preparation used to prepare the freeze-etch replica shown in Fig. 7. The protruding knob structures on the membranes can clearly be seen (arrows). (Reproduced with permission from Wrigglesworth *et al.* 1970).

(*b*) *Inner membrane.* When mitochondria are disrupted, the inner membranes thus exposed can be negatively stained, for electron microscopy. When examined under these conditions of staining, such membrane preparations are characteristically seen to be studded with large numbers of spherical particles about 8–9 nm in diameter (see Fig. 6). These particles ("head pieces") appear to be attached to the matrix-facing surface of the membrane continuum by a "stalk" about 4 nm long and 3–4 nm wide. The membrane continuum itself appears to be composed of a two-dimensional array of particles. There is some doubt as to the precise geometry of these latter particles but they may be conceived as approximating to cylinders about 6–7 nm in height (i.e. across the thickness of the membrane) and 10 nm in diameter (i.e. in the plane of the membrane). It frequently appears in negatively stained preparations as if each particle in the membrane continum bears a "stalk" with a "headpiece" attached, so that the entire membrane appears to be composed of repeating tripartite units. Each such unit comprises a "basepiece" (i.e. the particle in the membrane continuum plus a headpiece-stalk segment.*

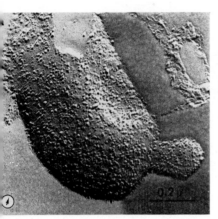

Fig. 7. Freeze-etch preparation of unfixed submitochondrial vesicles. The samples were suspended in distilled water for freezing. a. No etching, showing particle-covered fracture face b. 2 min etching, showing smooth etch face. (E) and particle-studded fracture face (F). A ridge (triple arrow) separates the fracture face from the etched face. Inset: Higher magnification view of the etch and fracture faces. (Reproduced with permission from Wrigglesworth et al. 1970).

There has been considerable controversy as to the physical significance of these observations and as to how the appearance of the membrane following negative staining correlates with that seen in conventionally sectioned material or in material examined by freeze-fracture or freeze-etching (Fig. 7). Any interpretation has to account for the fact that in conventionally sectioned material the electron microscopic appearance of the inner mitochondrial membrane is

* Since the headpiece-stalk segments are only clearly visualized at the edge of negatively stained preparations, it is possible that 1 : 1 association between basepieces and headpiece-stalks is illusory and that the precise arrangement, if it could be visualized in plan, would be seen to be more complicated, with headpiece-stalks attached to the interstices between three or four adjacent basepieces, depending on the precise mode of packing of the basepieces in the membrane continuum.

o

virtually unaltered when a substantial proportion of the phospholipid has been removed by extraction with aqueous acetone. Moreover, freeze-fracture of the inner membrane reveals a high density of globular particles embedded within the membrane continuum. Other relevant facts are:

(i) The removal of detachable protein (see above) from the membrane parallels the loss of "headpieces". (We shall subsequently see that this also correlates with the loss of certain enzymic functions from the inner-membrane preparation).

(ii) The residual protein in this "stripped" membrane is very firmly attached to lipid by what appear to be primarily hydrophobic bonds.

(iii) The negative charges of the phospholipid molecules in the membrane all appear to be exposed to the suspending medium.

(iv) Fragmentation of the membrane under controlled conditions, in the presence of detergent plus salt, yields lipoprotein complexes. Several of these complexes have been purified and have been shown to correspond to functional multi-component segments of the mitochondrial electron transfer chain. These complexes have molecular dimensions corresponding to those of the "basepiece" particles seen in negatively-stained preparations of the membrane.

(v) Removal of detergents from such dispersions of lipoprotein complexes results in the re-formation of membranous vesicles.

Fig. 8. Diagrammatic representation of the ultrastructure of the cristael membrane. ⠂, detachable protein; ▨ non-detachable protein; ●, polar end of phospholipid molecule; ⫙, fatty acyl chains of phospholipid.

The most satisfactory interpretation of these and other observations is that shown in Fig. 8. The membrane continuum is visualized as a biomolecular leaflet of phospholipid containing so much "endoprotein" that in effect it becomes a two-dimensional array of lipoprotein complexes. The protein species within any one complex are associated with one another and with their lipid components primarily by hydrophobic bonds. The lateral association between complexes is both hydrophobic and electrostatic. Any predominantly polar region on the surface of a protein component will be orientated away from the hydrophobic interior of the membrane. Such polar regions will be associated with the polar ends of lipid molecules, and with the surrounding medium and its polar solutes. The headpiece-stalk segment of the tripartite repeating units is "exoprotein" and contains no lipid. The precise chemical nature of its association with the membrane continuum is as yet not clearly defined, neither (as has already been mentioned) is the implied stoichiometry of one headpiece per basepiece a firmly established fact; indeed I tend to support the view that each headpiece is associated with a "domain" of several base-pieces. (The head-piece stalk illustrated in Fig. 8 is shown in the conformation as usually visualized by negative-staining, but the native situation of components of this detachable segment might not necessarily correspond with this simple representation). Since the base-pieces are unlikely to be completely cuboidal, (I have already described them in terms of cylinders), there will be interstices resulting from their packing in two dimensions. Such interstices would presumably be filled with pure bilayers of phospholipid.

This description of the ultrastructure of the inner membrane is a static one, but the true state of affairs is likely to be very dynamic. In particular, it is very likely that the geometry of the lipoprotein complexes, which, it must be remembered, are functional as well as structural elements, will change during cycles of catalytic activity. Moreover, it is also likely that the geometry of the headpiece-stalk segments and their conformation in relation to the corresponding base-pieces and the phospholipid continuum will change with the metabolic state of the mitochondrion.

(c) *Matrix.* In conventionally stained material, the matrix has a finely granular appearance with no apparent ultrastructure, although within the matrix a number of inclusions are frequently seen. Some of these are crystalline arrays of fibres or tubules of undefined composition, some are granules (e.g. of calcium phosphates) and some are fibrous, e.g. DNA filaments. When mitochondria are examined by the freeze-etching technique the appearance of the matrix depends upon the metabolic state of the mitochondrion prior to freezing. In the condensed state (see section on energized shape and volume changes) the protein components of the matrix appear to become organized into filamentous arrays. Indeed, there is growing evidence of structural organization in the matrix. This is hardly surprising; in the condensed state the concentration of protein in the matrix can approach 1 000 mg per ml!

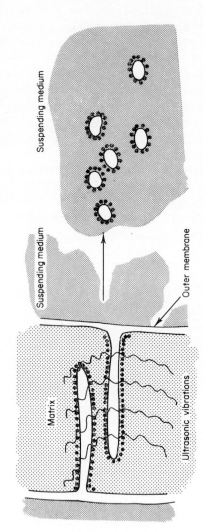

Fig. 9. Formation of submitochondrial particles by ultrasonic disruption of the cristael membrane. The outer membrane and inner boundary membrane are also disrupted, thus releasing the submitochondrial particles into the suspending medium.

Sub-mitochondrial particles

It is frequently very useful to study certain mitochondrial functions using fragmented rather than intact mitochondria. There are two principal advantages to this procedure. In the first place it simplifies the system and allows certain functions to be studied in the absence of soluble components. In the second place, fragmentation overcomes the permeability barrier, imposed by the mitochondrial membranes, between certain solute molecules and certain functional sites on the matrix-facing surface of the inner membrane. This latter advantage is the result of an effective topological inversion which occurs when fragmentation is achieved by procedures such as exposure to ultrasound. This is illustrated in Fig. 9. It should be stressed, however, that preparations of sub-mitochondrial particles obtained following such procedures are not homogeneous ones, and that moreover the properties of a given preparation of sub-mitochondrial particles depends very much upon the precise conditions of fragmentation.

B. FUNCTIONS

The functions of mitochondria may be described under four general headings: viz. the generation of reducing equivalents; the aerobic oxidation of these reducing equivalents; work performances (i.e. functions such as the synthesis of ATP), which are coupled to oxidation-reduction processes; and what I shall refer to as ancillary functions.

*Generation of reducing equivalents**

A source of the energy necessary for living processes is supplied to most cells as foodstuffs of photosynthetic origin. In the chloroplasts of green plants, water is oxidized, with the evolution of oxygen, and the transfer of electrons (with associated protons) to reduce carbon dioxide. The oxygen atom is very electronegative in comparison with the carbon atom, and so the transfer of bonding electrons (and associated protons) from the oxygen in water to the carbon in carbon dioxide requires a substantial input of energy. This is provided by photons trapped by the pigment systems of the photosynthetic apparatus.

Even if we take an over-simplified view of photosynthesis, it would be misleading to imply that the bonding electrons of water are transferred directly to the carbon skeleton that is being reduced. A key intermediate is $NADP^+$. This coenzyme collects the "reducing equiva-

---

* Readers who are not familiar with elementary principles of the energetics of biological oxidations should read a suitable treatment of this subject. e.g. in Appendix I of "Energy and the Mitochondrion", by D. E. Green and H. Baum, Academic Press, New York and London (1970).

lents" (i.e. the excited electrons generated in the photosystems) and in turn transfers them to the organic molecule being synthesized. This last transfer is generally a readily reversible one. In other words, the oxidation–reduction potential of the $NADP^+$: NADPH couple is not far removed from that of each of the reductive steps in the biosynthetic process. In addition to these simple reductive steps, photosynthesis also involves processes that are coupled to the cleavage of the terminal phosphoryl residue of ATP. We shall in due course be considering the mechanism whereby this ATP is regenerated in the chloroplast. In summary photosynthesis may be represented as in Fig. 10.

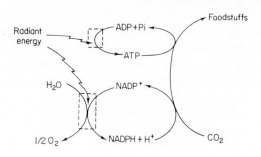

Fig. 10. Diagrammatic summary of photosynthesis.

When a cell catabolizes foodstuffs, cytoplasmic processes such as glycolysis effectively commence the reversal of photosynthesis (although the catabolic pathway followed is not a simple reversal of the corresponding biosynthetic one). A certain amount of conserved energy can be released at this stage, and used in the synthesis of ATP from ADP and P_i without any overall oxidation taking place. Such processes, for example the anaerobic conversion of glucose to lactate, may be looked upon as being formally equivalent to the recouping of a fraction of the energy put into the foodstuff in the non-reductive manipulations coupled to ATP-cleavage.

However, such sources of ATP are of minor importance to the energy economy of aerobic cells. As far as energy transfer is concerned, the key degradative processes take place in the mitochondrion and consist of the formal reversal of the reductive steps of biosynthesis. (Again, the actual pathway followed is different from the synthetic one). In the chloroplast, carbon dioxide is reduced by NADPH with only a relatively small expenditure of energy to drive the process in the synthetic direction. In the mitochondrion, foodstuffs are now oxidized, carbon dioxide is released, and NAD^+ (or $NADP^+$) is reduced. Again the overall change in free energy is, thus far, relatively small; but "reducing equivalents" (i.e. pairs of electrons on molecules of NADH) are now available for exergonic oxidation.

Certain oxidative processes, e.g. the pentose phosphate pathway, do take place in the cytosol but in such cases the reducing equivalents generated are (as NADPH) usually made available for the reductive syntheses of other cellular components. (The NADH generated in the Embden-Meyerhof pathway of glycolysis is a special case and will be considered later). However, where the degradation takes place in connection with the supply of energy for the cell, nearly all of the reducing equivalents are generated, mainly as NADH, in the mitochondrion. The key pathway for this degradation is the citric acid cycle and the mitochondrion houses all of the enzymes responsible for catalysing this cycle as well as all of the enzymes of fatty acid oxidation, (see Fig. 11 below).

Two oxidative steps in the above pathways represent anomalies; the oxidation of succinate to fumarate in the citric acid cycle, and the oxidation of fatty-acyl coenzyme A to the corresponding α,β-unsaturated derivative. In each case, the oxidation-reduction potential of the process is far more positive than that for the $NAD^+:NADH$ couple, so that the equilibrium constant for the coupled reduction of NAD^+ would be very unfavourable indeed. In the case of the fatty-acyl coenzyme A dehydrogenation step, an alternative electron acceptor is utilized, and so the "reducing equivalent" generated is not NADH, but a reduced "electron-transferring flavo-protein" with a more positive potential than NADH. Succinate is another special case, and for the moment we shall simply consider the "reducing equivalent" as being succinate itself, with an oxidation-reduction potential of around +30 mv.

In summary, the first general function of the mitochondrion is a manipulative one, the controlled and essentially reversible transfer of electrons from foodstuffs (i.e. the products of the reduction of carbon-dioxide) back onto adenine nucleotide coenzymes or other suitable carrier molecules.

Aerobic oxidation of reducing equivalents

(a) *The respiratory chain.* The mitochondrion contains a sequence of enzymes which catalyse the transfer of reducing equivalents from NADH to molecular oxygen via a corresponding sequence of prosthetic groups and coenzymes. The general features of this respiratory, or electron-transfer chain, may be summarized at this stage. Briefly, pairs of reducing equivalents are passed by way of carriers of relatively negative potential, such as flavoproteins, to carriers spanning a spectrum of more positive potentials (cytochromes, b, c_1, c, a and a_3, in that order), eventually to reduce an oxygen atom to give rise to water. In other words, the photosynthetic reduction of $NADP^+$ at the expense of the oxidation of water, is reversed in the respiratory chain.

(b) *Transfer of reducing equivalents from the cytosol.* Although details of the organization and functioning of the respiratory chain will be considered later, two points should be cleared up at this stage. The first is the fate of NADH generated in the cytosol. The mito-chondrion is not permeable to NADH. It is however permeable to the substrates for certain NAD^+-linked dehydrogenases such as malate dehydrogenase and $D(-)$ β-hydroxybutyrate dehydrogenase. Not only does the mitochondrion contain these enzymes, but corresponding enzymes are also localized in the cytosol. Thus, for example, cyto-plasmic malate dehydrogenase in the presence of NADH, will reduce oxaloacetate to malate. The latter may then enter the mito-chondrion where it is reoxidized to oxaloacetate with the generation of NADH. By this tactic therefore appropriate substrates can serve as a shuttle transferring reducing equivalent between the cytosol and the mitochondrion.

There are certain cases where a substrate can be reduced more or less irreversibly by cytoplasmic NADH. In other words the reduced substrate has an oxidation-reduction potential which is significantly more positive than that of NADH. Hence the transfer of such a reduced substrate into the mitochondrion does not allow for the generation of intra-mitchondrial NADH. One such case is the α-glycerophosphate : dihydroxyacetone phosphate shuttle, which is ex-tremely important in the oxidation of glycolytically-generated NADH in the insect flight muscle. Dihydroxyacetone phosphate is reduced to α-glycerophosphate in the cytoplasm. α-Glycerophosphate enters the mitochondrion where it represents a "special case", analogous to succinate and the "electron transferring flavoprotein".

(c) *Oxidation of reducing equivalents other than NADH.* The second point that needs to be cleared up is the fate of such "special cases", i.e. reducing equivalents whose oxidation-reduction potentials are more positive than that of NADH. These donors can feed their electrons directly into the respiratory chain at points along its length appropriate to their potential. Once in the chain, these electrons follow the same final pathway to oxygen as do electrons originating in NADH. The feeding-in of reducing equivalents is mediated by specific enzymes, usually flavoproteins, which are so intimately associated with the respiratory chain as to be considered part of it. The complement of such special enzymes will depend very much on the species and tissue of origin of the mitochondrion. Virtually all mitochondria however are rich in succinate dehydrogenase, and this is the only such enzyme that will be considered in the subsequent treatment of the structure of the respiratory chain.

The functional inter-relationships between the systems generating reducing equivalents and the respiratory chain is summarized in Fig. 11

Work performances

The difference in standard oxidation-reduction potential between

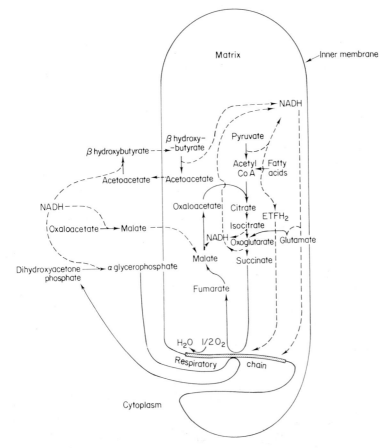

Fig. 11. The mitochondrion as the site of generation of NADH, and other donors of reducing equivalents to the respiratory chain. (ETFH$_2$ is the reduced form of the "electron transferring flavoprotein", a cofactor in the dehydrogenation of fatty acyl coenzyme A). The diagram also illustrates some of the shuttles whereby the mitochondrion is able indirectly to oxidize NADH generated in the cytoplasm.

the $NAD^+:NADH$ couple and the $\frac{1}{2}O_2:H_2O$ couple is 1·14 V. From this figure it may be calculated that the transfer of a pair of electrons from NADH to oxygen (under standard conditions) involves a free energy change of about −50 000 calories per mol, and hence might be expected to be an uncontrolled and irreversible process. In fact, however, most of the individual steps in the respiratory chain are carried out reversibily, because wherever a large fall in potential occurs the oxidation-reduction is coupled to another process which involves a corresponding gain in free energy. The precise nature of this latter, endergonic, energy-conserving process, is a key problem which we shall be considering later on. For the moment I shall refer to the form in which the energy is conserved (i.e. the "energy pressure", "energized state", or "high energy intermediate") as $\sim$, the elusive "squiggle" of the mitochondriacs.

The number of steps where $\sim$ can be generated is determined mechanistically, i.e. by the actual organization of the respiratory chain in relation to the coupling mechanism; but this number also has to satisfy the thermodynamic requirements imposed on the one hand by the free energy of the $\sim$-generating process, and on the other hand by the free energy changes in the respiratory chain. In the span from NADH to oxygen $\sim$ can be generated 3 times. Only $2\sim$ can be generated over the 0·8V potential span between succinate and oxygen.

(a) *Oxidative phosphorylation and its reversal.* In the presence of ADP and inorganic phosphate, (P_i), $\sim$ is discharged by the synthesis of ATP*. This process is reversible, as is the generation of $\sim$ during electron transfer. The overall process of the synthesis of ATP coupled to the respiratory chain is thus also reversible (see Fig. 13 below) and the accumulation of ATP not only inhibits respiration, but also, under appropriate experimental conditions, actually reverses the direction of electron flow. For example, in the presence of ATP, NAD^+ can be reduced by electrons originating in succinate. In the absence of a mechanism coupling the hydrolysis of ATP to this process the equilibrium constant for such a flow of electrons "up" the respiratory chain would be extremely unfavourable.

The synthesis of ATP coupled to respiration ("oxidative phosphorylation") and the reversal of the direction of electron flow, coupled to the hydrolysis of ATP, each constitute examples of "work perfor-

---

* The synthesis of ATP from ADP plus P_i with the discharge of $\sim$, as well as the generation of $\sim$ coupled to the hydrolysis of ATP are processes that always require the presence of Mg^{2+}. Whether this is a requirement of an enzyme that mediates these processes, or whether the true substrates for these processes are Mg^{2+}-complexes is not known. In subsequent discussion I shall assume that adequate amounts of Mg^{2+} are available.

mances" mediated through $\sim$. In a later section we shall be considering in more detail the mechanism of oxidative phosphorylation.

(b) *Energized transhydrogenation.* Mitochondria contain an enzyme system capable of catalysing the stereospecific transfer of a hydride ion, H^- (i.e. a hydrogen atom with a second electron) from NADPH to NAD^+. The process is reversible, and, since the oxidation-reduction potentials of the two pyridine nucleotides are almost identical, the equilibrium constant is very close to $1 \cdot 0$. However, when coupled to the discharge of $\sim$ this same system will catalyze the rapid and essentially irreversible transfer of hydride from NADH to $NADP^+$. This dramatic shifting of the equilibrium constant of the transhydrogenation represents another mitochondrial work performance energized by $\sim$; which can thus be driven either by the hydrolysis of ATP in the absence of electron transfer, or by electron transfer in the absence of ATP-synthesis. As with other mitochondrial work performances, the energised transhydrogenase is in fact reversible; in spite of its superficially irreversible characteristics. Thus, under conditions where $NADPH:NADP^+$ is high and $NADH:NAD^+$ is low, transfer of hydride from NADPH to NAD^+ can be coupled to the synthesis of ATP.

(c) *Accumulation of divalent cations.* Mitochondria have the capacity to bind small amounts (less than 10 nmol/mg protein) of Ca^{2+} with a very high affinity (Km $0 \cdot 1 - 1 \cdot 0$ μM). The capacity for this high-affinity binding of Ca^{2+} seems to reside in a specific Ca^{2+}-binding protein, which also possesses an affinity for Sr^{2+} and Mn^{2+}. Under conditions where $\sim$ can be generated, this protein can apparently act as a carrier, actively transporting Ca^{2+} (or Sr^{2+} or Mn^{2+}) into the inner mitochondrial membrane with the concomitant discharge of $\sim$.

In the absence of a permeant anion (see below) up to 80 nmol Ca^{2+} per mg protein can be accumulated by the membranes. This accumulation is paralleled by an ejection of protons into the suspending medium so that the mitochondrion itself becomes "alkalinized". In the presence of a permeant anion such as phosphate or acetate the accumulated Ca^{2+} appears in the matrix space as the corresponding soluble calcium salt. 200–300 nmol Ca^{2+} per mg protein can be accumulated in this manner. In the presence of phosphate *plus* ADP (or ATP) mitochondria can accumulate massive amounts of Ca^{2+} in the form of amorphous granules of calcium phosphate. These granules appear first on the inner surface of the inner membrane and finally accumulate in the matrix. Under appropriate conditions a mitochondrion can accumulate in this manner over 25% of its dry weight as calcium phosphate.

The ADP (or ATP) requirement for massive accumulation of Ca^{2+}, Sr^{2+} or Mn^{2+} is not understood, but it should be stressed that this requirement is apparently not concerned with the mode of generation of $\sim$, which may be either by ATP-hydrolysis or by respiration.

The process of active Ca^{2+} accumulation , at least in the presence of a permeant anion, is a strictly stoichiometric one, two Ca^{2+} ions being translocated per $\sim$ discharged. The Ca^{2+} transport system has in fact a very high affinity for $\sim$, higher than that of the ATP-synthesizing system. Hence, if $\sim$ is generated by respiration, no ATP is formed from ADP and P_i if Ca^{2+} is also present in the medium.

(d) *Accumulation of monovalent cations.* The inner mitochondrial membrane is relatively impermeable to monovalent cations unless they are present at rather high concentrations, nor does it appear to possess any specific binding sites for them. However, certain polypeptide antibiotics can act as ion transporters (ionophores) rendering the membrane permeable to specific cations. For example, valinomycin, (a compound containing twelve residues, three each of L-valine, D-valine, L-lactic acid and D-α-hydroxy isovaleric acid, linked cyclically by alternate amide and ester bonds,) renders the mitochondrion permeable to K^+. In the presence of valinomycin and an appropriate permeant anion, mitochondria can actively accumulate potassium salts (even when these are present in the suspending medium at low concentration), at the expense of the discharge of $\sim$.

(e) *Movement of protons.* It has been mentioned that the active uptake of Ca^{2+} in the absence of a permeant anion was associated with an ejection of protons. The movement of monovalent cations (in the presence of appropriate ionophores) may also be associated with a movement of protons in the opposite direction. Furthermore, the movement of certain anions into and out of the mitochondrion may be coupled to proton movements (see below), and the synthesis of ATP^{4-} from ADP^{3-} and P_i^{2-} at physiological pH involves the uptake of a proton. In view of these, and other related phenomena it is rather difficult to demonstrate unambiguously proton movements coupled directly to ATP hydrolysis or electron transfer. However, it is now generally accepted that under appropriate conditions two H^+ can be ejected from the mitochondrion per $\sim$ generated. What is still very much in dispute (as we shall be discussing in a later section) is whether this movement is incidental to some other process, whether it is the manifestation of a $\sim$-utilizing system for the active translocation of protons or whether the proton gradient set up (or the membrane potential which would result if the protons were allowed to exchange with other cations) actually represents the primary form of $\sim$.

(f) *Energized shape and volume changes.* The inner membrane of the mitochondrion, although permeable to water, is relatively impermeable to most solutes. Hence the mitochondrion behaves rather like an osmometer, the matrix space expanding and contracting in response to differences in osmolarity between the extramitochondrial and intramitochondrial solutes. The outer membrane in the isolated mitochondrion is rather permeable and inelastic, so that usually when we talk of changes in mitochondrial volume we are referring to changes in the relative volumes of the matrix space and the intermembrane space. Only when the outer membrane ruptures, as it does in the case of large amplitude swelling, is there a significant increase in total mitochondrial volume.

It follows from the above considerations that all net uptakes of salts, energized by respiration or the hydrolysis of ATP, will be accompanied by "mitochondrial swelling" (i.e. swelling of the matrix space). This swelling may eventually increase the rate of spontaneous dissipation of $\sim$ (see below) so that ion uptake energized by $\sim$ ceases. The accumulated ions then leak out of the mitochondrion which, under some conditions, spontaneously contracts again. This contraction may decrease the rate of spontaneous discharge of $\sim$ to a point where active ion uptake and associated swelling is re-established, thus giving rise to a cycle of volume oscillations.

Energized swelling associated with the uptake of salts can be inhibited or reversed by the addition of ADP plus Pi. The ATP-synthesizing system competes with the cation translocating system for $\sim$, and so inhibits swelling. Paradoxically however there are certain conditions where the hydrolysis of ATP can apparently be coupled to the contraction rather than the swelling of mitochondria. The so-called large-amplitude swelling of mitochondria induced by hypotonic media or by respiration in the presence of certain swelling agents such as thyroxin (see below) is reversed on addition of ATP. It has been suggested that this latter type of contraction represents a mechano-chemical process analogous to the contraction of muscle coupled to the hydrolysis of ATP.

Although we are primarily concerned here with volume changes coupled to the dissipation of $\sim$, it is important at this point to draw attention to certain changes in mitochondrial configuration that might not strictly fall within this category. As was mentioned in the section on mitochondrial morphology, the configuration of the inner mitochondrial membrane, both *in vitro* and *in vivo* depends upon its metabolic state. Unfortunately, although these changes are extremely significant, three factors have made it very difficult to define a systematic correlation between functional and configurational states. Firstly, it seems that a given set of changes at the molecular level might manifest themselves in a variety of gross morphological changes depending upon other parameters such as the composition and osmolarity of the suspending medium. Secondly, mitochondria from different sources with different numbers of cristae or with cristae of different configurations might exhibit what appear superficially to be quite different responses to a given set of changes in metabolic state, because of the superimposition of a variety of structural restraints upon a common set of fundamental responses. Thirdly, in order to demonstrate a meaningful correlation between metabolic state and mitochondrial configuration it is necessary to impose extreme conditions. The majority of the population of mitochondria must be in the same

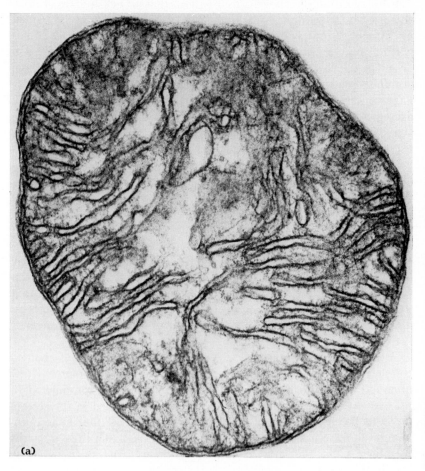

(a)

Fig. 12. a-d. Various configurational states of heart mitochondria. (Reproduced with permission from Korman *et al.* 1970).

state, metabolically and ultrastructurally, at the time of fixation for electronmicroscopic examination. Furthermore, the conditions must be so limiting that the majority of molecules of each of the catalytic components of these mitochondria must be in the same state.

Figures 12a-d illustrate the kind of changes that can reproducibly be observed in mitochondria from a particular source examined under a rigorously defined set of conditions. Figure 12a is an electron micrograph of an isolated mitochondrion from bovine heart, in the so-called orthodox configuration. This is the configuration usually seen in mitochondria *in situ* in tissue sections, and is considered to represent a "non-energized" state, i.e. a metabolic state where the rate of dissipation of $\sim$ greatly exceeds its rate of generation. The conditions usually employed in preparing tissue sections for microscopic examination tend to favour the non-energized state. A different non-energized state

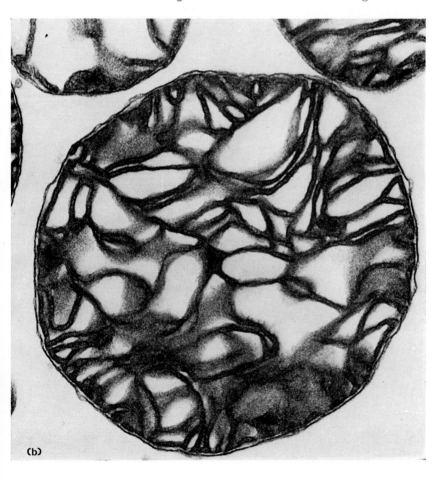

(b)

of an isolated mitochondrion from bovine heart is illustrated in Figure 12b. Here the intracristael space (although not the peripheral space) is seen to be very much enlarged so that there is very close apposition between neighbouring cristael membranes and hence a very condensed matrix space. This "aggregated" configuration of the non-energized state of heart mitochondria is the configuration most commonly seen in isolated preparations of such mitochondria under conditions favouring dissipation of $\sim$. The swelling of the intracristael space in isolated mitochondria might in part be due to the osmotic pressure of intermembrane proteins, as was discussed earlier. Other factors also appear to affect the transition from the orthodox to the aggregated configuration, e.g. the presence of Ca^{2+} ions or of free fatty acids.

Figures 12c and 12d illustrate two configurations characteristic of the "energized" state of isolated bovine heart mitochondria, i.e. of the

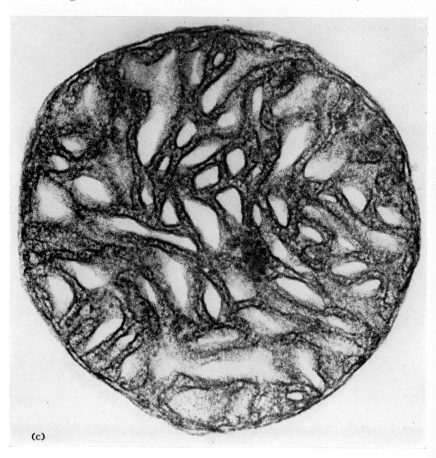

(c)

state where $\sim$ is being generated much faster than it is being dissipated. The configuration in Fig. 12b differs most characteristically from that in Fig. 12c in the closeness of apposition of neighbouring cristae. In the energized state (12c) the matrix is not as condensed as it is in the non-energized, aggregated state. The configuration in Fig. 12d (the "energized-twisted state") is seen whenever the energized state is generated in the presence of inorganic phosphate. Addition of ADP to mitochondria in this state results in the discharge of $\sim$, the synthesis of ATP and a reversion to the non-energized configuration of Fig. 12b.

The reader may very well come across other systemmatic descriptions of configurational changes related to metabolic states, in isolated mitochondria from, say, rat liver, or in mitochondria in cultured

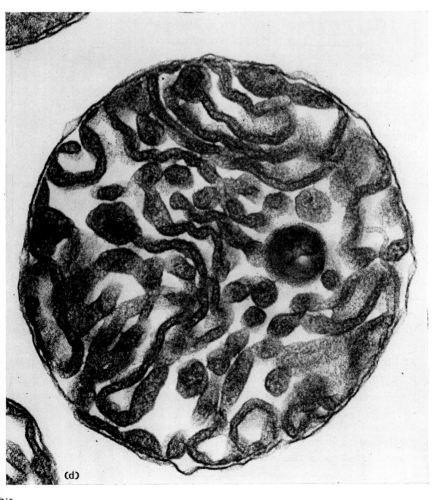

(d)

N*

cells *in vivo*, or indeed in isolated preparations of mitochondria from which the outer membranes have been removed (i.e. inner membrane-matrix preparations). Some of these descriptions and correlations might apparently conflict with the pattern of changes described above for mitochondria from bovine heart. What is common ground however is that under appropriate conditions a cycle of energizing and de-energizing of the mitochondrion (i.e. a cycle in which ∼ is generated and then utilized or dissipated) can be correlated with a cycle of configurational changes. It is possible that these changes are merely cycles of swelling and contraction associated with the uptake and release of salts. However, it is also possible that the configurational changes of the inner mitochondrial membrane are fundamental and represent relatively gross manifestations of changes of molecular conformation of components of the membrane itself and of the matrix. (It will be remembered that there is evidence of structural organization of the proteins in the latter compartment.) Such conformational changes might represent work performances coupled to the dissipation of ∼, or they might be incidental to but indicative of the generation and dissipation of ∼, or they might reflect the operation of a control mechanism concerned with the determination of the mode of utilization of ∼, or they might themselves represent a mode of conservation of energy. This latter possibility will be considered further in a later section.

(g) *Spontaneous dissipation.* In intact mitochondria ∼ has an inherent stability, but nonetheless in the absence of the conditions necessary for a work performance ∼ always dissipates spontaneously at a measurable rate. Some of this discharge of ∼ might in fact represent work performances concerned, say, with the active movement of endogenous ions; nonetheless, particularly in aged or damaged mitochondria, it is clear that there is a process whereby ∼ is spontaneously dissipated as heat.

The precise nature of this process clearly relates to the nature of ∼, which we shall subsequently be considering. Whatever the nature of the spontaneous dissipation, however, the existence of this process gives rise to two corollaries. The first is that mitochondrial respiration in the presence of respiratory substrates should always proceed at a measurable (even if greatly decreased) rate, in the absence of acceptor systems for the discharge of ∼ in the performance of work. The converse should also be true; in the absence of substrates for reversed electron flow or for other work performances, mitochondria should always be able to hydrolyse ATP at a finite rate, governed by the rate of spontaneous dissipation of ∼.*

*A complicating factor however is the presence of the "ATPase-inhibitor" (see p. 244), the inhibitory action of which varies with the functional state of the mitochondrian.

These points should be clear from Fig. 13 which summarizes the inter-relationship between the various processes by which $\sim$ is generated or discharged. The question of the discharge of $\sim$ as heat will be taken up again later on when we consider the action of uncouplers of oxidative phosphorylation, and when we consider further the phenomenon of "respiratory control", i.e. the extent to which respiration is controlled by the rate of discharge of $\sim$.

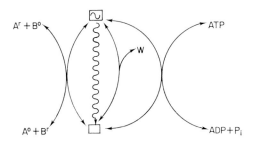

Fig. 13. Inter-relationship between systems coupled to the generation or dissipation of $\sim$. Contact between arrowed lines indicates coupling between systems, such that as one system undergoes a transition in one direction, (upwards or downwards), on the diagram, the other system undergoes a transition in the opposite direction. A^r and A^o, reduced and oxidised forms of respiratory carrier of relatively negative oxidation-reduction potential; B^r and B^o, reduced and oxidised forms of respiratory carrier of relatively positive oxidation-reduction potential; □, the non-energized state of the coupling system; ⧖, the energized state of the coupling system; W, work performance other than the synthesis of ATP; ⟿, spontaneous dissipation or uncoupling.

Ancillary functions

In addition to the systems which are concerned directly with the generation and oxidation of reducing equivalents, and with the coupled generation and utilization of $\sim$, the mitochondrion houses a number of other systems, concerned with general intermediary metabolism as well as with more specialized pathways. Among the former are enzymes concerned with the elongation of fatty acids, the synthesis of phospholipids, amino acid metabolism (e.g. aspartate aminotransferase), and gluconeogenesis (pyruvate carboxylase and phosphoenolpyruvate carboxykinase); among the latter are monoamine oxidase, and the steroid hydroxylation system of adrenal mitochondria.

In some cases the compartmentation of such systems within the mitochondrion can be seen to be relevant to other mitochondrial functions. For example, I have previously referred to the shuttle whereby NADH, generated in the cytosol, is utilized to reduce oxaloacetate to malate, which in turn enters the mitochondrion where it is oxidized back to oxaloacetate with the generation of intramitochondrial NADH. This shuttle is not usually completed simply by the passage of oxaloacetate out of the mitochondrion, because the inner membrane is relatively impermeable to oxaloacetate. Instead, the oxaloacetate undergoes a transamination reaction with glutamate. The products of this reaction, aspartate and α-oxo-glutarate, are readily transferred to the cytosol (see below), where they undergo a further transamination reaction to regenerate oxaloacetate and glutamate, ready for the next round of the shuttle. It can thus be seen that mitochondrial aspartate aminotransferase plays a specific role in the transfer of reducing equivalents between the mitochondrion and the cytosol.

In addition to such cases where the ancillary function is directly pertinent to key mitochondrial processes, there are also cases where the ancillary system might well be specifically concerned with the maintenance of mitochondrial integrity. An example of such a case is the existence in mitochondria of enzymes of phospholipid metabolism. In other cases there appears to be no obvious explanation for the localization of a particular function within the mitochondrion. It would be unwise however to assume that such localizations are without significance. Any compartmentation of a process within a particular subcellular domain automatically increases the possibilities for the control of that process.

Variations in functional complement

Just as mitochondria from different sources show morphological variety, so do they exhibit some variation in their enzymic complement. In the case of mitochondria from higher organisms such variations are usually restricted to those functions that I have classified as being ancillary. In this respect, not only is there predictable variation in the mitochondrial complement of enzymes concerned with specialized functions, but there is also some variety in the extent to which enzymes of general intermediary metabolism are confined to the mitochondria. Thus, phosphoenol-pyruvate carboxykinase is mitochondrial in the liver of the chicken, but is primarily cytoplasmic in rat liver and is more or less equally divided between mitochondria and the cytoplasm in the case of guinea pig liver.

Variation in the "minimal" mitochondrial systems, i.e. those for generating reducing equivalents, oxidizing them and performing work, are usually, in the case of higher organisms, only of a quantitative nature. Thus, the relative amounts of certain components of the respiratory chain is variable from one mitochondrial source to another. Also, the total complement of components directly involved in oxidative phosphorylation represents a variable fraction of the total mitrochondrial protein, as would be predicted on the basis of the known variability in the amount of cristael membrane in mitochondria from various sources. This latter variability, as has already been mentioned, is roughly correlated with the demand for ATP of the respective tissues of origin.

In addition to such quantitative variations in the capacities of the basic mitochondrial systems there are occasionally certain qualitative differences, particularly in regard to the permeability of the inner membranes to anionic substrate molecules. For example, in the case of the inner membrane of mitochondria from blowfly flight muscle, there appears to be a complete absence of any of the known specific carriers for anionic intermediary metabolites (see below). Moreover, these membranes are without the capacity either to bind calcium ions with high affinity or to transport them actively. In these latter respects they are like the mitochondria from certain yeasts. Indeed, this correlation between the absence of high-affinity calcium-binding and the inability to transport calcium ions actively is unlikely to be coincidental.

When one considers mitochondria from sources as primitive as yeasts, then more variety in the "minimal systems" become apparent. For example, mitochondria from certain yeasts, although able to oxidize NADH, do so with only the same yield of ATP as that obtained from succinate oxidation. In other words, they lack the first coupling site in the respiratory chain (see below). What is remarkable however, is not the extent of such variations, but the constancy of the structure and organization of the basic mitochondrial machinery.

C. PERMEABILITY OF MITOCHONDRIAL MEMBRANES

Virtually all studies on the permeability properties of mitochondrial membranes have been made using isolated mitochondria. It might therefore be wise to exercise some reservation before applying the conclusions reached from such studies to mitochondria in intact cells. This caution applies particularly in regard to conclusions concerning the rather fragile outer membrane. Another basis for reservation is the possibility that the peripheral space and the intra-cristal space are not always functionally continuous (see above). In interpreting data on mitochondrial permeability most workers have assumed that there are only two mitochondrial spaces under all experimental conditions. If this proved not to be the case then many findings would be open to radical reinterpretation.

Outer-Membrane

The outer-membrane of the isolated mitochondrion appears to be freely permeable to most molecules with molecular weights of less than 10 000. It is unusual that a membrane of homogeneous ultrastructure should allow free passage of molecules of such dimensions. It therefore seems possible either that there are well defined pores (possibly the hollow cylinders referred to in a previous section) or else that the membrane is damaged during the isolation procedure.*

Inner-Membrane

The inner mitochondrial membrane is freely permeable to water. it also appears to be permeable to small neutral solute molecules, but only to those with molecular weights up to about 150. It is thus impermeable to sucrose.

When we come to consider the permeability of the inner membrane to ions we immediately run into difficulties of interpretation. Many of these difficulties arise from the large numbers of fixed charges that are present, on the soluble proteins in the matrix compartment as well as on the inner membrane itself. These charges give rise to complex Donnan equilibria which depend to some extent upon the functional state of the mitochondrion. Such equilibria might inhibit the net movement of a given ionic species down a concentration gradient, whilst permitting a rapid exchange between ions of that species across the membrane.

In general, when it is stated that the inner-membrane is impermeable to an ionic species, what is implied is that, in the presence of a counter-ion to which the membrane is freely permeable, there is no rapid equilibration of concentrations between the suspending medium and the matrix space, i.e. there is no swelling of the matrix compartment when the salt in question is used as an osmotic support medium. However, in the case of complex anions such as ATP, permeability has been determined by direct analytical procedures.

The requirement for the membrane to be permeable to both ions of salt before swelling takes place is a corollary to the fact that electrical neutrality must be maintained. If a given ion is to enter the matrix it must either move with a counter-ion (to give rise to a net increase in the osmolarity of the matrix compartment) or else ions of equivalent charge must simultaneously move from the matrix into the intra-membrane space and hence, (the outer membrane being freely permeable), into the suspending medium. This latter process is referred to as "exchange diffusion".

There are certain ambiguities inherent in interpreting net salt movements in terms of ionic permeabilities. For example, if the membrane were permeable to NH_3, CH_3COOH and Cl^-, but not to NH_4^+ or CH_3COO^-, swelling might occur in ammonium acetate

*There is a recent report that cells may actually contain only one enormous mitochondrion and that isolated mitochondria are therefore re-sealed fragments.

but not in ammonium chloride. Only in the former case would the anion be sufficiently basic to carry across the membrane a proton originating in the NH_4^+, and hence permit the passage of uncharged NH_3.

(a) *Cations.* If permeability is judged by the rate of swelling of mitochondria suspended in salts of monovalent cations with an appropriate permeant anion, then the only cation to which the inner mitochondrial membrane appears to be freely permeable is NH_4^+. In fact the permeant species is probably uncharged NH_3 which is protonated again in the matrix compartment, and so, as discussed above, permeability to ammonium salts may be determined by the basicity of the associated anion.

By the criterion of salt-induced swelling, Na^+ may also penetrate the inner membrane, although less readily than NH_4^+. Na^+ in turn is a much better penetrant cation than is K^+. As we have mentioned previously, permeability of the inner membrane to monovalent cations is markedly increased in the presence of ionophorus antibiotics and there is now evidence of an endogenous ionophore for monovalent cations in mitochondria; this might account for their initial permeability. Most of the ionophores that have been studied fall into one of two classes, neutral ionophores such as valinomycin and ionophores, such as nigericin, that carry a net negative charge by virtue of an ionized carboxylate group. In the case of valinomycin, the penetrant species, in the presence of K^+ ions, is the positively charged valinomycin-K^+ complex. Therefore, for any net transport of K^+ in the presence of valinomycin it is also necessary for there to be present a permeant anion to achieve electrical neutrality. In the case of acidic ionophores such as nigericin, the penetrant species is a complex with K^+ and carries no net charge. Ionophores of the nigericin class therefore facilitate rapid exchange across the membranes of cations with which they can form permeant complexes. In particular, nigericin causes rapid exchange between K^+ and H^+ when there is an imbalance in the distribution of these species between the suspending medium and the matrix compartment.

In the case of divalent cations such as Ca^{2+} there is, in a sense, a natural ionophore, the specific calcium-binding protein to which I have referred previously. (There is recent evidence that mitochondria might, in addition, contain a lipid-soluble species of low molecular weight which can function as a specific ionophore for calcium.) It will be remembered that active uptake of Ca^{2+} into the membrane in the absence of a permeant anion was accompanied by an ejection of protons and that net uptake of calcium salts into the matrix compartment could be achieved in the presence of a permeant anion. At present however there is no complete understanding either of the role of the calcium-binding protein in the active accumulation and transport of Ca^{2+} or of the corresponding functions of the ionophorus antibiotics. It is therefore unwise to press too far the analogy between these two classes of compounds.

(b) *Anions.* Anions of weak, monobasic acids of low molecular weight readily penetrate the inner membrane.* Such anions include acetate and β-hydroxybutyrate. It is likely that they do so as undissociated acids. Thus, for example, mitochondria swell rapidly when suspended in ammonium acetate. The penetrant species in this case might well be NH_3 and CH_3COOH, with ionization of the latter and protonation of the former taking place in the matrix. It is however possible that the acetate ion itself crosses the membrane in exchange for an hydroxl ion (generated as the result of the protonation of NH_3). Indeed, it is very difficult to distinguish between transport mechanisms involving movement of an anion with a cation such as a proton, or in exchange for another anion such as an hydroxyl ion.

There is evidence that the inner-membrane might also be permeable to small anions, such as nitrate, which derive from strong acids. However, in general, it seems that the entry of most anions into the matrix is not a question of simple permeability but is controlled by specific carrier molecules each of which mediates either the facilitated transport of a given species of anion or specific exchange-diffusion between limited numbers of pairs of anionic species. These carriers, acting in concert, control the passage of metabolites into and out of the matrix compartment and hence contribute to the control of intermediary metabolism.

For example, phosphate may enter the mitochondrion on one carrier either on its own (the charge being balanced by other concommitant ion movements) or possibly directly in exchange for hydroxyl ions. In turn, and mediated by another carrier, malate or succinate may enter the mitochondrion in obligatory exchange for phosphate. A third carrier may then allow citrate to enter in exchange for malate or succinate. Similarly, a further carrier mediates exchange diffusion between malate and α-oxoglutarate whilst entry of glutamate is mediated by yet another carrier, this time in exchange for hydroxyl.

A very important exchange diffusion carrier is that which permits exchange of ADP for ATP across the inner membrane and hence allows rapid equilibration between ATP levels in the cytoplasm and the mitochondrion whilst ensuring that the net content of adenine

---

* The question of the entry of fatty acids of long chain lengths is a rather special and complex one and will only be dealt with here in connection with group transfer (see below).

nucleotides within the mitochondrion remains constant. There is some evidence that the ATP:ADP exchange carrier may be subject to long-chain fatty acids. The role of this particular carrier may dramatically be demonstrated by inhibiting it with a specific inhibitor, atractyloside. In the presence of atractyloside, intact mitochondria demonstrate neither oxidative phosphorylation nor any work performance mediated through externally added ATP. In sub-mitochondrial particles, where the inner surface of the inner membrane (see next section) has direct access to the suspending medium, atractyloside has no inhibitory effects.

When there is no net movement of salt into the mitochondrion then the total capacity of the matrix compartment for anions will be determined by simple considerations of electrical neutrality. Hence there will be conditions where the uptake of one anion will be in competition with that of another merely by virtue of the shortage of available cationic sites within the mitochondrion. When net movement of cations occurs, either passively or actively, then more anions will also enter the mitochondrion to maintain neutrality. It is likely that all anionic movement is essentially passive and that active uptake of salts is always dependent upon active movement of cations. However, cations themselves may exhibit exchange diffusion, for example of K^+ for H^+; and as we have already seen it is difficult to distinguish between an anion moving in its protonated form and one moving in exchange for an hydroxyl ion. It is therefore extraordinarily difficult to untangle the primary events in the active uptake of salts by mitochondria.

(c) Group transfer. Certain metabolites are transported across the inner membrane through the agency of carriers which are enzymes exhibiting the specific capacity to mediate group transfers between pairs of substrates. This is not a casual association of functions, the transport processes themselves are in fact group transfers rather than transfers of intact molecules.

One way in which long chain fatty acids enter the mitochondrial inner compartment is as thioesters with coenzyme A. However, the inner membrane is impermeable to acylcoenzyme A. Instead, the acyl group is first transferred to carnitine, and coenzyme A is released into the intermembrane space. This process is mediated by an enzyme centre localized on the outer-facing surface of the inner membrane. In turn, a catalytic centre of the same (or closely associated) carrier enzyme catalyzes the transfer of the acyl moiety from acyl carnitine to coenzyme A of the matrix compartment. This transfer takes place at the inward-facing surface of the inner membrane. The vectorial

nature of this second transfer is such that carnitine is released again
into the intermembrane space whilst acylcoenzyme A appears in the
matrix. Hence the carrier achieves the transfer of a long-chain fatty
acyl residue from external to internal coenzyme A via the formation
of the corresponding ester with carnitine.

Another group transfer is probably concerned with the transport of
oxalocetate out of the mitochondrion. In this case the carrier might
well be aspartate amino transferase, to which mention has already
been made in connection with the tactic whereby NADH generated
in the cytoplasm is effectively transferred to the mitochondrion. Glutamate may enter the mitochondrion on a specific carrier. Oxaloacetate
generated in the matrix compartment by the re-oxidation of malate
(the latter originating in the cytoplasm) undergoes transamination with
glutamate, the enzyme being vectorially orientated within the inner
membrane so that the products, aspartate and α-oxoglutarate are
released into the outer compartment. The vectorial nature of the
transamination is essential since the inner membrane appears to be
impermeable to aspartate as well as to oxaloacetate. Subsequently
there is a further transamination between aspartate and α-oxoglutarate
to regenerate oxaloactate and glutamate. This complex sequence of
events is summarized in Figure 14.

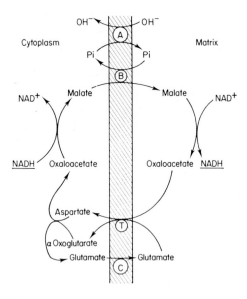

Fig. 14. The role of the inner-membrane-localized transaminase (T) in the transfer of
reducing equivalents from the cytoplasm to the mitochondrial inner compartment.
A, B and C represent specific anion porters, or exchange-diffusion carriers.

D. LOCALIZATION OF FUNCTIONS

Histochemical techniques are not yet sufficiently refined to determine unambiguously the precise location of most mitochondrial functions. The assignment of such functions to various compartments has therefore depended upon two other approaches, neither of which by itself is entirely satisfactory, but which in combination have allowed consistent conclusions to be drawn concerning functional localization.

One approach has been to consider the function of mitochondria in relation to the permeability of mitochondrial membranes. For example, intact mitochondria do not oxidize added NADH. The outer membrane is permeable to NADH but the inner membrane is not. Therefore, the NADH-generating dehydrogenases must be localized behind the inner-membrane barrier, i.e. in the matrix space. This approach is dependent upon unambiguous criteria for establishing permeability characteristics and also assumes that there are only two mitochondrial spaces. If the intra-cristael space were not functionally continuous with the peripheral space in the intact unswollen mitochondrion, then many assignments of localization might have to be reconsidered.

A second approach has been to "dissect" mitochondria chemically, using procedures such as osmotic shock and exposure to detergents. For example, the outer membrane may be removed, releasing the contents of the inter-membrane space and then the composition and properties of these two fractions and of the residual, inner membrane-matrix fraction may be examined. In turn, the latter fraction may be disrupted to release the components of the matrix compartment and leave an inner membrane fraction which may be subjected to further graded fragmentation and analysis. There are a number of inherent uncertainties in this approach.

The factors that give rise to these uncertainties include the following: possible contamination of the initial mitochondrial preparation with membranous material of non-mitochondrial origin, which might then be purified in the "dissection" process; difficulties in achieving sufficiently homogeneous sub-fractions at each step in the process of graded fragmentation; difficulties in characterizing sub-fractions as originating from a particular mitochondrial compartment without making unjustified assumptions as to what components constitute "markers" for that compartment; the possibility of modifying or redistributing components during the various fragmentation and fractionation procedures; and, inevitably, the necessity of making assumptions regarding the actual number of mitochondrial compartments. One must also remember that virtually all studies on localization of functions have been carried out using isolated mitochondria whose characteristics might differ significantly from those obtaining *in vivo*.

In spite of all of the above reservations, there is now a substantial measure of agreement as to where the various functional components of the mitochondrion are localized.

Matrix space

Within the matrix space are localized all of the components of the citric acid cycle with the exception of succinate dehydrogenase.* Also in this compartment are the enzymes and cofactors that mediate the pathway for the β-oxidation of fatty acids. In other words, the key function of the matrix is the generation of reducing equivalents in the form of NADH, succinate and the reduced electron-transferring flavoprotein of fatty acid oxidation. In addition, the matrix houses enzymes such as glutamate dehydrogenase, and probably most of the enzymes concerned with the replication, transcription and translation of the information encoded in mitochondrial DNA (see below).

Peripheral space

Any low molecular-weight components of the peripheral space will have been lost on isolation, in view of the permeability of the outer membrane. However, there are a number of enzymes which appear to be localized in this compartment and these represent a series of kinases, enzymes concerned with the transfer of phorphoryl residues between pairs of molecules, at least one of which is a nucleoside polyphosphate such as ATP. The enzymes, found in the peripheral space, include adenylate kinase, nucleoside diphosphate kinase, nucleoside monophosphate kinase and creatine kinase.

Outer-membrane

The outer-membrane appears to play a relatively insignificant role in mitochondrial metabolism. A few enzyme activities have been demonstrated to be associated with this membrane, most notably monoamine oxidase and a "rotenone-insensitive NADH-cytochrome c reductase". This latter system is quite distinct from the major mitochondrial pathway whereby electrons are transferred to cytochrome c from NADH; not only is it insensitive to the characteristic inhibitor, rotenone, (see below), but also the oxidation-reduction process cannot be coupled to ATP-synthesis. It has been suggested that the outer-

---

* In fact, it is possible that the complex enzymes of the cycle, pyruvate and α-oxoglutarate dehydrogenases, are attached to the inner surface of the inner membrane as "detachable" components.

membrane system might function *in vivo* as a reductant for some small carrier molecule such as a quinone which might shuttle between the outer membrane and the respiratory chain and thus provide an alternative link between cytoplasmically-generated NADH and the mitochondrial system for oxidative phosphorylation.

Certain other "ancillary" enzymes have also been shown to be associated with the outer membrane. These include kynurenine hydroxylase and some enzymes of lipid metabolism. Another enzyme, hexokinase, is frequently found to be associated with the outer-membrane, but in this case the enzyme is not a part of the membrane but is reversibly bound to it from the cytoplasm. The factors that might control this binding are not fully understood, but it is possible that the sequestration of hexokinase by mitochondria might serve to control glycolytic activity in some cells.

Inner-membrane

(*a*) *Functional complement.* The inner membrane houses the entire respiratory chain as well as the apparatus for transducing respiratory energy into $\sim$ and for mediating work performances energized by $\sim$. Thus, the inner membrane also contains the system that synthesizes ATP from ADP plus inorganic phosphate, and which, when acting in reverse generates $\sim$ at the expense of hydrolysis of ATP. Also associated with the inner membrane are the pyridine nucleotide transhydrogenase (which, it will be remembered, mediates the energized transfer of a hydride ion from NADH to $NADP^+$) and the specific calcium-binding protein (concerned with the energized accumulation of calcium salts).

It is obvious from what was discussed in the previous section that the inner membrane must in addition house all of the carriers mediating group transfer as well as the exchange-diffusion carriers concerned with the transport of anionic substrates between the matrix and the peripheral space. It is however not clear whether these carriers are dispersed homogeneously over the whole membrane, amongst the components concerned with oxidative phosphorylation, or whether the inner membrane is in fact a mosaic containing regions of quite different composition and function. It is tempting to speculate that the latter might be the case and that, moreover, these regions might correspond to morphologically distinct areas of the inner membrane. For example, that part of the inner membrane which lies parallel to the outer-membrane (the inner boundary membrane) might not carry the "head-pieces" which are characteristic of the cristael membrane, and which, as we shall see, have a specific role in oxidative phosphorylation. The suggestion has therefore been made that the anionic-carrier sites might be specifically localized in the inner boundary membrane, where indeed they would have the most direct access to the cytoplasm.

(b) *Vectorial aspects.* No membrane more complicated than a simple bimolecular leaflet of phospholipid is likely to be symmetrical across its thickness. In other words, the outward facing surface of a membrane is likely to be quite different from the inner facing surface particularly in regard to available functional groupings on proteins. If the membrane proteins are not free to rotate and move across the thickness of the membrane and if the membrane has a limited permeability to substrate molecules, then it follows that enzymic activities localized in such a membrane will have a vectorial character.

Such vectorial aspects are of particular importance in the case of the cristael membrane, a membrane whose sidedness is characterized by the headpieces which are unique to its matrix-facing aspect and also by the group transfer carriers which were discussed previously. The vectorial character of the membrane also applies to its function in respiration. For example, both NADH and succinate can only interact with the respiratory chain from the matrix facing surface of the membrane. (This restriction in so far as NADH is concerned, should already have become apparent from the fact that mitochondria are unable to oxidize externally added NADH). The matrix-facing surface is also the side of the inner-membrane where ATP is synthesized or utilized, (the site of interaction with ATP or with ADP plus inorganic phosphate is, in fact, the headpiece). Hence a functioning exchange-diffusion system is necessary for intact mitochondria to exhibit oxidative phosphorylation. (Again, this sidedness was predictable, since it has already been mentioned that oxidative phosphorylation is inhibited by inhibition of the ADP-ATP exchange-carrier with atractyloside).

Another example of the sidedness of the cristael membrane is the localization of cytochrome c. This cytochrome is localized on the outward-facing surface of the membrane. It is relatively easy to remove cytochrome c from the membrane, so that swollen mitochondria, (whose outer membranes are perforated), may readily be depleted of cytochrome c by salt extraction. Conversly, cytochrome c may readily be reintroduced into the membranes of such depleted mitochondria. However, cytochrome c cannot be extracted from certain sub-mitochondrial particles nor may cytochrome c be reintroduced into sub-mitochondrial particles prepared from cytochrome c-depleted mitochondria. The explanation of these phenomena is simply that the outward-facing surface of the mitochondrial inner membrane may become the inner-facing surface of a sub-mitochondrial particle (see Fig. 9).

Another outward-facing function of the inner membrane is α-glycerophosphate dehydrogenase. This fact is especially significant in the case of mitochondria from insect flight muscle where α-glycero-phosphate is a key source of reducing equivalents, and where the membrane is particularly impermeable to substrate anions. The vectorial aspect of this dehydrogenase is such that its substrate has no need to cross the inner membrane.

In general, one can say that any water-soluble molecule that inter-acts with functional components of the inner mitochondrial membrane does so in a strictly vectorial manner and that any products of such interactions are also released in a vectorial manner. Where the membrane is impermeable to such substrates or products then a functional sidedness will be demonstrable. In appropriate cases a sidedness demonstrable in intact mitochondria will be reversed in the case of a sub-mitochondrial particle. Such phenomena would be predictable for any highly structured system composed of complex, asymmetric, funtional units. Nonetheless, as we shall subsequently see, it is possible that the sidedness of the inner mitochondrial membrane has a unique functional significance, beyond that of compartmentation and control.

III. Oxidative Phosphorylation

A. THE RESPIRATORY CHAIN

Now that we have a general picture of the structural and functional framework within which the respiratory chain operates it seems appro-priate to consider the chain itself in more detail.

Components

As electrons (with or without associated protons) are passed from electron donors such as NADH or succinate to the final acceptor molecule, oxygen, a series of catalytic components associated with the cristael membrane undergo sequential reductions and oxidations. These components have generally been recognized by virtue of changes in their characteristic spectra attendant upon changes in the oxida-tion state of their respective prosthetic groups. Details of the chemistry of these components and of the spectroscopic techniques by which they have been investigated are beyond the scope of this chapter. However, we do need to consider the general nature of the various species which are known to be involved in the electron transfer process.

(*a*) *Flavins.* The respiratory chain contains at least two flavoproteins, one closely associated with the NADH-dehydrogenase function (i.e. the point of "feed-in" of reducing equivalents from NADH) and one associated with the succinate dehydrogenase function. The prosthetic group of the former is probably flavin mononucleotide (FMN) and is quite readily released from its apo-protein by treatment with dilute acid. The prosthetic group of the latter is flavin adenine dinucleotide (FAD) and is only released by digestion with proteolytic enzymes.

Mitochondria also contain other flavoprotein species, (some of which are concerned with processes such as the β-oxidation of fatty acids), but it is by no means clear how many of these additional flavoproteins are intimately associated with the cristael membrane. It is also not certain whether such firmly-bound flavoproteins function solely as the points of "feed-in" of reducing equivalents from other donors such as α-glycerophosphate, or whether there are any which function as additional components of the main respiratory chain itself.

The properties of a given flavoprotein; its oxidation-reduction potential, its characteristic absorption spectrum, its tendency to fluoresce, the firmness of binding of the flavin moiety, etc., are largely determined by the nature and environment of the particular apoprotein. The flavin moiety itself can only be FMN or FAD. In some flavoproteins the free-radical, half-reduced, form of the flavin is quite stable, and so it is possible for flavoproteins to act as one-electron redox component, cycling either between the fully oxidized and halfreduced states or between the half-reduced and fully-reduced states. It is more usual however to describe their functioning in the respiratory chain as simple, two-electron carriers cycling between the fullyoxidized and fully-reduced states.

(b) *Non-haem iron protein.* The cristael membrane contains significant amounts of protein-bound iron which is not associated with protoporphyrin rings. Some of this non-haem iron has been shown to be localized in the active centres of a number of specific non-haem iron proteins, which apparently act as one-electron redox carriers associated with the respiratory chain. A typical, oxidized, non-haem iron protein of the chain might contain two ferric ions associated with two sulphide ions, held in a specific geometric grouping within the tertiary structure of the protein itself.

Non-haem iron proteins are not unique to mitochondria but represent a large family of proteins some of which may contain selenide rather than sulphide, whereas others contain neither. They exhibit great variety in their oxidation-reduction potentials, the number of iron atoms which they contain, and their characteristic electron spin resonance spectra. This latter property has proved invaluable in characterizing certain of the non-haem iron proteins of mitochondria.

(c) *Ubiquinones*. It has been known for some years that the cristael membrane contains significant amounts of ubiquinone (coenzyme Q), a fully-substituted benzoquinone with a polyisoprenoid side chain. The amount of ubiquinone present varies among mitochondria from different sources and there is also some variability in the length of the side chain. In mitochondria from most mammalian cells (where ubiquinone is present in several-fold molar excess over other respiratory components) the side chain of the molecule is usually ten isoprenoid units in length. This particular quinone is frequently designated as UQ_{10}.

Ubiquinone undergoes oxidation and reduction during respiratory activity in mitochondria, cycling between the fully oxidized quinone and the fully reduced quinol, possibly via the free radical semi-quinone. However, there has been some dispute as to whether these oxidation-reduction processes are in the main pathway of the respiratory chain or represent a subsidiary, shunt pathway. It now seems most likely that the former is the case. Strong evidence for this has come from experiments where ubiquinone has been extracted from the cristael membrane to yield depleted preparations unable to oxidize NADH or succinate. Re-introduction of ubiquinone restored these respiratory activities.

(d) *Cytochromes*. The best known of all of the components of the mitochondrial respiratory chain are the pigments known as cytochromes. These are all haem proteins which, during respiration, undergo one-electron oxidation and reduction between the ferri- and ferro-forms. Each cytochrome species has a characteristic spectrum when in the reduced form. The spectrum, oxidation-reduction potential and other properties of a particular cytochrome are determined by three factors; the nature of the haem group itself, the properties of the apo-protein and the micro-environment of the cytochrome.

In mammalian mitochondria five different cytochromes have been recognized; cytochromes a an a_3 (each containing haem A), cytochrome b (whose haem moiety is identical to that of haemoglobin and myoglobin, i.e. protohaem IX) and cytochromes c and c_1 (each containing covalently linked haem C). Evidence is now accumulating that there are at least two forms of cytochrome b in mitochondria, probably identical in chemistry but differing in microenvironment and function (see below). It is possible that cytochromes a and a_3 are also identical in structure and that the characteristic differences between them (e.g. the binding of carbon monoxide by cytochrome a_3), derives entirely from their mode of organization within the cristael membrane.

Cytochrome c is a protein of low molecular weight (about 13 000) and is readily extracted from mitochondria, being localized (as we have already seen) on the outward facing surface of the cristael membrane. The other cytochromes have molecular weights between 30 000 and 75 000 and are very difficult indeed to isolate in soluble, monomeric form.

(e) *Copper proteins.* Two species of copper protein have been identified in mitchondria, in close association with cytochromes a and a_3 (see below). Each copper protein appears to act as a one-electron carrier in the respiratory chain, cycling between cupri and cupro-forms. They may be distinguished by the characteristic electron-spin-reasonance spectrum exhibited by the oxidized form of one of them. A derivative of one of the copper proteins has been obtained in soluble form and found to have a molecular weight of about 12 000, otherwise very little is yet known of their detailed chemistry.

(f) *Other components.* The components mentioned above do not necessarily exhaust the number of oxidation-reduction components involved in the respiratory chain. The species listed have been identified and characterized as oxidation-reduction carriers, almost exclusively by spectroscopic means. The only components which it has been possible to extract and study in soluble form without causing irreversible damage to the chain are ubiquinone and cytochrome c. Other components have had to be studied *in situ* in intact mitochondria, with sub-mitochondrial particles, or at best in functional fragments of the chain (see below). The controversies relating to the role of ubiquinone and the number of flavoproteins in the chain underline the enormous technical difficulties of resolving the various species involved when it is necessary to study them as minor components of insoluble complex, lipoprotein systems.

In view of these difficulties, it is quite possible that other, as yet unidentified oxidation-reduction processes might occur in the chain, particularly if they involved species with no strongly characteristic spectral properties. For example, the "core-protein" of the cristael membrane, to which I have previously referred, contains sulphydryl groups. In the functional chain these might undergo sequential oxiation and reduction. It would be very difficult to detect such a disulphide-thiol cycle in an intact particulate system. Other components might include iron proteins other than the well documented non-haem iron proteins, since not all of the iron of the cristael membrane can be accounted for unambiguously as proteins of the latter kind. Moreover, it is not impossible that metallo-proteins other than those of iron and copper might be involved in the respiratory chain. Indeed, there have been recent reports of molybdenum being associated with the flavoprotein of NADH dehydrogenase. There is also evidence of a lipid component, other than ubiquinone, whch might function as a redox component of the chain. The first of these observations now seems likely to have been an artefact, the second has yet to be substantiated; further evidence of the enormous technical difficulties associated with studies of the chemistry of the respiratory chain.

Oxidation-Reduction Potentials

Any proposed sequence of events in the respiratory chain, and any proposed mechanism for energy conservation coupled to respiration, must take account of and be consistent with the oxidation-reduction potentials of the individual components of the chain. These potentials will define the equilibrium constant for electron transfer between any pair of components as well as the change in free energy associated with any such electron transfer.

Several components of the chain take up protons as well as electrons on reduction. Moreover, most of the components are proteins whose properties will depend upon the pH of their environment. Therefore it is far more useful to define the potentials of chain components as standard oxidation-reduction potentials at pH 7 rather than as electrode potentials with all reactants (including H^+) at unit activity. The question of the dependence of potentials on hydrogen-ion concentration immediately underlines the difficulty that, when interpreting phenomena occurring in such a highly organized membraneous system as the respiratory chain, one has little idea of the effective pH to which a particular protein or prothetic group is exposed. Indeed a given component might be in a lipid milieu with little access to water. This last point raises in turn the more general problem that the potential of a component will depend on its environment and that this environment might well change with the functioning of the complex, energy-transducing systems of the cristael membrane. The very fact that, when $\sim$ is generated by the hydrolysis of ATP, electron transfer can proceed in the reverse direction, implies the oxidation-reduction potentials of some components must effectively become much more negative when the membrane is in the energized state. Whether this is because such components have been converted to modified chemical species or because their conformation, or local pH, or other aspect of their environment has been changed is a question directly related to the nature of $\sim$.

In the Table below are listed the approximate standard potentials of certain components of the chain. In most cases a range of potentials have been given since, by reason of theoretical and technical difficulties, the published data is conflicting. In the case of cytochrome b, the figure given is an approximate mean of a wide range of potentials cited for the two apparent forms of cytochrome b.

TABLE 1. Standard oxidation-reduction potentials*

System	E' (pH 7) volts
Oxygen/water	0·82
Cytochrome a_3; oxidized/reduced	0·36 – 0·53
Cytochrome a; oxidized/reduced	0·21 – 0·29
Cytochrome c; oxidized/reduced	0·23 – 0·28
Cytochrome c_1; oxidized/reduced	0·22
Cytochrome b; oxidized/reduced	0·00†
UQ_{10}; oxidized/reduced	0·00
Fumarate/succinate	0·03
Flavoprotein (NADH); oxidized/reduced	−0·22
Flavoprotein (succinate); oxidized/reduced	−0·22
NAD+/NADH	−0·32
NADP+/NADPH	−0·32

* Many of the values quoted must be treated as rough approximations to those obtaining *in situ* in mitochondria.
† See text.

Sequence

There have been several approaches to the determination of the sequence in which reducing equivalents are passed from NADH to oxygen. For example, sophisticated spectrophotometric apparatus has been used to determine the sequence in which components become reduced when NADH is added to a fully oxidized preparation of submitochondrial particles. Similar studies have also been made on the sequence of reoxidation when oxygen is added to a fully-reduced, anaerobic preparation. The general conclusion has been that the sequence, in terms of flavin and haem components, is: NADH $\rightarrow$ flavoprotein $\rightarrow$ Cyt.b $\rightarrow$ Cyt.c_1 $\rightarrow$ Cyt.c $\rightarrow$ Cyt.a $\rightarrow$ Cyt.a_3 $\rightarrow$ oxygen.

It is also accepted that succinate passes its reducing equivalents to the same cytochrome chain but via a different flavoprotein.

As was previously mentioned, the evidence now is that ubiquinone is essential for the oxidation of both NADH and succinate. Oxidation of reduced cytochrome c (or of artificial electron donors that "feed in" to the chain by reducing endogenous cytochrome c) does not require ubiquinone. Such studies clearly locate ubiquinone in the chain after the flavoprotein and before cytochrome c.

Further information on sequence has been obtained by using specific inhibitors of particular electron transfer steps, (see below), or by using fragments of the chain lacking certain components and only capable of catalyzing a limited span of electron transfer. As always with such complex system each approach has its disadvantages, but a synthesis of available information suggests that the general sequence to be that shown in Fig. 15. Other reducing equivalents, such as those supplied

NADH $\to$ flavoprotein $\to$ nonhaem-iron*

ubiquinone $\to$ cyt.b $\to$ cyt.c_1 $\to$ cyt.c $\to$ cyt.a $\to$ cyt.a_3 $\to$ oxygen
(nonhaem-iron) (copper proteins)

succinate $\to$ flavoprotein $\to$ nonhaem-iron

Fig. 15. Sequence of carriers in the respiratory chain. Each flavoprotein and nonhaem-iron protein indicated represents a different species. Carriers in parenthesis participate in electron transfer in the region of the chain below which they are indicated, but the precise sequence of electron transfer in these regions is uncertain. *It is likely that at least three different species of non-haem iron protein are involved in electron transfer over this span of the chain.

by α-glycerophosphate or reduced "electron-transferring flavoprotein" feed into the chain at the level of ubiquinone via specific flavoproteins, analogous to the succinate dehydrogenase system.

It will be seen that the sequence in Fig. 15 is consistent with the oxidation-reduction potentials listed in Table 1, in that at nearly every step reducing equivalents are passed to an electron acceptor whose potential is roughly equal to or more positive than that of the electron donor.

It will be noted that after ubiquinone there are no more two-electron carriers designated, so that for one molecule of NADH to be oxidized it might be expected that two molecules of each of the cytochromes would be involved. This is a question which will be taken up again later.

It will also be noted that between ubinquinone and oxygen there are no more hydrogen carriers, only electron carriers. In fact the non-haem iron proteins associated with the flavoprotein are also electron rather than hydrogen carriers. Indeed there is little evidence of direct transfer of hydrogen atoms anywhere in the chain and it is probably better to consider all oxidation-reductions in the chain as being essentially electron transfers, with protons being taken up from or released into the local environment or the suspending medium as required by the chemistry of the particular components involved. Eventually an oxygen atom is reduced and two protons are taken up; so that the overall oxidation of succinate to fumarate involves no net change of pH whilst that of NADH to NAD^+ involves a net uptake of one proton per molecule oxidized. (This proton is of course balanced by the proton released when NAD^+ is reduced by the substrates of the dehydrogenases; i.e. $XH_2 + NAD^+ \longrightarrow X + NADH + H^+$; $NADH + H^+ + \frac{1}{2}O_2 \longrightarrow NAD^+ + H_2O$.)

Respiratory Inhibitors

Our understanding of the sequence of events in electron transfer and of many other aspects of mitochondrial function owes much to the availability of inhibitors that are relatively specific for particular steps in the respiratory chain. Table 2 lists some of the more useful of these inhibitors. In some cases the precise site of inhibition is known (e.g. carbon monoxide, which combines directly with cytochrome a_3), in other cases (e.g. rotenone) the span of the chain that is inhibited is known for certain, but the precise binding site is not. Malonate is a competitive inhibitor for the succinate dehydrogenase step, all of the other inhibitors listed are noncompetitive.

TABLE 2. Some specific inhibitors of electron-transfer

Inhibitor	Span inhibited
Rotenone ⎫ Amytal ⎬ Piericidin ⎭	Flavoprotein (NADH) $\rightarrow UQ_{10}$
Malonate	Succinate $\rightarrow$ Flavoprotein(succinate)
Thenoyl trifluoroacetone	Flavoprotein(succinate) $\rightarrow UQ_{10}$
Antimycin A	Cyt.$b \rightarrow$ cyt.c_1
Carbon monoxide ⎱ Cyanide ⎰	Cyt.$a \rightarrow$ oxygen

Structural Organization

The electron transfer sequence shown in Fig. 12 might be considered to be an adequate description of the respiratory chain if electron transfer took place between reactants in solution. However, the components are organized within the structure of the cristael membrane. Are they arranged as single assemblies acting more or less independently and in parallel? Are there points of interaction between neighbouring assemblies? What is the stoichiometry of components within a given assembly, and does this allow for transitions between two-electron and one-electron carriers? What in fact is the structural organization of the chain within the membrane?

For a long time this general question was more or less ignored, and this is probably the chief reason for lack of progress in defining the nature of the energy conserving process. The generation of $\sim$ is coupled to the respiratory chain, and an understanding of the structure of the latter is a prerequisite for determining the nature of the former.

(a) *Mobile components.* The electron donors to the chain, NADH, succinate etc., are small molecules which presumably would have access to any point on the surface of the membrane at which they interact. The final electron acceptor, oxygen, is also freely mobile. Two other components of the chain are relatively mobile, ubiquinone and cytochrome *c*.

Ubiquinone is a small molecular present in relative molar excess over other components. For example, in mitochondria from bovine heart there are about eight molecules of UQ_{10} per molecule of cytochrome *c*. Ubiquinone is readily extracted from the membrane into solvents such as pentane and may be presumed to be localized with its polyisoprenoid chain buried deeply into the hydrophobic environment created by the hydrocarbon chains of the membrane phospholipid (see Fig. 8). One might not expect ubiquinone to be as mobile in this rather structured lipid phase as it would be in free solution, but there would be sufficient freedom of movement to allow a given molecule of ubiquinone to react with several neighbouring molecules, and thus give rise effectively to a rapidly equilibrating pool of ubiquinone over the whole membrane. Indeed, some movement of ubiquinone might be obligatory since it is more polar in its reduced than its oxidized form and the change in polarity might require movement to and from the aqueous interface.

Cytochrome *c*, as we have already seen, is readily extracted into salt solutions from the outward-facing surface of the inner membrane. It is a basic molecule carrying net positive charge at physiological pH, and hence is probably bound by simple ionic interactions to negatively charged groupings on the surface of the membrane (e.g. the polar residues of phospholipids such as cardiolipin).

Cytochrome *c* is a small protein and one might expect from the nature of its binding to the membrane that each molecule would have a certain mobility, (virtually rolling across the surface of the membrane), at least over a sufficient distance for it to interact with another molecule of cytochrome *c* and hence create another "mobile pool" analogous to that of ubiquinone.

It can thus be seen that if there are any structured respiratory assemblies they can only be of limited span; transferring electrons from NADH or succinate (or other donor) to the ubiquinone pool, transferring electrons from reduced ubiquinone to the pool of cytochrome *c*, and catalyzing the terminal oxidation of reduced cytochrome *c*. This concept immediately explains how it is that electrons from a variety of donors eventually follow the same terminal pathway to oxygen.

(*b*) *Complexes*. With the exception of ubiquinone and cytochrome *c* all of the other components of the respiratory chain are firmly attached to the membrane. At first sight the relative stoichiometry of these components is difficult to rationalize in terms of any meaningful organization. For example, according to some workers, the molar ratios; cyt.*b* : cyt.c_1 : cyt.*c* : cyt.*a* : cyt.a_3, are, in mitochondria from bovine heart; 2 : 1 : 2 : 3 : 3.* In mitochondria from liver the quoted ratios are quite different, for example the ratio cyt.c_1 : *a* is 1 : 0.7. In fact, within the same tissues at various stages in development or in various nutritional states the ratios may change, and yet all of the components appear to be functional. A clue to the explanation of this paradox is that certain ratios remain constant; cyt.*b* : cyt.c_1 is always (according to some authors) 2 : 1, cyt.*a* : cyt.a_3 is always 1 : 1. These constancies refer to components localized within the same span between "mobile pools", i.e. in the spans ubiquinone→cyt.*c* and cyt.*c*→ oxygen respectively. In other words, the stoichiometry is consistent with fixed electron-transfer assemblies of limited span but constant composition, variability being in the relative amount of each kind of such assemblies.

This line of reasoning has direct experimental support from studies on the isolation of components from fragmented cristael membranes. In the presence of suitable detergents and salts the cristael membrane can be dispersed into a heterogeneous mixture of lipoprotein subunits (see Fig. 8). Subfractionation of such dispersions has yielded four homogeneous fractions. Each fraction represents a preparation of a complex lipoprotein with a molecular weight of 300 000 and dimensions roughly corresponding to those of a base-piece subunit of the inner membrane, (see section on membrane ultrastructure). Moreover, each complex catalyzes an electron transfer span between a pair of mobile components: that designated as "Complex I" functions as an NADH-ubiquinone reductase, "Complex II" is a succinate-ubiquinone reductase, "Complex III" is a reduced ubiquinone-cytochrome *c* reductase, "Complex IV" is a cytochrome *c* oxidase.

The components of a given complex are the components already known to be associated with the corresponding span, and the enzymic activities are inhibited by the characteristic inhibitors of that span. For example, Complex III (reduced ubiquinone-cytochrome *c* reductase) contains cytochrome *b*, cytochrome c_1 and a specific

---

* Other workers claim that the ratios are 2 : 1 : 2 : 1 : 1. This represents a typical, and alarming, discrepancy. Fortunately, as we shall see, either set of ratios is compatible with our concept of the organization of the respiratory chain.

nonhaem-iron protein in the ratio $2:1:1$. Complex III is inhibited stoichiometrically by antimycin A, a specific and stoichiometric inhibitor of the corresponding span of the electron transfer chain in intact mitochondria. Complex III also contains phospholipid (up to 500 mg per g protein) and about half of its protein content can be accounted for as the "core protein" to which reference has already been made.

The individual components of a complex can be separated from one another, but this usually requires conditions far more drastic than those required to separate one complex from another. This lends support to the concept that each complex is a highly organized functional unit derived with little artifactual change from a corresponding unit in the native membrane.

Functionally the respiratory chain may thus be visualized as shown in Fig. 16. In ultrastructural terms we may consider the inner membrane continuum as being composed of a random, two-dimensional array of electron transfer complexes in functional contact with one another via mobile components. Thus, for example any Complex I in a given region of the membrane can supply electrons to any Complex III in that same region via the pool of ubiquinone. Hence, whatever the relative stoichiometry of the different species of complex in a given mitochondrion, all components can participate in respiration.

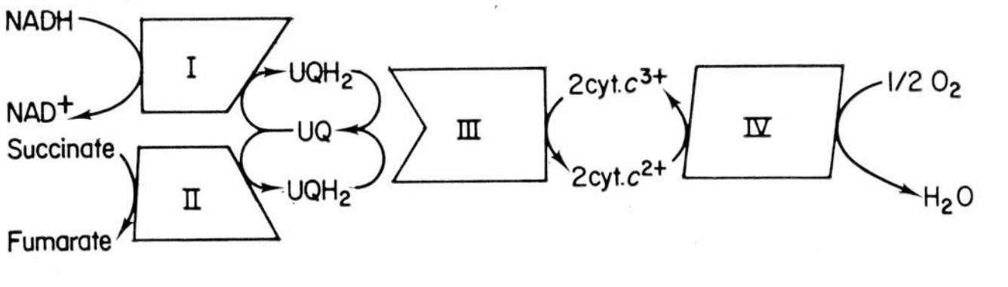

Fig. 16. Diagrammatic representation of the respiratory chain as a set of complexes functionally linked via mobile electron carriers.

(c) *Loops.* So far we have only considered the two-dimensional organization of the chain. However, as we have already seen, the cristael membrane exhibits a functional sidedness. Such a sidedness would be predicted from the chemical nature of the assymmetric, macromolecular assemblies of respiratory components which constitute the complexes of the chain. It would be extremely unlikely that such a highly organized structure as the cristael membrane, which exhibits a clear morphological sidedness with respect to the 8–9 nm headpieces, would be assembled from randomly orientated components. The actual orientations of many of the functional components of the chain are not known. Many of them are almost certainly buried within the hydrophobic core of the membrane and not directly accessible to externally added reagents. However, as we have already seen, the NADH- and succinate dehydrogenase functions are localized on the matrix-facing side of the membrane. It also seems probable that cytochrome a_3 interacts with oxygen on this side of the membrane. On the other hand, cytochrome c is localized on the side of the membrane that, in intact mitochondria, faces the cytoplasm. Thus, during their passage down the respiratory chain, electrons must "loop" across the thickness of the membrane from say, succinate, to cytochrome c and then back again to oxygen. How many such "loops" there are is a matter of dispute, related, as we shall see, to the nature of the energy-conserving processes. A complicating factor is that it is likely that ubiquinone can traverse the membrane and hence bring into functional contact components with opposite orientations.

Electron transfer within Complexes

It is beyond the scope of this Chapter to consider the mechanism of transfer of electrons between reactants that come into contact with one another in the appropriate orientation. Hence, we shall not discuss the mode of interaction between say, NADH and the corresponding dehydrogenase function of Complex I, or the ill understood, co-operative process whereby, (with no intermediary formation of free peroxide), a molecule of oxygen is reduced to two molecules of water by four electrons donated via reduced cytochrome a_3. What can be discussed is the mode of sequential interaction between relatively fixed components.

The mobile components, ubiquinone and cytochrome c present no problem in this respect. In each case they are relatively free to move between their respective electron donors and electron acceptors. (There is however evidence that the oxidized and reduced forms of ubiquinone not only function as redox components, but in some way influence electron transfer within Complexes I, II and III, and hence exercise control on the flux of reducing equivalents through the ubiquinone pool.) What is more difficult to account for is electron transfer within a complex. When cytochrome a is reduced by cytochrome c how is its reduced prosthetic group then able to come into contact with the haem moiety of cytochrome a_3, and how then is the latter to reorientate itself prior to interaction with oxygen?

The suggestion has been made that each fixed component of the chain retains the freedom to rotate about a fixed axis, so that it may interact sequentially between similarly fixed components localized on either side of it. This simple picture is inadequate. It has become abundantly evident that electron transfer within a Complex cannot be described in terms of a linear sequence with one electron following another. Transfer of a pair of electrons is discontinuous, each electron probably follows a different pathway, and the process is associated with major changes in the relative conformation of the components of the Complex.

In general terms, the kind of process that might occur is that a component, X, is reduced by one electron from some electron donor (W), thus altering the conformation of the Complex so that a previously inaccessible component, Y, becomes available to receive a second electron from W; the reduction of Y now causes another conformational rearrangement which allows reduced X to come into contact with and be oxidized by a third component, Z; a further electron transfer from Y to X may now occur, etc.

Convincing support for the existence of such an elaborate, co-operative networks of electron transfer within a given Complex comes not only from spectroscopic studies, but also from direct chemical evidence that major cyclical rearrangements occur when isolated Complexes are reduced and reoxidized.* The existence of pathways of this kind within the Complexes accounts for the mechanism by which electrons are transferred across the thickness of the membrane, and also for the mechanism by which pairs of electrons are handled within Complexes containing one-electron carriers. The conformational perturbation which is an integral part of the electron transfer process also has a direct significance in energy conservation, (see below).

B. ATP SYNTHESIS

Now that we have considered the respiratory chain where $\sim$ must first be generated, we have to consider the apparatus whereby $\sim$ is utilized to synthesize ATP from ADP plus inorganic phosphate.

P/O Ratios

The number of times that $\sim$ is generated as a pair of electrons traverses the chain is, as we have already mentioned, determined mechanistically. Changes in the ratio of succinate to fumarate for example, might alter the rate of generation of $\sim$ coupled to the oxidation of succinate, but not the stoichiometry of the process.

---

* These conformational changes are not at all surprising. The bond lengths and sometimes the entire geometry of simple co-ordination compounds of transition metal ions can be markedly changed on oxidation and reduction. Changes in the structure of a relatively simple redox protein such as cytochrome c on changing the valence state of the central iron atom extend over the entire tertiary structure of the molecule. Even more dramatic changes might be expected in more complex systems, particularly when there is a protonation step associated with the electron transfer process.

Similarly, the dissipation of $\sim$ is a process whose rate or even direction might be controlled by the relative concentrations of the reactants but whose stoichiometry remains constant. Therefore, for every pair of electrons traversing the chain there is a "theoretical" yield of ATP; i.e. there is an "ideal" P/O ratio, where this is defined as the number of molecules of ATP formed from ADP plus inorganic phosphate per atom of oxygen reduced.

Experimentally determined P/O ratios are usually significantly lower than this "ideal" since other $\sim$-dissipating processes compete with that for ATP synthesis. However, it is generally agreed that the "theoretical" yield of ATP coupled to the oxidation of NADH corresponds to a P/O ratio of $3 \cdot 0$. The corresponding P/O ratio for succinate is accepted as being $2 \cdot 0$.

Sites of coupling

The electrons from succinate follow the same pathway from ubiquinone to oxygen as do those from NADH. Moreover, there is little difference in oxidation-reduction potential between succinate and ubiquinone, and so there is no possibility of $\sim$ being generated over this latter span. It therefore may be concluded, from the respective P/O ratios for the oxidation of NADH and succinate, that $\sim$ is generated twice between ubiquinone and oxygen and once between NADH and ubiquinone. This latter conclusion is supported by the finding that, under appropriate conditions, the transfer of reducing equivalents from NADH to an externally added excess of ubiquinone, (in the presence of antimycin A to prevent the reoxidation of the reduced acceptor) can be coupled to the synthesis of one molecule of ATP per molecule of NADH oxidized.

Further confirmation that there is a single "coupling" site, (i.e. site where $\sim$ is generated) in the NADH-ubiquinone span has been obtained from studies on reversed electron transfer, in sub-mitochondrial particles. Electrons from succinate reduce ubiquinone and, (in the presence of inhibitors of terminal oxidation), the NADH-ubiquinone span may then be driven in reverse if $\sim$ is generated by the hydrolysis of ATP. Under ideal conditions the stoichiometry of this process approaches that of one ATP molecule being hydrolyzed per molecule of NAD^+ reduced by succinate.

Sub-mitochondrial particles will readily oxidize certain externally added electron donors whose oxidation-reduction potentials are close to that of cytochrome c. The oxidation of electrons "fed in" from such donors at the cytochrome c level in the chain may experimentally be shown to be coupled to ATP synthesis with a P/O ratio approaching 1·0. Hence, there appears to be one "coupling site" between cytochrome c and oxygen.

By difference one would expect there to be one coupling site between ubiquinone and cytochrome c. This conclusion has been confirmed directly by experiments in which NADH has been shown to reduce externally added cytochrome c (in the presence of cyanide to inhibit cytochrome oxidase) with the synthesis of two molecules of ATP per pair of electrons transferred, (i.e. a $P/2e^-$ ratio of 2·0). Moreover externally added, reduced ubiquinone has been shown to reduce cytochrome c with a $P/2e^-$ ratio of 1·0.

It can thus be seen that each of the Complexes I, III and IV traverse a single coupling site.* Many attempts have been made to identify the sites of generation of $\sim$ more closely , i.e. to locate them between given pairs of carriers whose oxidation-reduction potentials are sufficiently far apart to allow for the generation of $\sim$ when electrons are transferred between them. In particular, attempts have been made using spectroscopic techniques to identify points of respiratory control, i.e. pairs of adjacent carriers exhibiting anomalous degrees of reduction under conditions where respiration is inhibited by virtue of the accumulation of $\sim$, (see next section). If the generation of $\sim$ were coupled to the transfer of electrons between a carrier X and a carrier Y, whose oxidation-reduction potential was substantially more positive than that of X, then under conditions where $\sim$ was rapidly dissipated Y would tend to accept electrons from X. However, if $\sim$ were allowed to accumulate and exercise respiratory control at the site of its generation, then electron transfer between X and Y would be inhibited, and X would become significantly more reduced.

In spite of much effort of this kind directed at the identification of coupling sites in the chain there is very little agreement as to their precise localization. This is scarcely surprising in view of what we now know of the discontinuous nature of electron flow through the Complexes. In my opinion it is more constructive to look upon each Complex as an apparatus not only for transferring pairs of electrons between mobile carriers, but also for intergrating and conserving the free energy changes associated with each of the single-electron transfers taking place within that span.

Respiratory control

In the preceding section we have again come upon the phenomenon of respiratory control and the associated phenomenon of reversed electron transfer. The generation of $\sim$ is reversibly coupled to

---

* In mitochondria from certain yeasts where the P/O ratio for NADH oxidation is 2·0, it has been shown that the coupling site associated with Complex I is lacking.

respiration so that the accumulation of ∼ tends to inhibit respiration. The dissipation of ∼ is reversibly coupled to the synthesis of ATP. Therefore as the ratio of ATP to ADP in the mitochondrion increases, the free energy associated with the phosphorylation of ADP increases to the point where ATP synthesis can no longer be driven by the dissipation of ∼; indeed, if exogenous ATP is added the direction of the reaction is reversed with the hydrolysis of ATP and the net generation of ∼. Hence ATP accumulation tends to inhibit respiration, ATP addition tends to reverse respiration.

In vivo, these phenomena are reflected in a fundamental biological control mechanism; the respiration of a cell or of an animal is geared to the work load. If there is a demand for ATP for cellular activity then the cytoplasmic ratio of ATP to ADP falls. This fall is reflected (via the ADP:ATP exchange system) within the mitochondrion and respiratory control is released. When the work demand has been satisfied ATP accumulates and respiratory control is reimposed.

In vitro it is relatively easy to demonstrate the phenomenon of respiratory control. If intact mitochondria are suspended in a suitable medium containing substrate and inorganic phosphate the rate of respiration is relatively slow. The mitochondria under these conditions are said to be in State IV.* When ADP is added, ATP is synthesized, ∼ is dissipated and so respiratory control is released and rapid (State III) respiration ensues. If relatively small amounts of ADP are added, then State IV respiration is reestablished when most of the ADP added has been converted to ATP.†

If ∼ dissipates spontaneously in the absence of the substrates necessary for ATP synthesis then the extent of respiratory control is consequently decreased. Hence the ratio of the rate of respiration in State III to that in State IV (the "Respiratory Control Index") is a measure of the stability of ∼ in any particular preparation of mitochondria. It should be stressed that respiratory control is not necessarily an index of the capacity of a particle to synthesize ATP. Certain preparations of mitochondria, and most preparations of sub-mitochondrial particles, exhibit no respiratory control but may retain a

---

* The various States defined for mitochondria correspond to conditions where respiration is limited by different factors. The States are characterized both by relative rates of respiration and by the steady-state degree of reduction of the various carriers in the chain. In State I, levels of both ADP and substrate are low; in State II, there is adequate ADP but no substrate; in State III levels of both ADP and substrate are high; in State IV there is adequate substrate but ADP levels are low; in State V levels of ADP and substrate are high but the system has become anaerobic.

† It is possible from such an experiment to determine the P/O ratio of the mitochondria used, by dividing the number of moles of ADP added by the number of gram atoms of oxygen utilized during State III respiration, since at the point of re-establishment of State IV virtually all of the ADP added will have been converted to ATP.

capacity to synthesize ATP with quite high P/O ratios. In such preparations, the process of spontaneous dissipation of ~ is fast enough not to limit respiration, but this process has a lower affinity for ~ than has the system which synthesizes ATP when ADP and P_i are available.

The phenomenon of ADP-dependent respiratory control is not necessarily evidence that the generation of ~ during respiration is reversible. For example, the coupling mechanism might involve the conversion of a respiratory carrier to a derivative whose break-down was subsequently coupled to the synthesis of ATP (see below). In the absence of ADP, respiration would then be limited by the rate of spontaneous regeneration of unmodified carrier. Hence, respiratory control might to some extent be a mechanistic rather than a thermodynamic phenomenon. On the other hand, the control of respiratory rate by the ATP:ADP ratio, and the reversal of electron transfer in the presence of ATP are positive evidence of reversibility at the coupling site.

Uncoupling of oxidative phosphorylation

When mitochondria or submitochondrial particles respire but can-not carry out work performances such as the synthesis of ATP they are said to be "uncoupled". Uncoupling can be of three kinds: the result of lesions at the primary coupling site so that ~ is no longer generated; the manifestation of an overriding process whereby ~ is dissipated; or the result of damage to the system which synthesizes ATP. Particles exhibiting lesions of the third kind are not strictly uncoupled if they retain the capacity to perform other work func-tions such as energized transhydrogenation. In all cases the particles will exhibit no ADP-dependent respiratory control, but it should be remembered that lack of respiratory control is not necessarily an index of uncoupling. Moreover the performance of work functions which take precedence over ATP-synthesis in their affinity for ~ (e.g. accum-ulation of calcium salts), and hence which also abolish respiratory control, should not be confused with the truly uncoupled state.

Uncoupling can occur as a result of patent structural damage to the cristael membrane (e.g. by osmotic shock or by the addition of detergents) or as a result of specific interaction with compounds classi-fied under the general heading of "uncouplers". These include natur-ally occurring molecules such as long chain fatty acids, and many synthetic compounds. The best known of the latter is 2, 4-dinitro-phenol, but this is a relatively inefficient uncoupler by comparison with compounds such as certain benzimidazoles and carbonyl cyanide phenyl-hydrazones which can cause complete uncoupling when present at incredibly low concentrations. *In vivo* such uncouplers are thermogenic and highly toxic, giving rise to uncontrolled and fruitless cellular respiration.

The precise mechanism of action of uncouplers is unknown, since the nature of $\sim$ is unknown. Moreover, various classes of uncoupler might act in different ways, some causing ultra-structural lesions in the membrane, some acting as substrates for cyclic energy-dissipating processes. Many uncouplers do however share one common feature, they tend to increase proton conductance across lipid bilayers and consequently facilitate proton movement across the cristael membrane. The possible significance of this property will be discussed later.

As was mentioned earlier in this Chapter, any process (other than ATP synthesis) which increases the rate of dissipation of $\sim$ will also inevitably tend to increase the rate at which ATP is hydrolysed. Hence in any particle where the system for ATP synthesis is intact the addition of an uncoupler will tend to generate ATPase activity.

Inhibitors of oxidative phosphorylation

There is a set of compounds which prevent oxidative phosphorylation by inhibiting energy-transfer processes. Such compounds do not interact directly with the respiratory chain nor should they be confused with uncouplers. These inhibitors deserve mention because they have proved to be very valuable tools in mitochondrial research.

One class of such compounds is represented by atractyloside, the specific inhibitor of the ADP : ATP exchange reaction. When atractyloside is added to intact mitochondria, endogenous ATP levels rapidly rise and, since exchange with external ADP is prevented, respiratory control is imposed. As we have already seen, atractyloside is without effect on oxidative phosphorylation in submitochondrial particles. It is also without effect on uncoupled respiration or on other work performances energized by respiration in intact mitochondria.

Another class of inhibitor is represented by the antibiotic oligomycin. Oligomycin apparently inhibits the reversible system for the synthesis of ATP coupled to the dissipation of $\sim$. Hence, in intact mitochondria or in submitochondrial particles, oligomycin inhibits ATP synthesis coupled to respiration, and all work performances energized by ATP. It also inhibits spontaneous, or uncoupler-stimulated ATPase activity. Like atractyloside, oligomycin reimposes respiratory control on mitochondria respiring in the presence of ADP. Again, respiratory control is released on the addition of uncouplers.

Oligomycin has the additional and very interesting property of inhibiting the respiration of submitochondrial particles which are coupled but normally exhibit no ADP-dependent respiratory control.

Apparently therefore oligomycin stabilizes $\sim$ sufficiently in these particles to impose a kind of respiratory control. This "control" is in turn released by uncouplers. It is as though the $\sim$-dissipating process that is responsible for the lack of respiratory control in submitocondrial particles were the result of an abortive energy transformation between $\sim$ and the system for ATP synthesis. Such a transformation, like the corresponding (but effective) one which results in ATP synthesis, might be inhibited by oligomycin. In connection with this stablization of $\sim$ it is interesting to note that oligomycin not only permits work performances (other than ATP synthesis) energized by respiration, but in many cases actually stimulates them.

A third class of inhibitor is a group of fluorescent compounds of unknown structure called aurovertins. Aurovertin appears profoundly to decrease the affinity of the ATP-synthesizing apparatus of ADP, but not for ATP. Hence, in the presence of aurovertin, respiratory control is imposed and oxidative phosphorylation is inhibited. There is however no inhibition of other work performances energized by respiration or by ATP-hydrolysis.

Coupling factors

(a) *General considerations.* One of the standard procedures for resolving any complex biochemical system is that of fragmentation followed by functional reconstitution. Therefore much effort has been expended in resolving the system for oxidative phosphorylation by the controlled extraction of mitochondrial components so that the capacity for oxidative phosphorylation is lost, followed by attempts at reconstituting oxidative phosphorylation by the addition of "factors" isolated from the extract or from whole mitochondrial preparations.

There has been an unfortunate tendency to designate as a "coupling factor" any preparation that marginally improves the measured P/O ratio of damaged particles. Few such factors have anything to do with the primary energy-conservation process associated with the electron transfer chain. Those that are of any functional significance at all are primarily concerned with the terminal process of ATP synthesis. Regrettably, even factors of this kind have rarely been homogeneous and have usually been prepared and assayed by procedures that are difficult to reproduce outside the laboratories of their origin. In addition, many factors are only effective in combination with other equally ill-defined preparations.

Even the best defined "coupling factor", the macromolecular complex designated as F_1, (see below), can function not only in an enzymic capacity in ATP-synthesis, but also, when fragmented into its component protein molecules, can improve the P/O ratios of certain

S

particles, apparently by stabilizing $\sim$ in a manner similar to the action of oligomycin. Neither such conservatory action, nor for that matter recoupling dependent upon the removal of natural uncouplers can be deemed as true, reconstitutive coupling.

These cautionary comments on coupling factors in general do not detract from notable advances which have come from the reconstitutive approach. Such studies, although so far giving little insight into the nature of $\sim$ have revealed the site of ATP synthesis and many of the properties of ATP-synthetase system.

(b) *The ATP-synthetase system.* It has been convincingly shown by reconstitutive studies, supplemented by electron microscopy and immunological techniques, that the 8–9 nm headpieces which stud the matrix-facing surface of the cristael membrane represent the site of ATP synthesis. Isolated preparations of these particles as usually prepared, (in the form of the coupling factor designated "F_1") exhibit high ATPase activity. Presumably this is an abortive manifestation of the native function of this particle, to mediate the reversible process whereby ATP is generated at the expense of dissipation of $\sim$.

F_1 has a molecular weight of about 350 000. It is a complex which dissociates into inactive subunits on exposure to cold. The precise composition of the complex is not certain but it seems likely that there are a total of 9 subunits, not all of which are identical. In fact, the complex seems to contain 5 different protein species.

Another coupling factor, designated as "Factor A" behaves as though it were a more native form of the 8–9 nm particle, in that it restores the capacity of depleted sub-mitochondrial particles to synthesize or utilize ATP without itself acting as an ATPase. In this respect it behaves like a combination of F_1 with a specific inhibitory factor which has also been isolated from the detachable components of the cristael membrane. (However, there is no evidence that Factor A contains the inhibitory factor.) In the case of Factor A, or F_1 plus inhibitor, it is as though some active centre in the complex were now in a hydrophobic environment, thus inhibiting the ATPhase activity.

Structurally, the link between the site of generation of $\sim$ i.e. the respiratory chain, and the site of its utilization for ATP synthesis, i.e. the headpiece, is the "stalk". Presumably therefore the stalk, represents also a functional link and may thus be looked upon as part of the native ATP-synthetase. Two components have so far been isolated that could correspond to this link. One is a protein ("Factor B") with a molecular weight of about 30 000, which contains two sulphydryl groups whose integrity is essential for coupling. The other component is a "basic coupling factor" with a molecular weight of about 20 000.

If F_1 is attached to mitochondrial membranes from which the basic coupling factor has been removed, it is active as an ATPase, but is neither coupled nor sensitive to inhibition by oligomycin. The basic factor restores both of these properties. Moreover, if F_1 is attached to membranes depleted both in the basic factor and in phopholipid, the ATPase activity of the resulting preparation is completely inhibited by the addition of the basic factor and can then only be restored on addition of phospholipid. Thus the basic factor confers both phospholipid-dependence and oligomycin sensitivity; which suggests that oligomycin acts on the intact system by displacing phospholipid from a key site. This in turn suggests that the transfer of $\sim$ mediated by the functional stalk required a hydrophobic environment.

There is some evidence that the attachment of the stalk to the membrane continuum itself involves another, very hydrophobic, protein, but the nature of this has yet to be clearly defined. In addition, there is evidence for the existence in the membrane of a small proteolipid molecule which is concerned with energy transfer between the respiratory chain and the ATP synthetase system. As these factors become better defined in terms of structure, location and rôle in the energy transfer process, the reconstitutive approach will finally prove its worth.

Exchange reactions

Further evidence concerning the process whereby ATP is synthesized comes from a number of isotopic exchange reactions.

(*a*) *ADP:ATP exchange.* If non-respiring particles are incubated with unlabelled ATP and ^{14}C-labelled ADP, ^{14}C-ATP may subsequently be isolated from the mixture. This exchange is taken as evidence for a reaction sequence that involves a phosphorylated intermediate, i.e. : ATP $+$ X $\rightleftharpoons$ ADP $+$ X $-$ P.
Such a process would tend to equilibrate the label between the two nucleotides.

(*b*) $P_i:ATP$ *exchange.* If non-respiring particles are incubated with a mixture of ATP and ^{32}P-labelled inorganic phosphate (P_i), ^{32}P-labelled ATP is formed. This is usually interpreted as evidence for the sequence :

$$\text{(i) ATP } + \text{ X} \rightleftharpoons \text{ADP} + \text{X–P}$$
$$\text{(ii) X–P} + \text{Y} \rightleftharpoons \text{X–Y} + P_i$$

This sequence, acting in reverse, would tend to transfer a labelled phosphoryl residue from the $^{32}P_i$-pool onto the ADP generated in reaction (i), to give rise to labelled ATP. This overall exchange is inhibited by oligomycin.

(c) ^{18}O *exchanges.* Two other exchange reactions are of consider-able interest, that between $H_2{}^{18}O$ and P_i; and that between $H_2{}^{18}O$ and ATP. The former process is sensitive both to uncouplers and to oligomycin and is dependent upon the presence of ADP. It is how-ever independent of the state of oxidation of the components of the respiratory chain. In other words it is consistent with a direct dehydra-tion of ADP+P_i in the synthesis of ATP, the oxygen removed being that of the P_i, so that a phosphoryl transfer to ADP is achieved:

$$ATP + H_2{}^{18}O \rightleftharpoons ADP + H^{18}O—\overset{\displaystyle O}{\underset{\displaystyle O^-}{\overset{\|}{\underset{|}{P}}}}—O^- + (\sim)$$

(i.e. the bridging oxygen atom between the two terminal phosphorous atoms of ATP originates in ADP, not in P_i).

It will be noted that, a direct dehydration reversibly coupled to the dissipation of $\sim$, but without any chemical intermediates would also give rise to the other exchanges mentioned previously.

The oxygen exchanges would also be consistent with indirect dehdration via an anhydride of the X–Y type as postulated to explain the $ATP:P_i$ exchange, i.e.

$$\text{(i)} \quad ATP + X—OH \rightleftharpoons ADP + X—O—\overset{\displaystyle O}{\underset{\displaystyle O^-}{\overset{\|}{\underset{|}{P}}}}—O^-$$

$$\text{(ii)} \quad X—O—\overset{\displaystyle O}{\underset{\displaystyle O^-}{\overset{\|}{\underset{|}{P}}}}—O^- + YH \rightleftharpoons X—Y + HO—\overset{\displaystyle O}{\underset{\displaystyle O^-}{\overset{\|}{\underset{|}{P}}}}—O^-$$

$$\text{(iii)} \quad X—Y + H_2O \rightleftharpoons X—OH + YH + (\sim)$$

Reaction (iii) would exchange ^{18}O between water and XOH. Reactions (i) and (ii) would exchange ^{18}O between XOH and P_i.

The key restriction which the experimental evidence imposes on schemes of this kind is that reaction (iii) must not itself involve electron transfer, since the ^{18}O exchange is apparently independent of the functioning of the respiratory chain. Another important restriction is the fact that the $H_2{}^{18}O:P_i$ exchange is faster than the $P_i:ATP$ ex-change. This observation would be consistent with a direct union of ADP and P_i if, for example, there was at the energised active centre of the synthetase an exchange between bound and free P_i that was very rapid in comparison with the rate of dissociation of ATP.

The $H_2^{18}O$: ATP exchange poses even more serious problems. It is easy to see how such an exchange could arise, hydrolysis of ATP by $H_2^{18}O$ would insert an ^{18}O atom into the released P_i and one of the other oxygen atoms might be eliminated from this P_i on the reformation of ATP, so that the latter would now possess an ^{18}O atom on its terminal phosphoryl residue. However, such a mechanism would require that the $H_2^{18}O$: ATP exchange could not be faster than the $^{32}P_i$: ATP exchange. Paradoxically, (and, I suspect, uniquely to oxidative phosphorylation), the $H_2^{18}O$: ATP exchange is much faster than the $^{32}P_i$: ATP exchange. Mechanisms have been proposed which account for this apparent paradox, but all too often the problems of the relative rates of the exchange reaction are overlooked in models set up to explain the mechanism of exidative phosphorylation.

Photosynthetic phosphorylation

Before we come finally to consider the nature of $\sim$ we should tidy up one loose end which was left over from the early part of this chapter, the question of how ATP is synthesized during photo-synthesis. This is not irrelevant to our present considerations, since the chloroplant system may give an insight into the mitochondrial one.

Briefly, electrons from water, excited by a light reaction, reduce a component of relatively negative oxidation-reduction potential. In turn, electrons flow from this reduced component to another component of more positive potential. (The electrons of this second component are subsequently excited by a second light reaction).

Between these two components lies an electron transfer chain with components and characteristics which are in many ways analogous to those of the corresponding chain in mitochondria. The flow of electrons between the photo-reduced donor and the terminal component of the photosynthetic electron transfer chain is coupled to the synthesis of ATP. The ATP-synthetase apparatus of the chloroplast has structural and functional similarities to the 8–9 nm particles of the cristael membrane. Moreover the chloroplast may exhibit work performances (e.g. ion uptake) energized either by light-induced electron transfer or by the hydrolysis of ATP. However, all vectorial processes in the chloroplast membrane have a sidedness which is analogous to that of submitochondrial particles rather than that of intact mitochondria.

It seems very likely that oxidative phosphorylation and photosynthetic phosphorylation share a common principle of energy transduction. In view of this it is interesting to note that ATP may be generated in chloroplast membranes by artificially imposing a pH gradient between the inner and outer compartments. Similar experiments have been performed, although not so convincingly, using mitochondria, the pH gradient being in the opposite direction. This mode of ATP synthesis might be interpreted in three ways: as a reversal of a $\sim$-energized ion translocation, as a manifestation that $\sim$ is in fact a pH gradient (or more generally, a membrane potential) or merely as a generation of the energized state by artefactual means.

B. THE NATURE OF $\sim$

The meaning of $\sim$

The various functional properties of the cristael membrane which we have considered require that the membrane must be the locus of a "coupling system" which can exist in two states, a non-energized state (N) and an energized state (E). The difference in free energy between these two states is $\sim$. In an intact mitochondrion, when two electrons are transferred across a single coupling site of the respiratory chain (e.g. from NADH to ubiquinone), there is an obligatory and stoichiometric transition of the "coupling system" from the N to the E state. Similarly, in the presence of ADP plus inorganic phosphate the relaxation of E to N is stoichiometrically coupled to ATP synthesis.

If C_1 and C_2 are a pair of two-electron carriers at either side of a coupling site, then we may summarize oxidative phosphorylations by the two equations:

(i) C_1 (reduced) $+ C_2$ (oxidized) $+ N$
$\leftrightharpoons C_1$ (oxidized) $+ C_2$ (reduced) $+ E$

(ii) $ADP + P_i + E \leftrightharpoons ATP + H_2O + N$

What we now wish to identify is the nature of the coupling system, and the chemistry of N$\rightarrow$E transition which it undergoes.

The magnitude of $\sim$

The extent to which the position of equilibrium of each of the reactions (i) and (ii) above is displaced from the equilibrium state of the corresponding uncoupled reaction is an index of the free energy of the N$\rightarrow$E transition. Hence it should be possible to determine the magnitude of $\sim$either from measurements of the states of oxidation and reduction of carriers in the chain when the system is poised in respiratory control (reaction (i)), or from measurements of the ATP, ADP and P_i levels in the suspending medium when there is no net synthesis or hydrolysis of ATP (reaction (ii)).

The former approach is difficult, since it involves presumptions concerning the precise location of the coupling spans and the precise values of the standard oxidation-reduction potentials of the respective carriers. The latter approach is more useful, although care has to be taken to correct for factors, such as pH and Mg^{2+}-ion concentration, that affect the free energy of hydrolysis of ATP. Correction also has to be made for the free energy change associated with the transport of ATP out of the mitochondrion in exchange for ADP. (These species carry different amounts of net negative charge and so may require additional, energized ion movements to balance the exchange). On the basis of such studies it has been calculated that $\sim$ may correspond to a free energy change of up to 17 k cals per mol.

The standard free energy of hydrolysis of ATP is about 7·5 k cals per mol, but in the cell, under conditions where ATP is being synthe- sized, the actual energy change (taking account of actual concentra- tions of ADP, P_i, Mg^{2+} etc.) may be between 8 and 12 k cals per mole. The standard free energy change when two electrons are trans- ferred between a pair of carriers differing in standard oxidation- reduction potential by 0·25 V is 11·5 cals per mol, but the actual free energy change will depend upon the relative amounts of the oxidized and reduced forms of the carriers. Similarly, the actual free energy of the $N{\longrightarrow}E$ transition will depend upon the extent to which the system is in the N and E states respectively.

What we can conclude from such considerations is that the coupling system has the capacity to function as a stoichiometric transducer of energy between the respiratory chain and the ATP-synthetase and that its nature must be consistent with its being able to generate an "energy pressure" of up to 17 k cals per mol when fully promoted to the E state.

Chemical-bond hypotheses

Until recently most workers in the field assumed that $\sim$ corres- ponded to a chemical bond. This was an eminnently reasonable assumption for a stoichiometric and obligitarily-coupled system. An excellent analogy was the mechanism for "substrate-level phosphory- lation" associated with the oxidation of α-oxoglutarate to succinate in the citric acid cycle.

The actual oxidative step in the process catalysed by the α-oxo- glutarate dehydrogenase complex is not a direct transfer of reducing equivalents between α-oxoglutarate and NAD^+, but the oxidation of the decarboxylation product, succinic semialdehyde (R.CHO) by oxidized lipoic acid ($\begin{smallmatrix} S\diagdown \\ | \quad L \\ S\diagup \end{smallmatrix}$). There is a substantial difference between the oxidation-reduction potential of the R.CHO: R.COOH

couple and that of the $\begin{array}{c}HS\\ \diagup \\ HS\end{array}$ $L:$ $\begin{array}{c}S\\ \diagup \\ S\end{array}$ L couple; hence the reaction

$$RCHO + \begin{array}{c}S\\ | \\ S\end{array} L \rightarrow RCOOH + \begin{array}{c}HS\\ \diagup \\ HS\end{array} L \text{ would be exergonic and}$$

irreversible. In fact however, the reaction does not result in the formation of free RCOOH and $\begin{array}{c}HS\\ \diagup \\ HS\end{array}$ L. Instead, electron transfer is directly coupled to the formation of a thio-ester bond between the expected products, i.e: between the oxidized reductant and the reduced oxidant:

$$RCHO + \begin{array}{c}S\\ | \\ S\end{array} L \rightarrow R\!-\!\underset{\underset{O}{||}}{C}\!-\!S\diagdown \underset{HS\diagup}{\quad} L.$$

The free energy of hydrolysis of the bond formed is smilar to that for the terminal pyrophosphate bond in ATP. Hence a series of reversible group-transfer reactions becomes possible:

i) $R\!-\!\underset{\underset{O}{||}}{C}\!-\!S\diagdown \underset{HS\diagup}{\quad} L + \text{coenzymeA.SH} \rightleftharpoons R.\underset{\underset{O}{||}}{C}\!-\!S.\text{coenzymeA} + \begin{array}{c}HS\\ \diagup \\ HS\end{array} L$

ii) $R.\underset{\underset{O}{||}}{C}\!-\!S.\text{coenzymeA} + P_i \rightleftharpoons R.\underset{\underset{O}{||}}{C}\!-\!O\!-\!\underset{\underset{O^-}{|}}{\overset{\overset{O}{||}}{P}}\!-\!O^- + \text{coenzymeA.SH}$

iii) $R.\underset{\underset{O}{||}}{C}\!-\!O\!-\!\underset{\underset{O^-}{|}}{\overset{\overset{O}{||}}{P}}\!-\!O^- + GDP \rightleftharpoons R.\underset{\underset{O}{||}}{C}\!-\!OH + GTP$

iv) $GTP + ADP \rightleftharpoons GDP + ATP$

Meanwhile the reduced lipoic acid released at (i) is reoxidized by a flavoprotein which in turn, is reoxidized by NAD^+.

The analogy between such a process and that occurring, say, in the NADH : ubiquinone span of the respiratory chain is very tempting. Complex I has components that could be candidates for donor or acceptor roles in the bond-forming electron transfer, as well as components to reoxidize the reduced acceptor once it has been released from the energy-conserving bond by a group-transfer process. Moreover, the ADP:ATP and P_i:ATP exchange reactions mentioned previously are specifically consistent with a series of group transfers.

However, any chemical-bond hypothesis applied to the respiratory chain has to account for the phenomenon of uncoupling by damage to the membrane or by catalytic amounts of specific reagents. It has been suggested that the group-transfer chain can only function in a hydrophobic environment by virtue of the extreme lability (particularly in the presence of water) of one of the intermediate compounds. This lability is also held to account for failures to isolate any convincing "high-energy intermediate".

Another difficulty of the hypothesis is that of accounting for the apparently identical behaviour of all three coupling sites to most uncouplers, although the chemistry of the "high-energy intermediates'" formed at each site must be quite different, since these compounds must each contain different respiratory carriers. However, the chemistry of the formation of such intermediates when one-electron carriers are involved might not be strictly analogous to the chemistry of the formation of succinyl-lipoate, and might involve a class of bond that is always unusually labile, whatever the precise nature of the ligands involved.

A final difficulty is that of the observations concerning the exchange of ^{18}O between water and inorganic phosphate, which is dependent upon ADP but independent of the oxidation state of the chain. As we have already see, this requires that the primary dehydration reaction may not be an electron-transfer process. Even this criticism may be overcome by certain sophisticated models of the chemical-bond hypothesis. Moreover, there have now been a number of model experiments with relatively simple systems in free solution which illustrate that oxidation-reduction reactions involving carriers of the cytochrome type can be coupled to the formation of chemical bonds which in turn, by group transfer, can give rise to ATP.

Evidence that these interesting models might be directly relevant to mitochondrial oxidative phosphorylation comes from demonstations that during coupled respiration one of the cytochromes b apparently becomes converted into a new species with a new absorption spectrum and new oxidation-reduction potential. However the properties of a cytochrome depend upon its environment, and so these findings may mean that cytochrome b is a sensitive indicator of the generation of

$\sim$ and not that $\underset{.}{\sim}$ is a bond between the cytochrome and some new ligand.

Chemi-osmotic hypothesis

The limitations of the chemical-bond hypothesis and in particular the failure to isolate any high-energy intermediates, encouraged certain workers to abandon the first premise of this hypothesis and to reconsider the problem from first principles. The result has been the development of the so-called chemi-osmotic hypothesis, which is summarized in a simplified form in Fig. 17.

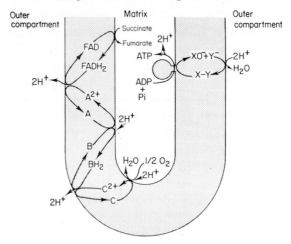

Fig. 17. A simplified scheme for oxidative phosphorylation.

The starting point for this hypothesis is the apparent dependence of the coupling phenomenon upon the integrity of the cristael membrane, a membrane of limited permeability and known vectorial character. It is postulated that the membrane is impermeable to protons but that the hydrolysis of ATP by the ATP-synthetase apparatus effectively results in the active translocation of protons from the matrix to the outer compartment. (In the case of sub-mito-chondrial particles such a "proton pump" would transfer protons from the suspending medium into the space enclosed by the membrane). As presently formulated the hypothesis requires that the hydrolysis of one molecule of ATP is coupled to the effective transfer of two protons. It is not considered that the protons actually traverse the membrane in this system. The hydrolysis of ATP is considered to be a complex, vectorial reaction involving a reversable condensation between two membrane-localized species, XOH and YH. XOH and YH tend to ionize to give the corresponding anions, XO^- and Y^-. The overall process for mitochondria might be formulated as follows, (the directionality being reversed in submitochondrial particles):

Inside: $ATP + XO^- + Y^- + 2H^+ \leftrightharpoons ADP + P + X—Y$

Outside: $X—Y + H_2O \leftrightharpoons XO^- + Y^- + 2H^+$

Sum: $ATP + H_2O + 2H^+ \text{ (in)} \leftrightharpoons ADP + P_i + 2H^+ \text{ (out)}$

It is considered that if the pH gradient across the membrane were sufficiently great, then the pump would be driven to operate in reverse and thus achieve the synthesis of ATP by the dehydration of ADP+ P_i. An alternative driving force to a pure pH gradient would be a membrane potential derived by the exchange diffusion of protons for other cations. In the chemi-osmotic hypothesis $\sim$ is a "proton motive force" (p.m.f.) across the membrane which is the sum of the contributions of a pH gradient and a membrane potential.

For this hypothesis to account for the mechanism of oxidative phosphorylation it would require that the transfer of a pair of electrons across each coupling span of the respiratory chain should be coupled to proton translocation with a stoichiometry corresponding to that for the ATP-energized proton pump. In other words, two protons per coupling site would have to be transferred from the matrix to the outer compartment. If this were achieved, then presumably the p.m.f. thus built up would eventually inhibit further coupled respiration unless it were discharged in the process of driving the ATP-energized proton pump in reverse.

We have already seen that there are "loops" in the respiratory chain, and, further, that some carriers (e.g. flavoproteins and ubiquinone) are protonated on reduction whereas others (e.g. cytochromes and nonhaem-iron porteins) are not. It is therefore possible to formulate a respiratory chain in which carriers of these two kinds alternate, and which is looped to and fro across the thickness of the membrane, so that respiration is coupled to the translocation of protons. For example, if A^{2+} :A, B:BH_2 and C^{2+} :C were a sequence of such alternating carriers, one might formulate the oxidation of succinate in mitochondria as :

Inside:　　succinate$+FAD \rightarrow$ fumarate$+FADH_2$

Outside:　$FADH_2+A^{2+} \rightarrow FAD+A+2H^+$

Inside:　　$A+B+2H^+ \rightarrow A^{2+}+BH_2$

Outside:　$BH_2+C^{2+} \rightarrow B+C+2H^+$

Inside:　　$C+\frac{1}{2}O_2+2H^+ \rightarrow C^{2+}+H_2O$

Sum:　　　succinate$+\frac{1}{2}O_2+4H^+\text{(in)} \rightarrow$ fumarate$+H_2O+4H^+\text{(out)}$

Such a formulation is consistent with the known sidedness of the succinate-dehydrogenase and cytochrome a_3-oxidase functions (both of which are matrix-facing in the intact mitochondrion), and also with the stoichiometry of two protons translocated per coupling span, (there being two coupling spans associated with the oxidation of succinate).

According to the chemi-osmotic hypothesis uncouplers act by rendering the membrane permeable to protons and hence discharging the p.m.f. (Certainly, uncouplers may be shown to increase proton conductivity in some synthetic membrane systems). Spontaneous leakage of the potential would correspond to loss of respiratory control. All energized ion movements would be driven by the membrane potential. Indeed, a great strength of the hypothesis is that it is sufficiently flexible to account for very many otherwise puzzling aspects of mitochondrial function, (in particular, ionophore-mediated cation accumulation). This flexibility however also means that it is very difficult indeed to put the hypothesis to any definitive test.

The chemi-osmotic hypothesis has been criticized on thermodynamic grounds, but these critical discussions are not yet resolved. It may also be criticized as presently formulated in that it is not consistent with evidence relating to the actual sequence of carriers and the actual coupling spans. However, since there is scope for other, as yet undefined, carriers, in the chain, it is possible that the scheme might be modified to account for such criticism. The hypothesis cannot as yet account satisfactorily for the phenomenon of energized transhydrogenation between NADH and $NADP^+$. It is also not consistent with evidence relating to certain allegedly site-specific uncouplers and inhibitors of oxidative phosphorylation, since the hypothesis predicts that $\sim$ has the same character at whatever site in the chain it is generated. This latter criticism may not be very serious since the evidence on such site-specificity is itself open to criticism.

The hypothesis has been directly tested by measurement of proton movement coupled to respiration in mitocohondria and in submitochondrial particles. Under carefully controlled conditions movements in the predicted direction and of predicted stoichiometry have been detected. However, all of these experimental findings are open to alternative explanations; proton translocation might be simply another possible work performance energized by the dissipation of $\sim$, or proton movement could be secondary and occur in exchange for some other cation whose translocation was a primary energized event, or the proton movement could be an indicator of the energized state, i.e. an ionization concomitant with an energized structural change in some component of the membrane. An additional criticism is that the time-course of the measured proton movements is claimed by some workers to be inconsistent with a primary coupling event.

Whatever the final fate of the chemiosmotic hypothesis, it has made a major contribution to mitochondrial research, since it has focused attention on aspects of the mitochondrion that were hitherto largely ignored; the vectorial character of the cristael membrane and the manner in which ion movement across this membrane may be directed and controlled.

The conformational hypothesis

Another way in which the synthesis of ATP (or of some prior anyhydride of the X-Y type), could be achieved is by the generation at a catlytic centre of conditions favouring the dehydration process. For example, the generation of a localized proton in a hydrophobic environment might be sufficient to facilitate the removal of water from P_i to give rise to a phosphoryl radical which could then attack ADP. This would be one of many alternative schemes involving structural rearrangement of the functional groupings at a catalytic site. In all cases, the free energy for driving ATP synthesis (i.e. $\sim$) would be the free energy of the conformational transition of the catalytic site. Such a transition might involve a massive rearrangement of the conformation of a complex of protein subunits, or simply the separation of a pair of unlike charges over a relatively small distance. The reverse of this process, the generation of a conformationally perturbed site coupled to the hydrolysis of ATP, would have analogies to the conformational change coupled to ATP-hydrolysis which is the molecular basis of muscular contraction.

There have been a number of suggestions that ATP-synthesis might indeed be mediated by such conformational changes, but it is only in recent years that the conformational theory of energy transduction has gained serious support. Two main lines of evidence contributed to the development of this hypothesis. The first was the correlation between certain kinds of configurational states of the cristael membrane with the relative rates at which $\sim$ was being generated and dissipated. The second was the demonstration of major conformational changes associated with electron transfer through the Complexes of the respiratory chain.

According to the hypothesis, a conformationally energized state is generated either when a pair of electrons traverse a Complex of the respiratory chain or when ATP is hydrolyzed. There are structural links between the 8–9 nm headpieces and the complex lipoproteins of the membrane continuum. It is thought that these structural links mediate the transmission of conformational energy between the two functional segments of the membrane (i.e. the ATPase and the electron-transfer Complexes).

In the conformationally energized state ATP synthesis in the head-pieces is now favoured. Also, the environment of the individual components of the electron-transfer Complexes is changed so that their effective oxidation-reduction potentials become such as to inhibit respiration and favour reversed electron transfer. The whole charge-distribution within the membrane might be altered in such an energized state, and this, together with the physical reorientation of membrane components dependent upon the subsequent discharge of this energized state, can be made to account for all energized move-ment of ions. Certainly, binding studies with fluorescence probes have suggested that the energized state is characterized by redistribution of charge within the membrane.

The various configurational states of the cristael membrane have been taken to be relatively macroscopic manifestations of changes in conformation of the individual Complexes of the membrane, possibly mediated by induced changes in the organization of components in the matrix. The observed configurational states depend on factors such as the presence or absence of inorganic phosphate, and this might imply that there are alternative conformations of the energized state of the Complexes. A certain conformation, favoured by a given set of condi-tions, might in turn favour a given work performance.

The related phenomena of uncoupling and loss of respiratory control could be explained in terms of the stability of the various conforma-tionally perturbed states. This stability might depend very much upon the constraints imposed by the mode of organization of the Complexes within the membrane. Any damage to the membrane might release these constraints and allow rapid, spontaneous relaxation of the energized state. If the energized state were a localized separation of change the uncouplers might function by increasing local proton conductance.

Distinction between hypotheses

The three hypotheses that we have considered are not as radically different from one another as they might seem at first sight. The formation of a bond between components each of which was attached to a complex structure within the membrane might very well require a major conformational change, and the free energy of formation of the bond might thus be largely conformational energy. Similarly, conformational energy might include a large component due to separa-tion of charge, and it is only a question of distance and direction of separation which determines whether a membrane potential is effectively generated. The orientation of the complexes of the respiratory chain might be such that separation of charge due to conformational change is always at right-angles to the plane of the membrane, and always in the same direction.

In my opinion, the only fundamental distinction between the coupling theories is the distance over which energy transduction occurs. In the chemical-bond hypothesis, the chain of group transfers, from the first compound involving respiratory components to the ATP-synthetase, is not likely to extend more than the distance between two ATP-synthetase units. Similarly, the simpler formulations of the conformational hypothesis would predict that ATP would be preferentially be synthesized by a headpiece within the same "conformational domain" as the electron-transfer complex where the primary coupling event occurred. On the other hand, according to the chemi-osmotic hypothesis as presently formulated, any ATP synthetase in a given membrane can be driven by the potential generated by any electron-transfer loop anywhere in the same membrane. In this case, the coupling is mediated by a potential due to differences in concentration of cations in solution in two homogenous compartments, i.e. the matrix and the outer compartment.

Such evidence as there is on this particular point seems to me very strongly to favour the concept that energy is preferentially transduced within restricted domains. If this is so, it constitutes a serious criticism of the chemi-osmotic hypothesis.

IV. Biogenesis of Mitochondria

A. ORIGIN OF MITOCHONDRIA

There is little evidence that mitochondria ever appear *de novo* in a cell. (Although there have from time to time been suggestions that mitochondria might arise from extension of the cytomembranes.) Mitochondria are partitioned between daughter cells on cell division. If the mitochondrial content of such daughter cells subsequently increases this is apparently achieved by an increase in the size of existing mitochondria followed by a rather random process of budding-off. (Present evidence suggests that in a zygote it is the mitochondria of maternal origin which are the precursors of the mitochondria of the new organism.) Even in the case of yeasts where respiratory metabolism is suppressed by anerobic growth in the presence of fermentable substrates, "promitochondria" (mitochondrial "ghosts" with little residual respiratory capacity) can still be detected within the cells. On derepression. it is these "promitochondria" which subsequently become modified to give rise to the fully functional mitochondria which appear as the cell assumes aerobic metabolism.

The fact that new mitochondria always appear to derive from pre-existing mitochondria (or promitochondria) might be taken to imply that mitochondria are autonomous and self-replicating. As we shall see below, the actual extent of mitochondrial autonomy is very limited. Moreover, it seems that the outer membrane is simply a cytoplasmic vacuolar membrane. Nonetheless, the possible existence of a continuous "line" of mitochondria (or at least of inner membrane-matrix) from one generation of an organism to the next does pose the interesting question of the origin of mitochondria. Present evidence points strikingly to their being derived from aerobic prokaryotes which, far back in evolution, adopted a symbiotic existence within primitive eukaryotic cells, whose energy requirements had previously been met by anaerobic metabolism.

B. MITOCHONDRIAL PROTEIN SYNTHESIS
Synthetic apparatus

Mitochondria contain a full complement of the apparatus necessary for protein synthesis: DNA, DNA-dependent DNA-polymerase, DNA-dependent RNA-polymerase, amino acid activating enzymes and corresponding transfer-RNAs, ribosomes, and all of the additional factors required for messenger RNA-directed ribosomal peptide synthesis. Hence, under appropriate conditions, isolated mitochondria incubated with radioactively labelled amino acids can incorporate them into specific protein species.

Mitochondrial DNA from any given cell is quite different from the corresponding nuclear DNA. It usually exists as a double-stranded circular form, strikingly similar to DNA of bacterial origin. There may be several DNA molecules per mitochondrion, but it seems likely that they are identical. The molecular weight is rather low; for DNA from mammalian mitochondria it is about 1×10^7 daltons, for mitochondria from more primitive cells the molecular weight is generally greater than this, but still low relative to nuclear DNA.

It is claimed by some workers that mitochondrial ribosomes tend to be smaller than corresponding cytoplasmic ribosomes, and fall into the category of "70 S" particles, i.e. they tend to have sedimentation characteristics closer to those of bacterial ribosomes than to those of the "80 S" ribosomes which are characteristic of the cytoplasmic protein-synthesizing system in higher organisms (see Chapter 5). Similar discrepancies are said to be apparent when the subunits of mitochondrial ribosomes, or their RNA and protein components, are compared with the corresponding components of cytoplasmic ribosomes. One of the complications in such studies is that mitochondrial ribosomes are very difficult to isolate and may well be bound to the cristael membrane, although it is usually assumed that they are free in the matrix.

Characteristics of mitochondrial protein synthesis

The incorporation of amino acids into protein by isolated mitochondria has certain characteristics that are much closer to those of bacterial systems than to those of the cytoribosomal system of eukaryotic cells. Thus, in mitochondria as in bacteria, the initiation of protein synthesis involves formyl methionyl *t*RNA, whereas cytoribosomes use some other initiator. Moreover, as is the case for the bacterial system, protein synthesis in mitochondria is inhibited by chloramphenicol and macrolide antibiotics (macrocyclic lactones such as erythromycin) and is insensitive to cycloheximide, a characteristic inhibitor of protein synthesis associated with the 80 S cytoribosomes.

These facts give persuasive support to the theory of the evolutionary relationship between mitochondria (or, more strictly, inner membrane-matrix) and bacteria. It is interesting to note that further evidence for this evolutionary relationship is the characteristic lack of cholesterol in both bacterial and cristael membranes.

Synthesis and assembly of mitochondrial components

It has become clear from many lines of evidence that the great majority of mitochondrial components are coded for by nuclear DNA and synthesized by the cytoribosome system. It is not surprising therefore that, when isotopically labelled amino acids are incorporated by isolated mitochondria, most of the label appears in a very limited number of uncharacterized and insoluble components of the cristael membrane. It is likely that these labelled components are coded for by mitochondrial DNA. The RNA of mitochondrial ribosomes and at least some species of mitochondrial *t*RNA are also probably coded for by mitochondrial DNA. These sets of minor components must represent very nearly the limit of the information that can be stored in the length of the DNA molecule of a mamalian mitochondrion. It is possible that one or two mitochondrial proteins might be coded for in the mitochondrion and synthesized in the cytoplasm, or coded for by nuclear DNA and synthesized in the mitochondrion. It is also just possible that a few non-mitochondrial proteins are coded for by mitochondrial DNA.

The limitations both in the amount of information in mitochondrial DNA and in the number of components synthesized by isolated mitochondria underlie the fact that, whatever their origin, mitochondria retain very little autonomy. The presence of relatively larger DNA molecules in mitochondria of more primitive organisms

T

perhaps is a pointer to the way in which mitochondria, during the course of evolution from their original autonomous form, gradually relinquished the responsibility of coding for their own components.

The lack of autonomy of the mitochondrion does not imply that its protein synthesizing apparatus is redundant. The appearance of fully functional mitochondria in derepressed yeast is blocked by inhibitors of mitochondrial protein synthesis such as chloramphenicol. Instead, all that appear are mitochondrial "ghosts" grossly depleted in the components of the cristael membrane. Moreover, there are a number of cases known where cytoplasmic mutations, presumably in mitochondria DNA, give rise to micro-organisms exhibiting various forms of respiratory deficiency. It may therefore be presumed that the protein synthesizing system of the mitochondrion is responsible for the synthesis of certain proteins which are essential for the organization of the cristael membrane, and possibly even of the soluble enzyme systems of the matrix. In the absence of these organizing proteins other mitochondrial components of cytoplasmic origin cannot be incorporated.

Under normal circumstances one must imagine that there is some liaison between the mitochondrial system for protein synthesis and that of the cytoplasm, so that there is synchrony in the production of mitochondrial components. This is not to imply that all of the components of mitochondrial membranes are synthesized at identical rates. Each component, protein or phospholipid, will have a characteristic rate of turnover which is controlled so as to maintain constancy of mitochondrial composition whilst allowing where appropriate an increase in the size or number of mitochondria in the cell. What is at present completely unknown is the mode of transport of components from their site of synthesis on cytoplasmic ribosomes to their eventual loci in the mitochondrion.

V. MITOCHONDRIA IN VIVO

A. PHYSIOLOGICAL FUNCTIONS

Primary functions

The primary function of mitochondria in cells is that of ATP-production coupled to the oxidative degradation of foodstuffs. In addition, mitochondria play an important role in the general metabolic traffic within cells (e.g. in effecting key steps in the pathway of gluconeogenis), as is apparent from the localization of various key enzymes within the mitochondrion. The question is whether mitochondria play any other specific role related to their energy-transducing capabilities.

Ion-movement

Apart from ATP-synthesis, the *in vitro* capability for work performance that is most likely to be of physiological significance is that for the accumulation of calcium salts. In particular, there is some evidence that, in those muscle fibres that are rich in mitochondria but have a rather sparse sarcoplasmic reticulum, it is the mitochondria rather than the sarcoplasmic reticulum that mediate the rapid changes in intracellular calcium levels that are associated with cycles of muscular contraction and relaxation.

Calcium ions are thought to be concerned in mediating many other physiological responses, from hormone release to blood platelet aggregation, and the mitochondrion might well play a key role in controlling these processes through the control of intracellular calcium levels. It has also been suggested that the mitochondrial capability for the massive accumulation of calcium phosphate might be central to the process of calcification of hard tissues.

There is a possibility that other energized ion movements (and also the energized redistribution of water between mitochondrial compartments,) might have physiological significance beyond that of the normal functioning of the mitochondrion in intermediary metabolism and ATP-synthesis. For example, it has been suggested that mitochondria may participate in the transcellular transport of certain solutes, particularly in cases such as the tubular cells of the kidney where there is a characteristic distribution and polarization of the mitochondrial population. Changes in the conformation of kidney mitochondria associated with changes in salt and water intake of test animals have also been taken to point to a role in ion movement.

Heat production

A further role for the mitochondrion is that of heat production. There are three major ways that the free energy of oxidation of foodstuffs can eventually give rise to heat. The first is by the hydrolysis of ATP, which may be direct or indirect. For example, muscular contraction energized by the hydrolysis of ATP is a major source of body heat in many species. Another mode of hydrolytic heat production is the ATP-dependent synthesis of triglycerides (from fatty acids and α-glycerophosphate) and their subsequent hydrolysis; which thus represents a metabolic "short-circuit" effecting the hydrolysis of ATP. There is some evidence that this might be a mode of heat production in brown fat during "non-shivering thermogenesis" in neonates of a number of mammalian species, (possibly including the human baby), in hibernating mammals and in cold-acclimatized adult rodents. In "non-shivering thermogenesis", adrenergic stimulation of brown fat, a specialized tissue, rich in mitochondria, gives rise to a prolonged increase in local heat production which can raise the body temperature of the animal by several degrees, or maintain a constant body temperature against a decrease in environmental temperature, without concomittant muscular activity.

A second possible mode of heat production is the spontaneous dissipation of ∼ under conditions of loss of respiratory control. (Actual uncoupling, where ∼ is dissipated as heat even in the presence of ADP, is probably not a normal physiological phenomenon). The extent to which respiratory control is ever lost under normal conditions is uncertain, but it now seems possible that non-shivering thermogenesis in brown fat (see above) involves the reversible release of respiratory control in response to appropriate autonomic stimuli. It is also possible that, even in euthyroid subjects, thyroid hormone (see below) might mediate a certain degree of loosening of respiratory control and thus increase heat production.

A third mode of heat production is oxidative phosphorylation itself, especially under conditions where ADP:ATP ratios are high. Under such circumstances the free energy of formation of ATP is low. Hence, not only can oxidative phosphorylation proceed at a maximal rate, (dependent upon availability of substrate), but also, a significant proportion of the respiratory energy becomes available as heat. It therefore follows that any process that makes heavy demands on cellular ATP will tend to give rise to direct heat production by mitochondria.

B. EFFECTS OF HORMONES

Thyroid hormone

Mention has already been made of the fact that, under appropriate conditions, thyroxin will cause extensive mitochondrial swelling *in vitro*. Since such swelling is associated with uncoupling, there have been suggestions that this *in vitro* phenomenon may be related to the effect of thyroid hormone upon the basal metabolic rate *in vivo*. This seems unlikely since the amounts of thyroxin required to produce mitochondrial swelling are far above physiological levels.

Administration of thyroid hormone to hypothyroid animals does however result in morphological changes in the mitochondria of certain tissues. In general there is a relative increase in the amount of cristael membrane. This is consistent with the concept that the primary effect of thyroxin is upon protein synthesis, and indeed there is evidence that thyroxin does increase the rate of synthesis of mitochondrial components.

Mitochondria isolated from thyrotoxic animals tend to exhibit very little respiratory control. This lesion might reflect an imbalance in the process of assembly of the cristael membrane. It will be remembered that two distinct systems for protein synthesis are concerned in the biogenesis of mitochondria. Presumably thyroid hormone exerts direct or indirect control over both of them. Under conditions of excessive biosynthesis, the liaison between the systems may function imperfectly giving rise to imperfectly assembled membranes in which $\sim$ has a greater tendency to dissipate spontaneously. However, this is mere speculation.

It is not known whether other hormones also affect the biogenesis of mitochondria. It has been shown that the amount of mitochondrial material in the muscles of experimental animals can increase very significantly following prolonged exercise, but it is not known how this effect is mediated.

Other hormones

Quite a number of hormones have been shown to affect mitochondria morphology *in vitro*. Thus, mitochondrial swelling has been induced by oxytocin, vasopression, insulin and growth hormone. However, there is little evidence to relate these effects to any *in vivo* function. Perhaps the only *in vitro* phenomenon of this kind that might be of physiological significance is the effect of parathyroid hormone, which has been shown under certain conditions significantly to increase the uptake of inorganic phosphate by mitochondria. One reason why it is very difficult to interpret such *in vitro* affects is that many hormones contain readily accessible disulphide bonds which might interact rather nonspecifically with sulphydryl groupings of the mitochondrial membrane.

There have been reports that, during normal pregnancy, changes occur in the morphology of liver mitochondria, associated with the appearance of crystalloid inclusions. Oral oestrogens apparently give rise to similar effects. There is also some evidence that progesterone might affect the movement of substrate anions into mitochondria. But, in general, there is little systematic information on how such hormones might control mitochondrial function or whether such control is relevant to the mode of action of these hormones.

In addition to morpholgical effects on mitochondria, evidence is now accumulating that points to direct hormonal effects on mitochondrial metabolism. For example, pyruvate dehydrogenase has been shown to exist in two interconvertible forms, the interconversions being mediated by a protein kinase and a phosphatase respectively. These latter enzymes are themselves subject to further control, so that a system appears to be operative analogous to the well known control of glycogen phosphorylase. Not only has a protein kinase been demonstrated in mitochrondia, there is also evidence of a lipid kinase which might affect ion movements into mitochrondia. In turn it is possible that these systems will prove to be controlled directly by hormones such as insulin, or indirectly by cytoplasmic levels of cyclic AMP, or by cyclic AMP which might possibly be generated in the mitochondrion. Another direct effect of a hormone has been demonstrated with prostoglandin E_1. We have recently found that very low levels of this local hormone greatly affect the binding of calcium to the mitochondrion. Moreover, anti-inflammatory agents such as aspirin and indomethacin, which are known to affect prostaglandin synthesis also alter the calcium-binding characteristics of the mitochondrion. These findings might merely indicate that the mitochondrion is a useful model system for studying membrane-active agents. They might however implicate the mitochondrion as a target organelle for prostaglandins and/or for pharmacological agents.

C. PATHOLOGY

Mitochondrial pathology

Cell damage in many gross pathological states involves obvious changes in mitochondrial morphology. It may be presumed that in many cases these effects are secondary to some other cellular lesion, or else manifestations of general damage to the membranous sytems of the cell.

There are however certain examples of unambiguously specific mitochondrial pathology. Mention has already been made of the loss of respiratory control in mitochondria from thyrotoxic animals. There is also at least one case on record of a patient who was not thyrotoxic, the mitochondria of whose tissues were without respiratory control. This defect was apparently a genetically determined lesion. Primary biliary cirrhosis is, in a sense, another "mitochondrial disease", since it represents an auto-immune response to an antigen of mitochondrial origin which has yet to be fully characterized.

A number of toxic agents are known to affect mitochondrial function. These include uncouplers and certain bacterial toxins. It is also very likely that membrane-active agents such as anaesthetics could interfere with oxidative phosphorylation. For example, genetically-determined sensitivity to halothane anaesthesia which manifests itself as a malignant hyperpyrexia appears to be a mitochondrial lesion of some kind. In fact we have recently found that even normal mitochondria may be uncoupled in the presence of anaesthetic levels of halothane, provided that low levels of calcium are also added. Moreover, it also appears that a number of inhalation anaesthetics have the common property of specifically inhibiting the NADH:UQ span of the respiratory chain. It therefore seems that, we may now be entering a phase of mitochondrial research where great emphasis will be laid on possible correlations between clinical observations and mitochondrial ultrastructure and function.

Mitochondria and malignancy

It is perhaps appropriate to end this section with some speculation on the possible involvement of mitochondria in the transformation of cells to malignant forms. At one time it was quite fashionable to believe that malignant transformations arose as a result of loss of respiratory capacity. Such a simple view is probably no longer tenable. However, there is one aspect of the mitochondrion that might deserve further consideration in this respect, and that is its capacity to code for a limited number of specific proteins, at least some of which appear to play an essential role in the assembly of a functional membrane system.

Each mitochondrion has several copies of the same DNA molecule, and this might give some stability against mutations. However, any mutation that increased the rate of replication of a particular DNA molecule would have a selective advantage, so that some time after such a mutation a "clone" of mitochondria would arise in which it was primarily the mutant DNA that was expressed.

In micro-organisms mutations of this kind are generally characterized by respiratory deficiency. However, it is by no means beyond the bounds of possibility that mitochondrial DNA codes for proteins with functions other than those concerned with the assembly of the cristael membrane. If this were the case then some other aspects of cellular organization might become disturbed. Such a lesion could be in the control of a metabolic pathway or in the ultrastructure of another membrane. In any event, it is intriguing to note that abnormal forms of mitochondrial DNA have been reported in leukaemic cells.

VI. PEROXISOMES

A. INTRODUCTION

Under appropriate conditions more than 35% of the respiration of an isolated liver slice is non-mitochondrial. The organelle primarily responsible for this additional consumption of oxygen is the peroxisome. It is this respiratory function which accounts for the inclusion in this Chapter of a short section on peroxisomes.

It should be stressed at the outset that, unlike mitochondria, peroxisomes from various sources have no common, unique and clearly defined morphological characteristic. The term peroxisome is a biochemical definition. It is applied to any particle which contains, in association with the enzyme catalase at least one oxidase which produces hydrogen peroxide.

Peroxisomes occur in the cells of mammalian liver and kidney, and possibly in bone. They also occur in certain protozoa, in germinating fatty seedlings and in the leaves of several varieties of plant. Only in the case of peroxisomes from the mammalian tissues has any extensive study been made of the morphology of the organelle.

B. MORPHOLOGY

In the cells of kidney and liver, peroxisomes correspond to the morphological entity designated as microbodies. These are spherical particles with a diameter of about 0·5 nm. They are bounded by a single membrane and possess a fine, granular matrix. In the case of liver peroxisomes there is also a dense core with a crystalloid appearance (see Fig. 18). It seems likely that this crystalloid is actually an ordered array of tubules. This polytubular core is usually only seen in cases where the complement of peroxisomal enzymes includes urate oxidase. In human liver where urate oxidase is absent the peroxisomes have no dense core.

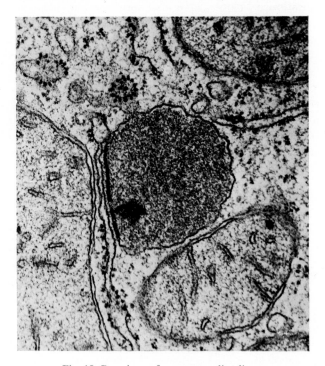

Fig. 18. Peroxisome from mammalian liver.

The number of peroxisomes in a liver cell is about a quarter of the number of mitochondria. In liver, peroxisomes frequently appear to be in the process of budding off from the smooth endoplasmic reticulum, with which the peroxisomal membrane sometimes seems to be continuous. Indeed it is possible that, in the intact cells, peroxisomes always remain attached to the endoplasmic reticulum by a stalk-like connection.

Isolated peroxisomes have sedimentation characteristics similar to those of lysosomes, although they may readily be separated from the latter by centrifugation through appropriate density-gradients. Unlike lysosomes or mitochondria, peroxisomes are insensitive to osmotic shock, since the peroxisomal membrane is freely permeable to solutes such as sucrose. Furthermore this membrane appears to be more resistant to fragmentation by detergents than is the lysosmal membrane.

C. ENZYMIC COMPLEMENT

Catalase

All peroxisomes, by definition, contain the haemoprotein catalase. This enzyme is usually present in very large amounts, and constitutes about 40% of the total protein of the peroxisome. In spite of the permeability of the peroxisomal membrane, the organelle exhibits latency with respect to catalase activity. This is because the enzyme is present at such a high concentration that the diffusion of substrate becomes a rate-limiting process. Catalase can be released into solution from peroxisomes, and *in vivo* the enzyme has a very short half-life. It seems unlikely however that the peroxisome actually functions as an intracellular secretion granule for catalase.

It is quite well known that catalase levels are depressed in the livers of tumour-bearing animals. This may be correlated with the finding that typical microbodies appear to be absent in cells of certain rapidly growing hepatomas. The significance of such changes is not known.

Catalase can mediate two kinds of reaction involving hydrogen peroxide. One reaction is the reduction to water of one molecule of hydrogen peroxide, the electron donor being a second molecule of hydrogen peroxide which itself is oxidized :

$$H_2O_2 + H_2O_2 \rightarrow 2H_2O + O_2$$

This reaction is favoured by low levels of catalase, high levels of hydrogen peroxide and low levels of alternative electron donors.

In the presence of suitable alternative electron donors such as phenols or aliphatic alcohols, catalase mediates a peroxidation reaction :

$$H_2O_2 + RH_2 \rightarrow 2H_2O + R$$

This reaction is favoured by high levels of catalase and low levels of hydrogen peroxide.

Urate oxidase

Urate oxidase has so far only been found in hepatic peroxisomes, and in the particles of one species of protozoon. Unlike catalase or the other oxidases that may be present in peroxisomes, urate oxidase is not soluble even upon disruption of the peroxisomal membrane. The enzyme is probably firmly associated with the polytubular structure of the peroxisomal core. It appears to be a copper-containing enzyme which catalyzes the oxidation of urate to yield an unstable product which breaks down spontaneously to allantoin and carbon dioxide. The oxidant in the reaction is oxygen which is itself reduced to hydrogen peroxide.

Other oxidases

Peroxisomes are, by definition, characterized by the association of catalase with one or more hydrogen peroxide-producing oxidases. Apart from urate oxidase, the other oxidases which are found in peroxisomes are readily soluble flavoproteins. There is some overlap in the specificities of certain of these oxidases. It is therefore simpler to define the kind of substrate which might be oxidized by peroxisomes from various sources, without defining the actual enzyme responsible in each case.

Substrates for peroxisomeal oxidases include: D-amino acids (liver, kidney, protozoa); L-amino acids (liver, kidney); glycolate (liver, kidney, protozoa, seeds, leaves); L-α-hydroxy acids (liver, kidney, protozoa, leaves); and glyoxylate (protozoa). In most cases the various oxidases have a rather low affinity for oxygen and this may be a factor of some physiological significance.

Other enzymes

Apart from catalase and various oxidases, only one other enzyme has yet been demonstrated to be present in peroxisomes from any mammalian tissue. This enzyme is an $NADP^+$-linked isocitrate dehydrogenase, which has been found in peroxisomes from rat liver. In the peroxisomes from certain leaves, two other oxido-reductases have been identified, malate dehydrogenase and an NADH-linked glyoxylate reductase.

Malate dehydrogenase is also present in peroxisomes from germinating fatty seedlings, but, far more significantly, peroxisomes from this source and from protozoa contain a complement of other enzymes of the glyoxylate cycle. The glyoxylate cycle can be looked upon as a by-pass across the citric acid cycle. In the key reaction of the glyoxylate pathway, the enzyme isocitrate lyase cleaves isocitrate to give rise to one molecule of succinate plus one molecule of glyoxylate. The latter condenses with a molecule of acetyl coenzyme A to give rise to malate. A molecule of malate is also produced from the succinate molecule (via the citric acid cycle). Hence there are now two molecules of malate, which, (by way of the normal citric acid pathway), will be converted to oxaloacetate, condense with acetyl coenzyme A and eventually give rise to two molecules of isocitrate (see Fig. 19).

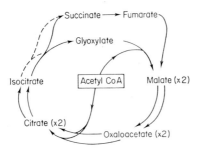

Fig. 19. The glyoxylate cycle. For each turn of the cycle three molecules of acetyl-coenzyme A are taken up, and two molecules of isocitrate are generated for every molecule of isocitrate entering the cycle.

It can thus be seen that, unlike the ctiric acid cycle where the number of intermediate molecules remains constant, the glyoxylate cycle is "anaplerotic", i.e. there is a replenishment of the number of intermediate molecules. This permits the net synthesis of carbohydrate from acetyl coenzyme A derived from the metabolism of fatty acids. It will be remembered that in mammalian cells, which do not contain the enzymes of the glyoxylate by-pass, net gluconeogenesis from fatty acids is not possible.

D. FUNCTION OF PEROXISOMES

Not all cells contain peroxisomes. Moreover, many hydrogen peroxide-producing oxidases, (e.g. monoamine oxidase and xanthine oxidase) are not peroxisomal in origin. There also seems to be sufficient catalase in the cell sap rapidly to destroy any hydrogen peroxide produced by such oxidases. It is therefore difficult to maintain that peroxisomes are essential either for cellular metabolism or to protect cells from hydrogen peroxide of metabolic origin. What then is the function of the peroxisome?

It is now generally believed, that, in a sense, peroxisomes are the vestigial remains of organelles which once played a key role in the adaptation of primitive anaerobic cells to the appearance of oxygen in the atmosphere. The new aerobic conditions arose when photosynthetic organisms achieved the ability to utilize water as an electron donor. Oxygen was toxic to anaerobic cells, both because it spontaneously oxidized cellular components and because a product of such oxidations was hydrogen peroxide, which was also toxic. The development of an organelle with specific oxidases in association with catalase permitted cells to remove both oxygen and the toxic product of oxidase activity, i.e. hydrogen peroxide. The substrates for the oxidases were usually fermentation products such as hydroxy acids, and these were thus removed simultaneously with oxygen. In addition, other fermentation products such as alcohols could be removed by catalase acting in its role as a peroxidase. Hence the process of protection of the cell against oxygen was turned to metabolic advantage.

There were also obvious advantages when the enzymes of the relatively complex glyoxylate by-pass were organized within an organelle. The association of these enzymes with the peroxisome in particular rather than with any other membranous sytem might have been merely coincidental. However, it might be relevant that the glyoxylate cycle involves the NAD^+-dependent oxidation of malate to oxaloacetate, and this process is controlled by the rate of reoxidation of NADH. Now, the peroxisome can function as an indirect NADH-oxidase. For example, plant peroxisomes contain both an NADH-linked glyoxylate reductase (which generates glycolate and NAD^+) and glycolate oxidase (which regenerates glyoxyate). The net effect of the functioning of these two systems is the oxidation of NADH by oxygen, which in turn allows further oxidation of malate to oxaloacetate. The original association of the glyoxylate bypass with the peroxisome might thus have been more than fortuitous. Some reactions of the overall pathway have now been taken over by the predominant mitochondrion but some enzymes of the cycle remain associated with the peroxisome. This would not be surprising, particularly in cases such as germinating fatty seedlings where there is extensive gluconeogenesis from fat, and little oxidation of carbohydrate.

What, however, is the reason for the survival of the peroxisome in the cells of higher organisms? In the leaves of plants it would appear actually to be disadvantageous, since in the presence of a trace amount of glycolate, the peroxisome as we have seen could function as an NADH-oxidase, giving rise to "photosynthetic respiration" which would direct reducing equivalents away from photosynthetic pathways. Perhaps this system has survival value as a "safety valve" for disposing of surplus reducing power generated under conditions of high light intensity.

In mammalian cells there might still be a specific metabolic role for urate oxidase (although this has disappeared with man and the higher apes) and possibly an as yet undetermined role for D-amino acid oxidase. However, their survival as part of an integrated system would be surprising if the integration served no function. There are four clues as to what such a function might be. Firstly, liver and kidney are tissues which are exceptionally concerned with gluconeo-genesis rather than with the oxidation of carbohydrate. Secondly, the affinity of the oxidases for oxygen is, as we have seen, rather low. Thirdly, the conditions in the peroxisome tend to favour a peroxi-dative function for catalase. Finally, there are cytoplasmic enzymes, e.g. lactate dehydrogenase, (acting on pyruvate), and alcohol dehydro-genase, (acting on acetyldehyde), which can oxidize NADH and at the same time give reaction products (lactate and ethanol) which may be re-oxidized by the peroxisome. This re-oxidation may be mediated either by an appropriate oxidase or by catalase acting as a peroxidase. In each case, substrates for the cytoplasmic enzymes are regenerated (pyruvate and acetaldehyde respectively in the cases cited). Hence, again, the peroxisome may effect the net oxidation of NADH.

In such cells, when oxidation of carbohydrate does become neces-sary, i.e. when cytoplasmic NADH has to be oxidized rather than utilized in gluconeogenesis, there may be inadequate shuttles to trans-fer reducing equivalents from the cytoplasm to the mitochondrion. Under such circumstances, cellular oxygen levels might tend to rise, and hence "switch on" the NADH-oxidase system which is provided by the peroxisome functioning in conjunction with cytoplasmic dehydrogenases. No ATP would be generated in such a process, but at least glycolysis would be allowed to proceed and pyruvate would be made available for oxidative phosphorylation in the mitochon-drion.

If this hypothesis were valid, then there would be, in some cells at least, a collaboration between mitochondria and peroxisomes in cellular respiration.

REFERENCES AND RECOMMENDED READING ON MITOCHONDRIA

Allman et al. (1970). J. Bioenergetics 1, 75, Fig 2A.

Ashwell, M. and Work, T. S. (1970). The biogenesis of mitochondria. Ann. Rev. Biochem. 39, 251–290.

Fawcett, D.W. (1966). "The Cell: Organelles and Inclusions: An Atlas of Fine Structure." W. B. Saunders, Philadelphia.

Green, D. E. and Baum, H. (1970). "Energy and the Mitochondrion". Academic Press, New York and London.

Green, D. E. *et al.* (1970). Conformational Basis of Energy Transduction in Membrane Systems, (several original papers) in *J. Bioenergetics* **1**.

Greville, G. D. (1969). A scrutiny of Mitchell's chemiosmotic hypothesis of respiratory chain and photosynthetic phosphorylation. In "Current Topics in Bioenergetics", (Sanadi, D. R. ed), Vol. 3, pp. 1–78, Academic Press, New York and London.

Klingenberg, M. (1970). Metabolite transport in mitochondria. An example for intracellular membrane functions. In "Essays in Biochemistry", (Campbell, P. N., and Dickens, F. eds), Vol. 6, pp. 119–159, Academic Press, London and New York.

Korman *et al.* (1970). *J. Bioenergetics* **1**, 10–14, Fig 1A-D.

Lehninger, A. L. (1970). Mitochondria and calcium ion transport, *Biochem. J.* **119**, 129–138.

Lehninger, A. L. (1964). "The Mitochondrion". John Wiley and Sons Inc., New York.

Munn, E. A. (1969). Ultrastructure of mitochondria. In "Handbook of Molecular Cytology", (Lima-de-Faria, A. ed), pp. 875–913. North-Holland Publishing Company, Amsterdam.

Racker, E. (ed), (1970). "Membranes of Mitochondria and Chloroplasts", Van Nostrand, Reinhold, New York.

Racker, E. (1970). The two faces of the inner mitochondrial membrane. In "Essays in Biochemistry". (Campbell, P. N. and Dickens, F., eds). Vol. 6, pp. 1–22. Academic Press, London and New York.

Singer, T. P. (1968). "Biological Oxidations". Interscience Publishers, New York.

Slater, E. C. (1971). The coupling between energy-yielding and energy-utilizing reactions in mitochondria. *Q. Rev. Biophys.* **4** (No. 1), 35–71.

Wrigglesworth, J., Packer, L., and Branton, D., (1970). *Biochim Biophys. Acta.* **205**, 125–135.

Williams, R. J. P. (1969). Electron transfer and energy conservation. In "Current Topics in Bioenergetics". (Sanadi, D. R. ed), **Vol. 3,** pp. 79–156, Academic Press, New York and London.

ON PEROXISOMES

de Duve, C. (1969). The peroxisome: A new cytoplasmic organelle. *Proc. Roy. Soc. Ser. B.* **173**, 71–83.

7. Lysosomes

J. B. LLOYD
Biochemistry Research Unit, University of Keele

and

F. BECK
Anatomy Department, University of Leicester

I. Introduction

Lysosomes are unusual among the cellular organelles in that their existence was initially inferred entirely from biochemical studies. These studies were a painstaking series of experiments on the fractions obtained by differential centrifugation of rat liver homogenates and were made by Professor C. de Duve and his colleagues at the University of Louvain, Belgium, in the early nineteen fifties. De Duve reached the conclusion that his experimental results could only be explained if rat liver cells possessed a type of organelle distinct from the mitochondrion and containing the five acid hydrolases acid phosphatase, ribonuclease, deoxyribonuclease, cathepsin and β–glucuronidase. His data indicated that these organelles were approximately $0\cdot4$ μm in diameter and had a limiting membrane that was permeable to glucose but impermeable to larger molecules such as sucrose or the contained enzymes. In 1955 De Duve was sufficiently confident in his findings to suggest a name for the new particle and his choice (from the Greek λυσις loosing, and σωμα body, reflecting the characteristic hydrolytic enzymes) has been accepted ever since. (The only disadvantage of lysosome is its similarity to lysozyme, and generations of students have made, and still make the mistake of confusing the two.) In 1956 De Duve and Novikoff published an electron micrograph of a lysosome-rich centrifugal fraction of rat liver and made the tentative (in fact, correct) identification of lysosomes with the pericanalicular dense bodies of the parenchymal cells. From that time the understanding of lysosomes has been furthered by both biochemical and morphological techniques, and this chapter attempts to present the integrated view that has resulted. The early work leading to the discovery of lysosomes remains, however, a most interesting and instructive study; it has been recorded in fascinating detail by De Duve himself (1969).

Students of lysosomes have been well-served by reviews of their subject. Extensive reviews were published by De Duve (1959) and De Duve and Wattiaux (1966). Lysosomes were the subject of a Ciba Foundation Symposium (1963) and more recently of a comprehensive three-volume multi-author compendium (Dingle and Fell, 1969; Dingle, 1973) and a laboratory handbook (Dingle, 1972). Books and published symposia have appeared on aspects of lysosome function and malfunction, and these will be mentioned in the appropriate sections below. In contrast to specialist publications, student text books, particularly of biochemistry, frequently give a very inadequate account of lysosomes. This is exemplified by one recently published and deservedly popular textbook of biochemistry, which accords lysosomes a single index entry, referring to a few lines of text on one page, while giving mitochondria over 50 index entries. We believe that this neglect is unjustified and hope to show, in this chapter, that the study of lysosomes is both intrinsically interesting and also highly relevant to medical science.

II. The Vacuolar System

Lysosomes may be defined as cytoplasmic vacuoles containing acid hydrolases and bounded by a trilaminar "unit" membrane. They probably originate in association with the Golgi apparatus (see below), but are usually destined to fuse with other vacuoles of various origins so that their contained hydrolases can meet and digest substrates without ever being released into the surrounding cytoplasm. Lysosomes can also fuse with the cell membrane, releasing their contents extracellularly. It is thus impossible to consider the functions of lysosomes in isolation from certain other cellular phenomena, in particular the generation of vacuoles with which lysosomes may fuse. Lysosomes must therefore be seen as part of a family of related particles, collectively known as the vacuolar system, or vacuolar apparatus. The interrelationship of its various parts will be considered in the remainder of this section.

A. NOMENCLATURE

Lysosomes are termed primary lysosomes from the moment of their origin up to the point when they fuse with another vacuole. From that point on, whatever their subsequent history, they are known as secondary lysosomes. This distinction is a very useful one for theoretical purposes, though it is often not possible in practice to be sure whether any particular profile seen in an electron micrograph is a primary or a secondary lysosome.

Lysosomes may fuse with vacuoles that arise from at least three sources and, in each case, these vacuoles bear substrates for digestion. The first type of vacuole is that arising by pinocytosis or phagocytosis. Its membrane is derived from the plasma membrane and it carries into the cell material of extracellular origin. Such a vacuole is known as a heterophagosome and the secondary lysosome formed by its fusion with a lysosome or lysosomes (primary or secondary) is sometimes called a heterolysosome. The second type of vacuole is the storage granule of certain secretory cells, such as those of the anterior pituitary, where fusion with lysosomes provides a mechanism for the destruction of excess secretory product. This process has been termed crinophagy or granulolysis and the membranes that fuse appear to originate from the Golgi apparatus. Indeed there is even evidence that in certain physiological states hydrolytic enzymes and the specific cellular secretory products are packaged into the same Golgi vacuoles (see below). Thirdly there is the phenomenon of autophagy; it is possible for areas of cytoplasm to become sequestered within a vacuole, an autophagosome, whose membranes may be derived from the smooth endoplasmic reticulum (see below). After fusion with lysosomes, the autophagosome becomes an autolysosome. The term digestive vacuole is often used to designate any secondary lysosome in which digestion is progressing. A lysosome that has digested its substrate, but is left containing indigestible residues, is a telolysosome and a telolysosome that has lost its enzymes and contains only residues is a post-lysosome.

B. ORIGIN OF THE PRIMARY LYOSOMES

An account of protein synthesis at the molecular level has been given in Chapter 34 and in Chapter 5 it has been explained that different cellular machinery is brought into play depending on whether the protein synthesized is structural or secretory. In other

S*

words, although the molecular basis of protein synthesis in eukaryotes is invariably based upon transcription of DNA followed by translation of RNA on ribosomes, the cellular events depend very much on the type of protein produced. In general terms proteins that contribute to the fabric of the cell are formed in association with polyribosomes lying free in the cell cytoplasm. On the other hand, proteins required for export out of the cell, or those with a specific structure-based function to perform within the cell, are usually secreted into the granular endoplasmic reticulum and may be "packaged" by the Golgi apparatus into Golgi vacuoles for transport to the cell surface or to specific intracellular sites. An excellent example of this mode of protein secretion is given in Chapter 31 and concerns the export of enzymes from the cells of the exocrine pancreas. The catabolic enzymes within primary lysosomes are directed to various components of the vacuolar system (i.e. heterophagosomes, autophagosomes, excess granules of specific cellular secretions) and are therefore conveniently handled in a similar fashion. Very many ultrastructural studies, e.g. Smith and Farquhar (1966), performed on a variety of cells have indicated that Golgi cisternae—but possibly in certain cells only those at the mature face—are frequently acid phosphatase-positive. Associated with the phosphatase-positive cisternae small "vesicles" also containing acid phosphatase, the primary lysosomes, are usually demonstrable (Fig. 1); and they represent the physiological equivalent of the condensing vacuoles in the cells of the exocrine pancreas.

A comprehensive view of the formation of primary lysosomes is proposed by Novikoff et al. (1971). These workers have postulated a complex relationship of Golgi-associated endoplasmic reticulum and lysosomes (GERL); it is suggested that lysosomal enzymes are elaborated in the granular endoplasmic reticulum and pass to a specialized portion of the smooth endoplasmic reticulum associated with the mature (concave) face of the Golgi apparatus. Here primary lysosomes are budded off (Fig. 2). The GERL hypothesis of primary lysosome formation has been prompted by the frequent observation that only cisternae in the region of the concave face of the Golgi complex are acid phosphatase-positive. These cisternae may represent smooth endoplasmic reticulum rather than true Golgi apparatus (if indeed a real distinction exists) and the Golgi vesicles would then, at least as regards origin, correspond physiologically to those described by Caro and Palade (1964) in the pancreas (Chapter 31). The fact that many lysosomal enzymes are glycoproteins suggests that the Golgi complex has a part to play in their formation (see Chapter 5). It is probably too restrictive to define "Golgi vesicles" only as organelles that transport material from the endoplasmic reticulum to the Golgi apparatus although they

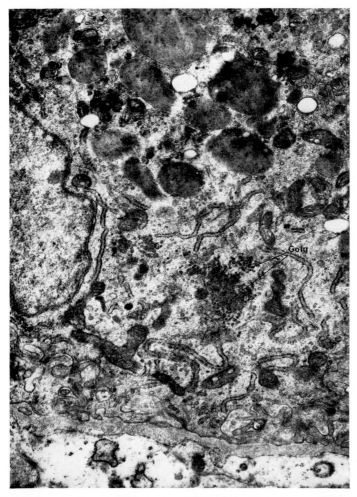

Fig. 1. Rat visceral yolk sac epithelium stained for acid phosphates by a modified Gomori technique. Phosphatase-positive Golgi regions (Golg) seem to be giving rise to primary lysosomes which are passing into the cell cytoplasm. × 17 500. (From Beck *et al.*, in Methods in Mammalian Embryology, ed. J. C. Daniel, with permission).

undoubtedly perform this function in the cells of the exocrine pancreas. Novikoff *et al.* (1971) refer to such structures as "transitional vesicles" or "transitional sheets" (Fig. 2). There is however evidence that in certain cells organelles having the dimensions of Golgi vesicles are budded from the Golgi cisternae (Nichols *et al.*, 1971) or from GERL. So-called "coated" vesicles which are often found in the region of the Golgi apparatus may fall into this category though they may also be composed of plasma membrane that has been interiorized and is possibly in the process of incorporation into the Golgi apparatus.

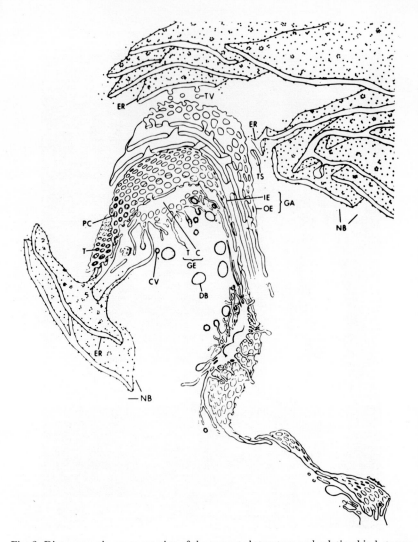

Fig. 2. Diagrammatic representation of the suggested structure and relationship between endoplasmic reticulum, GERL and Golgi apparatus in small neurones from adult and fetal rat dorsal root ganglion. From Nissl bodies (NB at the outer aspect of the Golgi stack) endoplasmic reticulum (ER) is thought to carry materials in transitional vesicles and sheets (TV and TS) to the outer element (OE) of the Golgi. The outer Golgi element is composed of irregularly anastomising tubules or small saccules. Two fenestrated elements of Golgi separate the outer and inner elements (IE) of the stack. The innermost element consists of a hexagonal array of tubules surrounding polygonal compartments. Tubules of smooth ER at left which are coming directly from rough ER or from GERL enter each polygonal compartment (PC). From a Nissl body (NB at left) endoplasmic reticulum (ER) presumably transports material including acid phosphatase to GERL (GE) which consists of cisternal portions (C) and tubules (T). The tubules of GERL form more or less regular anastomoses which are shown within the concavity of the Golgi apparatus in the centre of the figure. Towards the right the Golgi stack is shown twisted so that GERL lies above the innermost element and the other elements are not drawn. In the centre of the diagram the origin of dense bodies (DB) and coated vesicles (CV) from GERL is shown. The Golgi apparatus forms a continuous network coursing through the cytoplasm. (From Novikoff *et al.*, 1971, *J. Cell Biol.* **50**, with permission).

Dense Bodies

Many cells (e.g. those of the neurons of some rat dorsal root ganglia, heterophils in the peripheral blood etc.) contain "dense bodies" which are acid phosphatase-positive and of larger dimension than the "Golgi vesicles" described above (Fig. 3). They do not contain material suggestive of previous participation in either heterophagy or autophagy and the available evidence is consistent with their classification as primary lysosomes. Sometimes they may be storage granules of lysosomal enzymes in cells where the latter are utilized intermittently electron microscope they are membrane-bounded and their dimensions vary. According to the GERL concepts of Novikoff *et al.* (1971) dense bodies may be budded directly from Golgi-associated smooth endoplasmic reticulum (Fig. 2). In the admittedly somewhat special case of the heterophils in peripheral blood, however, Cohn and Fedorko (1969) have summarized evidence indicating the formation of the specific storage granules (i.e. the primary lysosomes) in these cells from Golgi vacuoles that separate from cisternae of the Golgi apparatus, while Bainton and Farquhar (1966) have demonstrated that the origin of granules in polymorphonuclear leucocytes occurs from *opposite* faces of the Golgi complex in mature and developing granulocytes. Nichols *et al.* (1971) have shown that monocytes make two different kinds of primary lysosomes, each from the Golgi apparatus. During cell differentiation large azurophilic granules are produced and serve as storage granules utilized during the first phases of phagocytosis. Thereafter, in mature cells lysosomal enzymes are packaged into small "Golgi vesicle" type of primary lysosomes.

Multivesicular Bodies

Multivesicular bodies consisting of many small vesicles surrounded by a single unit membrane are frequently found in mammalian cells (Fig. 4). It is likely that profiles of this nature are of various origins but some undoubtedly contain histochemically demonstrable lysosomal enzymes. The latter possibly result from the penetration of intact primary lysosomes of the "Golgi vesicle" type *through* heterophagosome membranes to lie within the substance of the latter. Such multivesicular bodies may therefore be a form of heterolysosome *inside* which the primary lysosomes still exist as discrete structures An alternative suggestions to account for the formation of multivesicular bodies is reviewed in the GERL concept as described by Novikoff *et al.* (1971).

We may summarize by saying that primary lysosomes appear to be

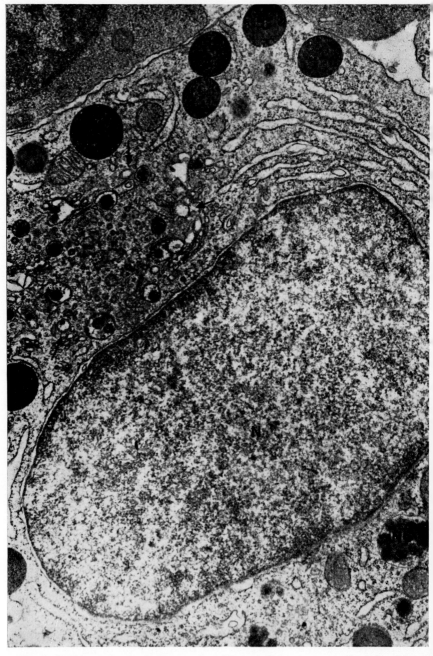

Fig. 3. Rabbit myelocyte showing nucleus (N) and Golgi complex (at arrows). The latter consists of lamellae, vesicles and vacuoles containing osmiophilic spherules. Nascent granules (NG) and granules (G) are present at cell periphery. These are primary lysosomes in the form of storage granules. × 22 500. (From Cohn and Fedorko, *In:* "Lysosomes in Biology and Pathology", Eds. J. T. Dingle and H. B. Fell, North Holland, with permission).

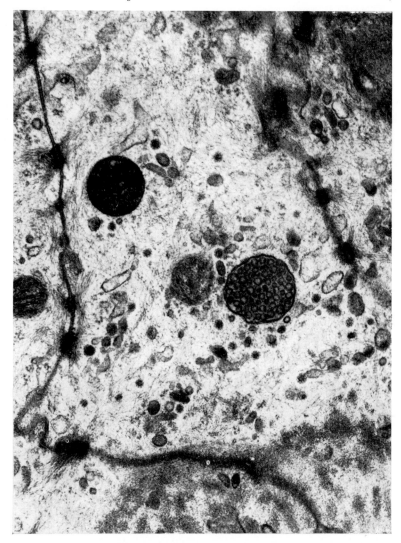

Fig. 4. Multivesicular bodies from rat ileum. Cytochemical evidence is required to establish the lysosomal nature of these organelles (× 30 000). (By courtesy of Prof. F. R. Johnson from Beck *et al.* In "Lysosomes, A Laboratory Handbook" (Ed. J. T. Dingle, North Holland, with permission).

formed in the region of the Golgi complex. Whether they are subject to precisely the same cellular processes as the zymogen granules of the exocrine pancreas (a classical model of cellular secretion) is still in doubt. It is important to bear in mind that primary lysosomes cannot be completely equated with Golgi vesicles in the functional sense tentatively suggested for these structures in the exocrine pancreas by Caro and Palade (1964). Indeed it is probable that primary lysosomes arise in a number of distinct ways (Bainton and Farquhar, 1966); this is also reflected in the many forms they take—ranging from small "Golgi vesicles" found in cultured blood monocytes to the large dense bodies of mature peripheral blood heterophils.

C. ENDOCYTOSIS

Endocytosis is a term coined to cover the two related phenomena of pinocytosis and phagocytosis. In endocytosis substances present in the environment of a cell are captured and internalized without any permeation through the cell membrane. The mechanism involves an invagination of the plasma membrane followed by a constriction of the channel thus formed, fusion of membranes and the separation of an independent vacuole. The contents of the vacuole are still separated from the cytoplasm by membrane and therefore, in a sense, may be thought of as an extension of the extracellular space.

Endocytosis has been the subject of recent reviews by Jacques (1969, 1973).

Occurrence

Although endocytosis may well occur in all animal cells, with the exception of the anucleate mature erythrocyte, it is a much more prominent activity in some cells than in others. In mammals the cells of the reticuloendothelial system, the renal proximal tubular epithelium and the thyroid follicular epithelium (Chapter 26) are among those with strong endocytic capacity, while the endothelial cells of the capillary walls are active in a specialized form of endocytosis termed diacytosis (see p. 283).

Morphological Considerations

Endocytosis refers to the capacity of a cell to ingest extracellular material into membrane-bounded spaces located within its substance. Certain cells have the capacity to take up both solid objects and quantities of extracellular fluid while others are incapable of interiorizing solids but retain the power to imbibe fluid to a greater or lesser extent. Endocytosis may serve numerous functions in the economy of the organism; it may be an essential step in cellular nutrition especially in unicellular organisms, it can be used by the cell to dispose of noxious agents such as bacteria or toxins in its microenvironment or it may merely serve as a convenient means of allowing macromolecules to traverse cells. The multiplicity of purposes served by endo-cytosis is reflected in its varying morphology and the simplest classification of the phenomenon is a morphological one. Thus phago-cytosis, a term introduced by Metchinkoff in 1892, refers to the ability of a cell to take up solid particles such as those of colloidal carbon. Pinocytosis (cell drinking) was first described by Lewis in 1931 in cultured cells; like phagocytosis it is observable by light microscopy. In both processes ingestion involves the projection of cytoplasmic processes from the cell surface which, in his cultured cells, Lewis described as the "ruffled membrane". When cells were first examined by electron microscopy it soon became clear that many more cell types were capable of endocytosis than had been supposed and the term micropinocytosis was used to describe the uptake of small quantities of extracellular fluid in vacuoles that are smaller than the resolving power of the light microscope. Ultrastructural studies have shown that micropinocytosis can take at least two main forms. The first appears to be principally associated with the passage of fluid *through* the cell. A characteristic of this process is that vesicles with a smooth inner surface are seen in the process of budding from the plasma membrane on one surface of the cell and of fusing with the membrane at the other. A large number of completed vesicles are seen still attached to the plasma membranes, while relatively few are seen either in the process of formation or free within the cell cytoplasm (Fig. 5). It appears therefore that in the process—which has been named diacytosis—there is rapid formation of micropinocytic vesicles together with a relatively rapid passage through the cell but a comparatively prolonged period during which the completed vesicle remains attached to the plasmalemma. Diacytosis is classically seen in capillary walls where it provides a convenient method for the transport of macromolecules out of the vessel into the extracellular space. The direction of flow of the vesicles cannot, of course, be determined simply by examination of electronmicrographs but the use of electron-dense markers (Daems *et al.*, 1972) combined with timed studies has allowed definite conclusions to be drawn in a variety of situations. A puzzling feature of diacytosis is the apparent ability of vesicles to cross cells without fusing with lysosomes. It could be that diacytosis is more than a specialized form of endocytosis and constitutes a separate class of cell activity.

The second form of endocytosis is one clearly associated with lyso-somes and usually occurs in relation to a special external coating on the plasma membrane. This coating is a mucopolysaccharide and may provide special receptor sites for compounds prior to their ingestion by the cell. It is undoubtedly true that certain substances are

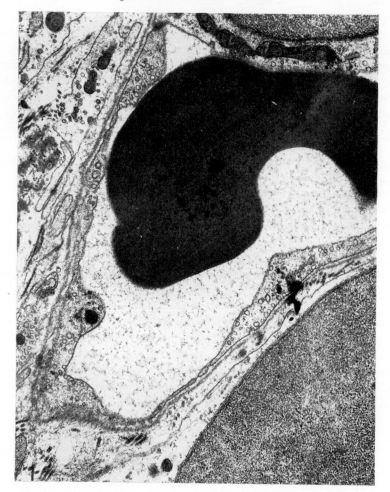

Fig. 5. Capillary endothelium from ferret endometrium showing clear evidence of diacytosis.

selectively endocytized (see below) and it is quite likely that this may be due to a selective affinity for the surface coating on the plasma membrane. Under the electron microscope the surface coating has a fuzzy appearance (Fig. 6); it may form a continuous layer around a cell actively engaged in endocytosis and may be increased in quantity in areas on the cell surface that are about to be interiorized (Jollie and Triche, 1971). Further surface specializations are frequently seen in pinocytizing cells. The surface membrane may be thrown up into a "brush border" made up of microvilli (Fig. 6) (as in the proximal kidney tubules) while the apical region of the cell may contain a complex system of interconnecting channels communicating, on the one hand, with the spaces between the microvilli and budding off, on the other, numerous micropinocytic vesicles into the interior of the cell (Jollie and Triche, 1971). Regular surface specializations in the form of discrete geometrically arranged plaques have been described in the surface coating of the apical canalicular system in certain cells (Wissig and Graney, 1968). As already indicated, micropinocytic vesicles are bounded by unit membrane and lined by the fuzzy coating typical of the plasmalemma; in common with the pinocytizing surface they also display bristle-like projections into the surrounding cytoplasm from which the term "coated vesicle" arises. In the substance of the cell, micropinocytic vesicles frequently fuse to form larger heterophagosomes which sometimes (for example in the ileal cells of the neonate rat) are large enough to be seen with the light microscope. At the same time electron-microscopic appearances suggest that a condensation of material is occurring within these organelles with loss of water into the surrounding cytoplasm.

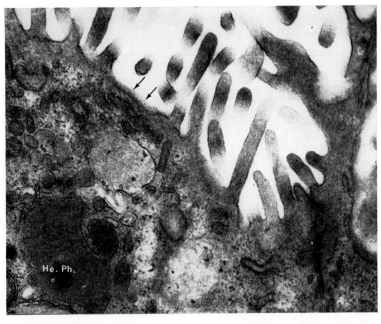

Fig. 6. Apical portion of two adjacent cells of the rat visceral yolk sac. The actively pinocytizing cells have well-developed microvilli; an electron-dense external coating (arrows) is seen at the surface of the plasmalemma and lining the micropinocytic vesicles. More deeply a heterophagosome He. Ph. is forming by fusion of such vesicles. × 36 000 (From Beck and Lloyd, In "Lysosomes in Biology and Pathology", (Eds. Dingle and Fell), North Holland, with permission).

Mechanism

Endocytosis has been shown in many cell systems to be an energy-dependent process, sometimes linked obligatorily with either anaerobic or aerobic metabolic pathways. Thus it is reported that phagocytosis in alveolar macrophages is impaired by inhibitors of oxidative phosphorylation but unaffected by glycolytic poisons; peritoneal macrophages react in the converse manner. The ATP requirement of endocytosis is probably expended in the membrane movements involved, and recent evidence suggests that these arise from contractions of the actomyosin-like proteins of the microfilaments which form a "cage" in the cytoplasm of some cells, just below the plasma membrane. Much of the evidence for this conclusion comes from experiments using a mould metabolite, cytochalasin B, which specifically, but reversibly, inactivates microfilaments and paralyses phagocytosis in macrophages (see Allison, 1972).

There remains the question of whether the various types of endocytosis share a common mechanism. There is, in fact, little to suggest that they do and A. C. Allison (personal communication) has recently shown that cytochalasin B does not inhibit micropinocytosis in macrophages. Earlier Z. A. Cohn (reported by Jacques, 1969) had shown that a number of metabolic inhibitors depressed pinocytosis but not phagocytosis in mouse macrophages.

Specificity

In endocytosis substances may enter cells by a non-specific and specific mechanisms. Non-specific entry results because the process involves entrapment of liquid and consequently of any solute or particle that liquid may contain. Specific entry occurs when a solute or particle interacts with the plasma membrane being internalized. A substrate that is attracted to the membrane will be incorporated into the cell in large amounts relative to other species present that are not so attracted. In theory at least, substrates could be repelled from the membrane and thus incorporated by endocytosis less efficiently than other solutes that entered simply in the liquid phase.

The size and shape of the endocytic invagination are also determinants of specificity. The geometry of a vesicle determines the ratio of its volume to its surface and therefore the relative importance of the two mechanisms of uptake in endocytosis. Moreover in phagocytosis the size of the endocytic vacuoles formed will impose an upper limit on the dimensions of the particles they contain.

There is strong evidence that specific mechanisms are of great importance in uptake by phagocytic cells. Much of this evidence comes from studies on the clearance of various substances from the bloodstream, where it is apparent that cells can discriminate between one macromolecule and another and even between native and denatured molecules of the same macromolecule, such as a protein. Recognition of particular molecules is presumably a function of receptors on the plasma membrane, but no detail of these interactions is known. In a recent review (Winterburn and Phelps, 1972) a role in recognition by such receptors prior to pinocytosis has been proposed for the glycosyl residues that are present on many proteins.

Kinetics

The kinetics of endocytosis will be rather complex in the case of insoluble particulate substrates or substrates that are repelled by the cell. Such cases are discussed briefly by Jacques (1969). The treatment that follows here is a simple one that should apply to soluble substrates taken in in the liquid phase and/or as a monolayer of molecules attached to the plasma membrane.

The rate at which a soluble substance will enter a cell by pinocytosis will be the sum of two terms, the first representing uptake in the liquid phase, the second uptake by adsorption on the membrane. Mathematically this is expressed by the quotation :

$$Q = Fc + \frac{SR \ c}{K+c}$$

where Q is the rate of uptake of the substance, c its concentration in the liquid surrounding the cell and K the dissociation content of the substrate—surface complex. F is the rate of pinocytic uptake of liquid, S the rate of internalization of cell surface and R the maximum amount of substrate that can be adsorbed per unit area of cell surface. In this equation all the terms are potential variables, if one changes the nature or concentration of the substrate. Each substrate will have its own distinctive K and R; if no surface absorption occurs, R will be zero, thus making the whole second term zero. But F and S, which are of course related to each other, may also vary, because substrates may stimulate or depress the rate of internalization of plasma membrane or alter the geometry of the pinocytic invaginations. In consequence the relationship between the rate of the uptake of a solute and its concentrations in the cell's liquid environment may be far from simple. From the equation given above, it can be deduced that uptake will be proportional to concentration only if two separate criteria are both met. The first of these is that changing c does not change F or S, i.e. the substrate itself must not stimulate or depress pinocytosis. The second is that either the second term must be zero or c must be insignificantly small in comparison with K.

Except in a very few cases, mentioned by Jacques (1969, 1973), even the above simple treatment of the kinetics of endocytosis has not been tested in an experimental system. This is largely because of technical difficulties: experiments on whole animals are subject to too many unmeasurable physiological variables and experiments on cells or organs in culture rarely yield sufficiently reproducible data for kinetic treatment.

Stimulation and Inhibition

Much has been written about the stimulation and inhibition of endocytosis by various substances, and a good deal more is implied in phrases such as "reticulo-endothelial blockade". However, the experimental evidence behind these statements is not always fully convincing. Phagocytosis of solid objects by macrophages etc., appears to be a phenomenon that requires an inducer, frequently the substrate particle itself. To be efficiently ingested by macrophages, bacteria must, in most circumstances, be coated with "opsonin", serum components that probably interact with specific receptor sites on the phagocytic cell (see Chapter 37 for a full discussion). In contrast, micropinocytosis probably occurs in many cells without any requirement for a specific inducer, although the rate of the process may be increased or decreased by appropriate stimuli. Thus in mouse peritoneal macrophages in culture pinocytosis is reported to be stimulated by a variety of substances, most of them anionic, and depressed by certain metabolic poisons (see above). However, it is important to note that the rate of pinocytosis was inferred from the number of pinocytic vacuoles seen in the cells, a parameter that may not correlate with the rate of internalization of membrane or of substrate. Even when the entry of substrate into cells is monitored, it is not always clear whether an inhibitor acts by decreasing the rate of internalization of membrane or by competing with substrate for substrate binding sites on the membrane. The effect of a substance on endocytosis may vary with its concentration: in at least one case, trypan blue, low concentrations stimulate while high doses inhibit pinocytosis (Williams *et al.*, 1973).

D. AUTOPHAGY

Profiles may be seen in a variety of cells (particularly in quiescent or involuting tissues) in which a portion of the cytoplasm with its contained organelles has apparently become sequestered in a membrane-bounded structure (Fig. 7). Sometimes though not invari-

ably the limiting membrane is double and, when the specimen is
stained for electronmicroscopically demonstrable lysosomal enzymes
such as acid phosphatase, the sequestered region usually gives a
positive reaction. These regions of the cell are called autolysosomes
and their precise method of formation is still in some doubt. It is
possible that a portion of the cell cytoplasm may be isolated as a
result of being surrounded by endoplasmic reticulum or Golgi
membrane, the inner membrane then either disappears or else fuses
with the outer membrane so the mature "autophagosome" has only a

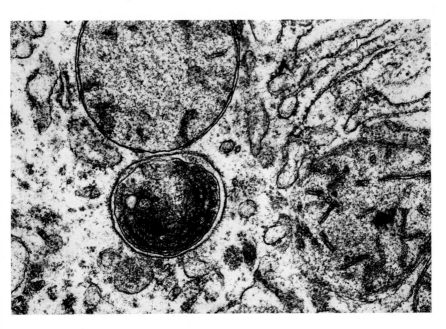

Fig. 7. Autolysosome from the hepatic parenchyma of a glucagon treated rat. The
structure is bounded by a double layer of unit membrane and contains degenerating
elements of cytomembrane × 70 000.

single membrane surrounding it. Some evidence in support of the
theory that the membranes fuse is given by the fact that the final
autophagosome membrane is about 10 nm thick while the constituent
cytomembranes of the Golgi or endoplasmic reticulum are only 6 nm.
Furthermore, when autophagic vacuoles are induced to form experimentally in the liver (by the injection of glucagon), the double
membrane surrounding them in the early stages of their formation
shows fine structural evidence of the presence of glucose–6–
phosphatase—an enzyme characteristic of the endoplasmic reticulum
(Chapter 5). Alternative (or possibly additional) mechanisms for the

formation of autophagic vacuoles are by the distortion of an endocytic vacuole to surround an area of cytoplasm or by a number of endocytic vacuoles surrounding and sequestering a portion of cell cytoplasm. In either case, the 10 nm membrane surrounding the autophagosome would originate in plasmalemma. Lysosomal enzymes may enter autophagosomes to form autolysomes as a result of fusion with primary lysosomes, a method similar to that whereby the hetero-phagosomes are converted into heterolysosomes. Alternatively, lysosomal enzymes may be present in autophagic vacuoles from their inception by virtue of their formation from Golgi cisternae or perhaps even because autophagic vacuoles may also form as a result of portions of cytoplasm becoming trapped between primary or secondary lysosomes. Clearly a number of the possible methods of autolysosome formation involve endocytic vacuoles and indeed endocytic markers (see below) are frequently found in the same vacuoles as the dis-integrating cell organelles characteristic of autophagosomes.

E. CRINOPHAGY

Crinophagy is a process observed in some secretory cells that store specific product in the form of membrane-bounded secretory granules. The process has been described in a number of endocrine cells and is well summarized by Farquhar (1969) for various cell types in the anterior pituitary gland. The first examples of crinophagy were observed in prolactin-secreting cells from rat pituitaries prepared at various times after removal of pups from the breast. It was observed that shortly after removal of the young, mature secretory granules fused directly with acid phosphatase-positive dense bodies (primary lyso-somes). Immature secretory granules appeared to fuse more readily with multivesicular bodies but the significance of this observation is not understood. A positive acid phosphatase reaction could also be obtained from the mature (concave) face of the Golgi apparatus, often from the same cisternae as those in which secretory granules were forming. About 72 h after the cessation of lactation morphological evidence of dissolution and degradation of the secretory products became apparent in the dense bodies with the formation of membranes and vacuolar residues. At the same time autophagic vacuoles became common as the protein synthesizing machinery of the cell was dis-mantled. Often autophagosomes and secretory granules undergoing crinophagy fused. In even later stages the "vacuoles" representing degraded cellular secretion often separated from their associated electron dense bodies.

Whatever the ultimate fate of the secretion, the process that distinguishes crinophagy is the initial fusion of a secretory granule with a primary lysosome, leading to the former's selective destruction.

III. Methods of Study and Properties of Lysosomes

A. BIOCHEMICAL METHODS

Differential Centrifugation

The original discovery of lysosomes resulted from work on the differential centrifugation of homogenates of rat liver, and this is still the technique most widely used for their biochemical study. The detail of the method has not changed significantly in the intervening years. Typically tissue is disrupted in an ice-cold isotonic solution of sucrose or KCl using gentle homogenization conditions that do not impose unduly large shearing forces. The homogenate is then subjected to successive centrifugations at increasing centrifugal forces, the pellet of precipitated material being removed and reserved after each spin. The concentrations of various enzymes may be measured in each fraction and will indicate, if each enzyme is known to occur in only one organelle or cell compartment, the relative concentration of each organelle in each fraction. The uses and pitfalls of this method are discussed at length by De Duve (1965, 1967).

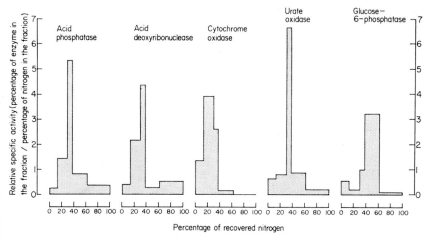

Fig. 8. Data of Table 1 presented in graphical form.

The scheme regularly used for rat liver uses four centrifugations (800 **g** for 10 min., 3 300 **g** for 10 min., 16 500 **g** for 20 min., 100 000 **g** for 3 min.) and thus yields five fractions—four pellets and a final supernate. Table 1 and Fig. 8 show the distribution of several enzymes

T*

TABLE 1. Distribution of reference enzymes after fractionation of the liver by differential centrifugation in 0.25 M sucrose.[1]

	Recovery (%)	% of recovered amount					Relative specific activity[2]				
		N	M	L	P	S	N	M	L	P	S
Nitrogen	98.9	13.4	16.5	7.5	24.7	37.9	1.00	1.00	1.00	1.00	1.00
Enzymes											
Acid phosphatase	101.8	3.5	23.8	40.0	19.7	13.0	0.26	1.44	5.33	0.80	0.34
Acid deoxyribonuclease	95.3	5.6	35.3	32.7	6.6	19.8	0.42	2.14	4.36	0.27	0.52
Cytochrome oxidase	88.9	11.4	64.8	19.6	4.2	—	0.85	3.93	2.61	0.17	—
Urate oxidase	94.9	8.6	13.5	49.8	21.3	6.8	0.64	0.82	6.64	0.86	0.18
Glucose-6-phosphatase	92.8	7.3	2.9	7.5	79.4	2.9	0.54	0.18	1.00	3.21	0.08

[1] Results of de Duve et al. (1955). Table 1 is reproduced with permission from Beaufay (1972).

[2] Relative specific activity = percentage of enzyme in the fraction/percentage of nitrogen (or protein) in the fraction.

among fractions isolated in this way. The lysosomal enzymes occur in highest concentration in the third pellet (the light mitochondrial, "L" fraction), although they are by no means confined to this fraction. The heterogeneity of lysosome size in rat liver is sufficient explanation of this distribution throughout several centrifugal fractions. Similarly the L fraction does not contain only lysosomes. It contains most of the peroxisomes, some mitochondria and some "microsomes" (p. 149). Attempts to prepare "pure lysosomes" from liver by differential centrifugation have generally been disappointing, although many schemes, often involving several successive steps and usually also including the use of density gradient centrifugation, have been reported. The yield of lysosomes is always low, making it likely that a particular selected sample of lysosomes, perhaps not typical of the population, has been isolated. Many other tissues than rat liver have been subjected to differential and density gradient centrifugation, and it is usually possible to obtain a fraction enriched in lysosomes, but impossible to obtain one free from major contamination with other organelles. In the special case of polymorphonuclear leycocytes, however, a granule fraction almost free from contamination can be obtained by these methods. There is another approach, however, that has been widely used, chiefly with rat liver. It involves altering the lysosomes *in vivo*, by the administration of the detergent Triton WR–1339, so that their size is increased and their density decreased (the detergent congregates specifically in the lysosomes). The specific decrease in density makes it possible to separate lysosomes from the other organelles by density gradient centrifugation (Leighton *et al.*, 1968). The lysosomes thus prepared are, however, considerably modified and care must be taken in extrapolating from their properties to those of normal lysosomes. An excellent recent summary of the methods available for isolation of lysosomes is given by Beaufay (1972).

Enzyme Complement and Digestive Capacity

The original description of lysosomes saw them as containing five enzymes, all hydrolases with an optimum pH on the acidic side of neutrality. Since that time the number of enzymes found to be associated with lysosomes has increased, but, with a very few exceptions, they all conform to the description "acid hydrolases". The exceptions include myeloperoxidase, which is present in the lysosomes of the neutrophil leucocyte, and some amidases reported to be maximally active at neutral pH or above.

Identification of an enzymic activity as lysosomal is usually by one of two methods. In the first the distribution of the activity among the various fractions obtained by a scheme of a differential centrifugation is studied, a similar "profile" to that for known lysosomal enzymes in the tissue in question being taken as evidence of a lysosomal localization. In the second method a cell fraction containing only lysosomes is prepared and its enzymic activities examined. As stated above, the only method by which a sufficiently pure fraction has been obtained from rat liver is by the use of Triton WR–1339; in most other tissues no adequate method has been described. By means of these methods lysosomes have been shown to have hydrolytic activity towards a wide variety of substrates. Peptide, glycosidic and ester linkages are all susceptible to hydrolysis and many distinct patterns of specificity are discernible within each of these large categories. Thus lysosomes display α–glucosidase, β–glucosidase, α–galactosidase, β–galactosidase, α–mannosidase, α–acetylglucosaminidase, α–acetylgalactosaminidase, β–acetylglucosaminidase, and β–glucuronidase activities among their glycoside hydrolases. Within the group of lysosomal peptidases, endo-proteinase activities are found (cathepsin B1, which resembles papain in its specificity, and cathepsin D, which resembles pepsin) and also exo-peptidases, such as cathepsin A (a carboxypeptidase) and cathepsin C, an aminopeptidase that removes dipeptide moieties from the N-terminus of polypeptides.

In the above discussion care has been taken to speak of enzymic activities rather than enzymes. Very few enzymes known to be lysosomal have been purified sufficiently for the extent of their catalytic activity to be studied. It is thus not possible at the present time to know how many distinct enzyme species are present in lysosomes. A few lysosomal enzymes have been purified. One of these is α–glucosidase, which was found to hydrolyse both 1,4– and 1,6–glucosidic linkages. Thus lysosomes need one enzyme only to achieve the complete catabolism of glycogen. Indirect evidence about the individual lysosomal enzymes sometimes comes from the study of inborn errors of metabolism. In type II (Pompe's) glycogen storage disease (see section V) α–1,4– and α–1,6–glucosidase activities are both absent, making it extremely likely that one enzyme species is responsible for both. Further the galactosidase and other glycosidase activities that are present in tissues from Pompe's disease cannot be due to the same enzyme as is responsible for the α–glucosidase activity.

It seems probable that the lysosomal hydrolases can digest the whole range of biopolymers synthesized by cells. For a few such substances (e.g. glycogen, triglycerides) only one enzyme is needed, but more commonly (e.g. proteins, heteropolysaccharides, glycolipids), several enzymes must act sequentially to achieve complete hydrolysis. The extent to which the mixed lysosomal enzymes can degrade biopolymers has been investigated by *in vitro* experiments with lysosomal extracts prepared usually by the Triton WR–1339 technique. In this way it has been shown that proteins are degraded to a mixture of free amino acids and apparently resistant oligopeptides (chiefly dipeptides), glycogen to glucose, nucleic acids to nucleosides and orthophosphate, and triglycerides to glycerol and fatty acids. The degradation of the many types of heteropolysaccharides and glycolipids has not been so thoroughly investigated and many such substances appear to be only very slowly attacked *in vitro* systems. It tends to be assumed, however, that *in vivo* they are degraded within lysosomes at a rate adequate to account for their turnover times and to prevent their intracellular accumulation (see Section IV).

In cases where several enzymes acting sequentially are needed for the breakdown of a macromolecule, it is interesting to know the order in which they act. This information is becoming available as individual lysosomal enzymes are purified to homogeneity. Lysosomes contain a number of peptidases, but probably only two of these are endo-enzymes capable of hydrolysing internal linkages in intact proteins and large polypeptides. It is therefore probable that intralysosomal proteolysis takes place by hydrolysis of internal peptide linkages by cathepsins B1 and D and the subsequent digestion of the resultant fragments by exo-peptidases such as cathepsins A and C. Information on the sequence of enzymic attack can also be inferred from the consequences of certain human genetic disorders. Thus metachromatic leucodystrophy, in which there is accumulation of cerebroside sulphate within lysosomes, is characterized by an absent lyosomal sulphatase (see Section V). It may be reasonably deduced that desulphation is the first step in the catabolism of cerebroside sulphate.

A detailed account of the enzymic activities found in lysosomes, together with suitable assay methods, is given by Barrett (1972).

Latency and Membrane Permeability

The enzymes of the lyosomes are associated with the lyosomal interior, thus being separated from the general cytoplasm by unit membrane. Therefore, if an attempt is made to assay a lysosomal enzyme in a tissue homogenate, the only activity that will be measured will be that due to lysosomes whose membrane has been broken or damaged sufficiently to allow enzyme and substrate to meet. The total enzyme can be measured only if all the lysosomal membranes are disrupted prior to assay. The enzymes of intact lysosomes are thus said to exhibit structure-linked latency.

Lysosomes are easily induced to break and thus expose their enzymes. Lysosomes suspended in water or any hypotonic solution do so and this is taken to be an osmotic effect, the consequence of lysosomes having a membrane permeable to water but impermeable to solutes (enzymes etc.) present in the lysosomal interior. Lysosomes suspended in isotonic salts or sucrose retain their integrity, and thus the latency of their enzymes, for longer periods, presumably because osmotic balance is achieved between inside and outside, as it must be in the living cell. Lysosomes also break, even if osmotically protected, if treated by substances, such as detergents and certain enzymes, that attack their membranes. Broken lysosomes do not necessarily release their enzymes in soluble form and in practice the extent of release varies from one enzymic activity to another and with the method used to induce breakage. Thus, two biochemical criteria can be used to assess the integrity of a preparation of lysosomes, the latency of their enzymes and the sedimentability of their enzymes on centrifugation, the former criterion being a more sensitive indicator of damage. Measurements of latency must be carried out under conditions that do not themselves disrupt the lysosomes : these will include short assay periods (10 minutes is the usually recommended maximum), a low assay temperature, if possible, and the presence in the assay medium of sufficient of an inert solute to achieve osmotic protection. The total activity in the lysosomes is measured under identical conditions but including also a substance (almost universally now, $0 \cdot 1\%$ Triton X–100) that ensures complete and immediate disruption of the lysosomal membrane. The difference between the first value (the free activity) and the second (the total activity) is the latent activity; in a carefully prepared suspension of lysosomes from rat liver the latent activity of most enzymes is approximately 90% of the total activity.

The permeability properties of the lysosomal membrane cannot be measured in any direct fashion, but a considerable body of indirect evidence suggests that substances whose molecular weight is below 200 traverse the membrane with ease. Many substances of molecular weight above 200 apparently cannot cross the lysosome membrane, but the available information largely relates to highly hydrophilic substances such as carbohydrates and peptides and it is likely that a higher maximum molecular weight barrier operates for substances with substantial hydrophobic character. It seems certain that all the biological macromolecules are unable to penetrate across the lysosome membrane. These properties of the membrane are important to lysosomal function. The membrane is an effective barrier to enzymes, so that the potentially destructive lysosomal hydrolases are confined to the lysosomes and denied access to substrates in the surrounding cytoplasm. Conversely, substrates undergoing digestion within lysosomes are retained in the lysosomes until digestion is complete. The end-products of digestion are, however, able to escape to participate in metabolic processes elsewhere in the cell. For a fuller discussion of lysosomal membrane permeability and the methods available to investigate it, see Lloyd (1973).

Intralysosomal pH

The acidic optimum pH exhibited by almost all the lysosomal enzymes is suggestive of an acidic pH within the secondary lysosomes in which they act and there is indeed some long-standing support for this suggestion from experiments on the uptake of indicators by phago-cytic cells. How the intralysosomal pH is maintained at a lower value than that of the surrounding cytoplasm is unknown, but it could simply reflect a Donnan equilibrium consequent upon the semi-permeable nature of the lysosomal membrane. Alternatively there could be an ATP–driven pump in the membrane. As yet there is no strong evidence for either mechanism.

B. MORPHOLOGICAL METHODS

Most basic cytological techniques have been applied to the study of lysosomes and the vacuolar system. They include light and electron microscopy, enzyme histochemistry, immunofluorescence and *in vivo* staining methods.

Light and electron microscopy of the vacuolar system

The nature of the various processes associated with the vacuolar system have been described above and illustrations of electron micro-scopic appearances during endocytosis (Fig. 6) diacytosis (Fig. 5) heterophagy (Fig. 9) and autophagy (Fig. 7) are given. Routine techniques of transmission electron microscopy are entirely adequate to demonstrate these. It is obvious, however, that fixed preparations of cells will only give a static picture of an essentially dynamic process. To overcome this serious drawback attempts have been made to introduce microscopically traceable "markers" that are taken up by endocytizing cells *in vivo* or *in vitro* and whose position in hetero-phagosomes and heterolysosomes can be demonstrated at timed intervals after the administration of a pulse dose. These techniques are only of value for actively endocytizing cells to which the markers have ready access. Often cells either do not pinocytize sufficiently for the methods to be used or else their topographical location *in vivo* is such that they are inaccessible to parenterally introduced marker molecules.

Fig. 9. Rat visceral yolk-sac epithelial cell stained for acid phosphatase by a modified Gomori technique. Apically located phosphatase-negative heterophagosomes may be distinguished from more deeply placed heterolysosomes. × 18 000. (From Beck *et al.*, *In:* "Methods in Mammalian Embryology". J. C. Daniel, ed. Freeman and Co., with permission).

At the level of the light microscope the simplest techniques involve the exposure of living tissue to certain vital dyes, such as trypan blue, that bind strongly to protein. Under these circumstances the cells of actively pinocytizing tissues are found to contain granules of the dye probably because they have taken up molecules of protein to which the dye was attached (Fig. 10). It is a pre-requisite that vital stains used for this purpose are not rapidly degraded and do not leave the vacuolar system. Their gradual concentration, however, could lead to changes—if only subtle ones—in the normal functions of cells containing them. A somewhat more physiological marker is a histochemically demonstrable protein such as horseradish peroxidase. This material is taken up by endocytizing cells, and its intracellular fate can be followed by specific staining techniques (Straus, 1964). Staining with either vital dyes or horseradish peroxidase can be combined with histochemical demonstration of certain lysosomal enzymes and a dynamic assessment of the process of heterophagy can thus be made.

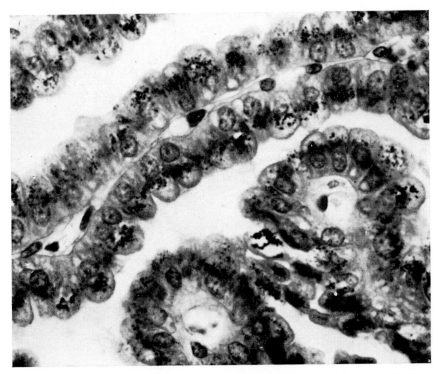

Fig. 10. Trypan blue granules present in heterolysosomes of the rat visceral sac epithelial cells 24 hours after subcutaneous injection of dye into the mother × 750. (From Beck and Lloyd, In: "Lysosomes in Biology and Pathology", J. T. Dingle and H. B. Fell, Eds North Holland, with permission).

By electron microscopy endocytosis can be followed using a number of electron-dense marker materials. Once again the most physiological techniques involve the use of naturally occurring macromolecules such as ferritin. Provided it is first purified, this iron-containing protein can be injected *in vivo* or introduced into culture media in high concentrations. It can subsequently be visualized with ease in the vacuolar system of actively pinocytizing cells with which it has had contact. By appropriate histochemical techniques horseradish peroxidase may be traced electronmicroscopically (Fig. 11) and is consequently also an extremely useful marker compound. Other particulate substances which are taken up by cells are frequently used as markers. Like the vital dyes they are not materials of types that the cell usually meets and consequently they may well have toxic effects. They are, however, easy to use and for this reason (again like vital dyes) retain a value in the study of the vacuolar system. Examples are colloidal suspensions of iron, gold and silver, latex and thorotrast.

Pinocytosis and the fusion of heterophagosomes with lysosomes can often be observed in cultured cells by phase-contrast microscopy.

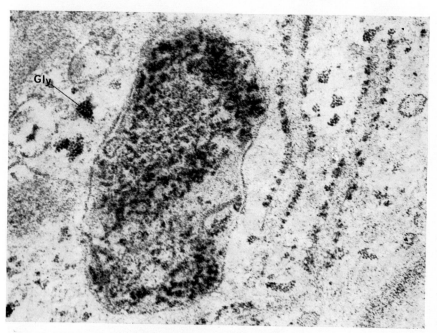

Fig. 11. Electron micrograph of heterophagosome from visceral yolk sac epithelium of rat one hour after intravenous injection of horseradish peroxidase. Stained for peroxidase. The enzyme reaction is membrane bounded. Patches of glycogen (Gly.) are seen scattered in the cytoplasm. × 10 0000. (From Beck and Lloyd, *In:* "Lysosomes in Biology and Pathology", J. T. Dingle and H. B. Fell, Eds North Holland, with permission).

The Histochemical Demonstration of Lysosomal Enzymes

Histochemical techniques for the light microscopic demonstration of acid phosphatase (Fig. 12), β–glucuronidase, arylsulphatase, N–acetyl–β–glucosaminidase, and certain non-specific esterases of lysosomal location are now matters of laboratory routine. A method for dipeptidyl aminopeptidase I has also been developed. With the electron microscope satisfactory methods for acid phosphatase (Fig. 9) and to a lesser extent arylsulphatase, dipeptidyl aminopeptidase I and non-specific (lysosomally localized) esterases are also available. Special attention must be paid to the preparation of the tissues, to fixation and to the staining procedures used and these are summarized by Beck *et al.* (1972). These methods are fundamental in the identification of both primary and secondary lysosomes.

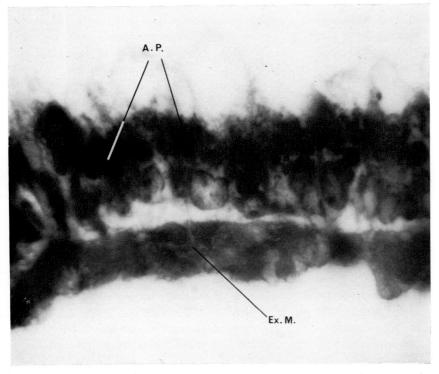

Fig. 12. Visceral yolk sac epithelium of rat stained for acid phosphatase. The enzyme is clearly localized in the supranuclear parts of the epithelial cells (A.P.). The underlying extraembryonic mesoderm (Ex.M.) is phosphatase-negative. × 1400. (From *J. Anat. Lond.* (1967) with permission).

A number of individual lysosomal enzymes in specific tissues of certain animals have now been purified and it has thus become possible to prepare antisera against these enzymes and then fluorescent labelled antibodies can be used to localize lysosomes by conventional methods of immunohistochemistry. For electron microscopy a method has been designed in which the antibody is conjugated with horse-radish peroxidase. It is possible to "stain" horseradish peroxidase for ultrastructural studies (see p. 300) and therefore the enzymes against which the antibodies have been prepared can be localized. Clearly the preparation of suitable antibodies and their utilization are highly specific and complex research methods that are not comparable in scope with routine histochemical methods; their use, however, makes large areas of research possible for the first time.

Other Methods for the Demonstration of Lysosomes

Relatively non-specific histological methods for the demonstration of lysosomes depend upon the staining of some of their non-enzymic components. These include staining for phospholipids and for acid and neutral mucopolysaccharides. Autofluorescence of lysosomes in tissue sections and in pellets prepared by ultracentrifugation has also been described, as has vital staining by a number of fluorescent and other dyes (Allison and Young, 1969). Non-enzymic demonstration of lysosomes because of its lack of specificity is in general of no great importance but some of the *in vivo* methods for vital staining represent unique techniques for visualizing primary lysosomes in living cells.

Conclusions

Electron microscopy, histochemistry, the use of "marker" macro-molecules and observations on living tissue have thrown a great deal of light on the main morphological events concerned in heterophagy. The processes of autophagy and crinophagy are far more controversial because most conclusions have to be based upon static pictures obtained from fixed material. It is difficult, therefore, to be certain of the nature and sequence of all of the events concerned as well as of their relative timing. It does seem clear, however, that heterophagy and autophagy are interrelated and experimental evidence leads one to be fairly confident in putting forward the scheme outlined on p. 275.

IV. Functions of Lysosomes

Thus far this article has been concerned to describe the properties of lysosomes and the cellular phenomena in which they participate.

In this section it is proposed to outline the importance of lysosomes in some normal physiological processes.

A. TURNOVER OF CELL AND TISSUE COMPONENTS

Turnover of Cell Constituents

In a cell that is not growing in size or changing significantly in content the rate of synthesis of macromolecules must equal the rate at which they are degraded. For many macromolecules, although not for all (glycogen is an obvious exception), the lysosomes are the major or sole location of appropriate degradative enzymes and it seems reasonable to assume that the regular turnover of enzymes, membranes and organelles takes place within lysosomes. Furthermore, in autophagy a mechanism exists for the sequestration of areas of cytoplasm large enough to include organelles such as mitochondria.

In spite of the attractions of this hypothesis there is remarkably little direct evidence that autophagy is responsible for the regular turnover of intracellular constituents and this view may be a considerable oversimplification. Although specific sequestration of particular organelles would seem feasible, and indeed micrographs showing autophagy often show mitochondria being enveloped by endoplasmic reticulum, autophagy seems to allow little scope for the selection of particular molecules for degradation. Furthermore there are intracellular hydrolases not associated with lysosomes or any other organelle, and these may have important but as yet undiscovered functions in catabolism of macromolecules. However, the ubiquitous presence of lysosomes in cells, including those with no significant capacity for endocytosis, suggests they play an important role; a role in the turnover of cell constituents currently seems most likely.

Turnover of Extracellular Constituents

The turnover of many extracellular substances such as those of the matrix of connective tissue and bone has been shown to involve lysosomal enzymes, the process apparently including both extracellular and intracellular events. The initial phase of digestion probably occurs extracellularly, by exocytosis of lysosomal enzymes, and is then completed within lysosomes after endocytosis of the partial degradation products. It has been suggested that exocytosis of lysosomal enzymes may occur by the fusion of a lysosome with an endocytic invagination that is still in continuity with the extracellular space. It is envisaged that lysosomal contents secreted into a "stagnant" environment such as the matrix of connective tissue or bone would not quickly diffuse away but might create local conditions of pH and enzyme concentration in which enzyme activity might continue. The administration of parathyroid hormone to an embryonic bone in organ culture causes bone resorption and is accompanied by a marked release of lysosomal enzymes into the medium.

The turnover of plasma proteins and of red blood cells is also achieved by the activity of lysosomes. Since damaged or malformed red blood cells are cleared from the circulation more rapidly than normal cells, it is assumed that the reticuloendothelial cells, especially in the spleen, are selective when engaged in phagocytosis. Similarly, denatured proteins are selectively removed by the liver, indicating that uptake is not a random process.

B. CELL DEATH AND TISSUE INVOLUTION

Programmed cell death is an important morphogenetic mechanism, occuring for example in the tadpole's tail at metamorphosis, and in the interdigital region and the urogenital system during the development of the mammalian embryo. The finding that lysosomal enzymes were histochemically prominent in such areas of necrosis led to an early suggestion that lysosomes might act as "suicide bags" within cells, rupturing in selected cells in response to some metabolic trigger. This view is not now accepted and it seems unlikely that lysosomes are the cause of cell death in physiological tissue involution. But when a cell dies, its autolysis is achieved by the secondary rupture of lysosomes. Also in many cases, after cell death has occurred, phagocytic cells migrate into the area of necrosis to remove and digest the resultant cellular debris: this process involves lysosomes, of course, and it is usually these immigrant cells that account for the raised levels of lysosomal enzymes seen.

The uterus undergoes very rapid involution post-partum, with destruction of connective tissue matrix and a decrease of cell volume in smooth muscle. This appears to be achieved by a combination of autophagy and the influx of phagocytic cells which engulf and degrade extracellular material. Similar mechanisms appear to be operative during involution of the mammary gland.

As far as is known, lysosomes never behave as "suicide bags". But certain substances, e.g. silica and sodium urate, if taken into lysosomes by endocytosis, can cause membrane rupture and consequent cell death (see Section V).

C. SPECIALIZED FUNCTIONS INVOLVING HETEROPHAGY

Defence Mechanisms

Foreign macromolecules and particles that may find their way into the bloodstream are effectively removed and destroyed by specialized cell active in endocytosis. Thus human neutrophil leucocytes engulf bacteria and fungal hyphae by phagocytosis, and then kill and digest them within lysosomes. The phagocytic event often requires that the prey is coated with opsonin (see Section II and Chapter 37). The killing process appears to involve a lysosomal peroxidase whose substrates are chloride ion and hydrogen peroxide, but the precise mechanism is not understood. Foreign proteins introduced into the bloodstream are rapidly removed by the cells of the reticuloendothelial system, in a similar way to denatured plasma proteins of the host.

The involvement of lysosomes in immunological processes has been discussed by Bowers (1969), who concludes that there is no evidence for a specific role for lysosomes in the primary response. Antigens are handled by the macrophages, whose function is twofold, the engulfment and digestion of the antigen and the transmission of specific information to the lymphocyte, the cell responsible for antibody syntheisis. The second function presumably involves the antigen prior to its digestion and indeed some antigen may permanently avoid endocytosis. The possible ways in which macrophages could pass information to lymphocytes are discussed in Chapter 37.

Nutrition

Heterophagy is a potentially useful nutritional device, being a mechanism for the capture of macromolecules of biological origin and for their conversion into simple substances which can then enter either anabolic or further catabolic pathways. Presumably the products of intralysosomal catabolism regularly supplement the supply of nutrient substances to cells, but in certain cells, this activity may be particularly important. In protozoa the "food vacuoles" are heterolysosomes and it is probable that the organisms' nutritional demands are met almost exclusively by this route. Higher organisms rely chiefly on extracellular sites for the digestion of macromolecules, but in many mammals there is a short period after birth before the digestive mechanisms of the gut become operative. During this period the digestion of orally ingested macromolecules is by heterophagy within the epithelial cells of the gut. In some species passive immunity is acquired during this period by the transfer across the enterocyte of antibodies imbibed in the colostrum. How these molecules escape degradation within lysosomes is not known (Brambell, 1970). A role for heterophagy has also been postulated (Beck and Lloyd, 1968) in the nutrition of the mammalian embryo at early stages before a chorioallantoic placenta is formed. In many mammalian species certain extraembryonic membranes are found to be active in endocytosis and to contain high levels of lysosomal enzymes; it is supposed that these membranes engulf and digest macromolecules, passing on the products of digestion to the growing embryo.

Hormone Release

In the thyroid the hormone thyroxine is stored in an extracellular depot in the form of a covalently-bound complex with protein, the complex being known as thyroglobulin. The effect of thyroid stimulating hormone (TSH) is to stimulate the endocytosis of thyroglobulin into the cells of the thyroid follicular epithelium. Within lysosomes the thyroglobulin is digested by the lysosomal proteinases, thyroxine being liberated to enter the bloodstream. This process is discussed more fully in Chapter 26. The mechanism by which TSH stimulates endocytosis is not known, nor is it understood how the action is specific, affecting one type of cell only.

D. CRINOPHAGY

Crinophagy is a specialized cell process for disposing of excess secretion by means of the lysosomal system. Many cells are able to respond very rapidly to an appropriate secretory stimulus; they can do this by modulating the rate of *release* of their specific product which must therefore be allowed to accumulate to some extent within the cell—usually in the form of membrane-bounded secretory granules. Alteration in the rate of *production* of a cell secretion, though it can take place very quickly under certain circumstances, cannot in general enable a cell greatly to increase its output almost instantaneously. Within this framework, circumstances can arise in which the stimulus to secretion is suddenly withdrawn and the cells concerned must then dispose of their accumulated secretory granules. This can be achieved by crinophagy or autophagy.

Apart from the prolactin secreting cells, crinophagy has also been observed in thyrotrophin and somatotrophin secreting cells, in neurosecretory cells and in the alpha cells of the endocrine pancreas. In the adrenal medulla and the parathyroid glands excess secretions appear to be disposed of by autophagy.

V. Lysosome Pathology, Pharmacology and Toxicology

Lysosomes are implicated in a wide range of pathological, pharmacological and toxicological phenomena. This for three chief reasons. Firstly their normal function, the disposal of waste substances, though important to cells, is not essential to their viability. The contrast here is with other organelles (nucleus, mitochondria, ribosomes), some of whose functions are of such importance that disturbances in metabolism are often incompatible with life. Thus inborn errors of metabolism in which lysosomal enzymes are absent are not immediately lethal to affected cells and individuals survive long enough to be born and present as clinical conditions. Secondly lysosomes by their very nature are potentially dangerous to cells, since their enzymes are likely to cause damage if they reach the cytoplasm or the extracellular space. Thirdly lysosomes are peculiarly accessible to exogenous substances, whether drugs or toxins, by the process of endocytosis. The various ways in which lysosomal functions can be deranged have been summarized by De Duve (1964, 1968).

A. CONGENITAL ENZYME DEFICIENCIES

The first human genetic disease to be identified as resulting from an absent lysosomal enzyme was Type II (Pompe's) glycogen storage disease. This condition is due to an autosomal recessive gene and is characterized by the accumulation of glycogen (of normal structure) within abnormally large lysosomes in many tissues, but chiefly in liver and muscle. The disease is progressive and affected children usually die, before the age of two, from the disruptive effects of stored glycogen on the function of heart muscle. The absent enzyme is lysosomal α–glucosidase, whose specificity allows it to degrade glycogen completely to glucose : the enzyme's chief function is probably the degradation of glycogen that unavoidably finds its way into secondary lysosomes during autophagy. The tissues that contain high concentrations of glycogen, namely liver and muscle, are naturally those most severely affected by the enzymes absence.

Shortly after the elucidation of the metabolic defect in Pompe's disease, Hers (1965) summarized the common features that might be expected to characterize conditions resulting from an absent or defective lysosomal enzyme. These are, in abbreviated form, (1) the replacement of normal lysosomes by large vacuoles, (2) the possibility of heterogeneity of the accumulated material, since a single lysosomal enzyme might be responsible for the degradation of several substances, (3) the variation from one cell type to another of the severity of the effects, (4) the diseases would be progressive in nature, and (5) the correlation between enzyme deficit and clinical manifestation might sometimes be difficult to establish. Hers also drew attention to a sixth feature, the existence of a theoretical basis for replacement therapy in inborn lysosomal diseases; this will receive additional comment below.

In the past decade many different storage diseases of previously unknown aetiology have been shown to be lysosomal in origin. Two examples are Tay-Sachs disease, a disease in which a naturally occur-

w

ring glycolipid, GM_2 ganglioside, is progressively stored in neuronal lysosomes, is due to an absent isoenzyme of β–N–acetylhexosaminidase and metachromatic leucodystrophy, in which sulphatides are accumulated in the brain, is due to an absent cerebroside sulphatase (otherwise known as arylsulphatase A from its long-known ability to hydrolyse phenolic esters such as nitrocatechol sulphate). Many other examples could be given and, in addition, several more storage diseases are suspected to result from absent lysosomal hydrolases although it is possible that some of these may be caused by an over-production of rather than a failure to degrade the characteristic storage material. It is already clear, however, that a high proportion of the lysosomal hydrolases, including almost all the known carbo-hydrases, are affected in one or other human disease. As stated above, this high proportion no doubt reflects the relatively peripheral role that lysosomes play in the cellular economy.

An interesting feature of the lysosomal storage diseases is the possibility of replacing the missing enzyme by injecting the patient with an extraneous enzyme of appropriate substrate specificity. If an exogenous enzyme is injected intravenously, it is rapidly removed from the circulation by endocytic cells and taken into their lysosomes for degradation. This effectively prevents enzyme therapy from being effective in most inborn errors of metabolism, as the administered enzyme fails to reach the subcellular site at which it is needed. For lysosomal enzyme deficiences, however, it facilitates this end, as the administered enzyme is directed specifically to lysosomes. If the enzyme is active under intralysosomal conditions and can resist degradation long enough, it should be able to effect some breakdown of the stored material. Attempts to use replacement therapy have been made in a number of the human lysosomal enzyme deficiencies. Unfortunately those cells most severely affected by the disease usually have a poor endocytic capacity. Thus a fungal glucosidase administered to patients with Pompe's disease was effective in depleting intralysosomal glycogen in the liver, but had no discernible effect in muscle.

The lysosomal storage diseases are the subject of a recent multi-author treatise (Hers and Van Hoof, 1973).

B. INTRALYSOSOMAL MODIFIERS

Many substances, if administered to an experimental animal, are found to congregate specifically in the lysosomes. Their route there is in most cases assumed to involve endocytosis, but positive evidence is usually lacking. (However, the alternative route, passage across the plasma and lysosomal membranes, seems implausible except for some small or lipophilic molecules.) Foreign substances in lysosomes may modify lysosomal function in a number of ways. If the effect is not to be ephemeral, it is necessary for the agent to be resistant to degradation by the lysosomal enzymes.

Modifiers of Lysosomal Morphology

The accumulation of certain substances within lysosomes causes changes similar to those seen in the storage diseases described above The average size of lysosomes increases up to a maximum of approximately 1·5 μm, and these enlarged lysosomes come to occupy a large part of the extranuclear cytoplasm. Substances that accumulate in this way include sucrose, dextran and polyvinylpyrrolidone. The morphological changes are not known to be accompanied by changes in lysosome function, although an increase in extra-cellular release of lysosomal enzymes has been reported in some cases. The phenomenon is discussed more fully by Lloyd (1973).

Modifiers of Enzyme Activity

The presence within lysosomes of some substance not usually found there could activate or inhibit one or more enzymes, and evidence is available that the teratogenic dyestuff trypan blue, the trypanocide suramin and the anti-rheumatic aurothiomalate can inhibit lysosomal enzymes in this way. (Davies *et al.*, 1971). Hypotheses have been proposed that link this effect with the substances' biological activities (see references in Davies *et al.*, 1971), but in each case the evidence is as yet inconclusive. Recently Dingle *et al.* (1973) have demonstrated intralysosmal inhibition of proteolysis by an antibody prepared against cathepsin D.

Intralysosomal enzyme inhibition could also explain the observation that in certain storage diseases there is accumulation not only of the substance predicted from the enzyme deficit but also of other biopolymers. Another similar explanation is possible if the intralysosomal pH is maintained by a Donnan equilibrium effect, for the accumulation of ionic substances too large to penetrate through the lysosomal membrane could change the pH to values that depressed enzymic activity. A polyanionic primary storage product such as sulphated polysaccharide could thus so alter the intralysosomal environment that enzymes degrading other substances were unable to function adequately.

The possibility of supplementing the normal complement of lysosomal enzymes has been discussed already as an approach to the therapy of inborn lysosomal disease. It has been shown experimentally that exogenous enzymes with an appropriate optimum pH and some resistance to the action of lysosomal proteinases can function within lysosomes after uptake by endocytosis. Thus yeast invertase after up-take into lysosomes can effectively hydrolyse endocytized sucrose, a substance that cannot be degraded by normal lysosomes and that, if ingested, simply accumulates.

Modifiers of Membrane Stability

When macrophages in culture are exposed to particles of silica, the mineral is phagocytized and appears in the cells' lysosomes. After a short period there is a breakdown of the membranes of affected lysosomes, a concomitant disorganization of the cellular ultrastructure and cell death. It is believed that silica disrupts membranes by specific hydrogen bond formation with membrane components and that the resultant leakage of lysosomal enzymes into the cytoplasm is responsible for the cellular necrosis. There is good reason to suppose that the industrial disease of silicosis has its aetiology in similar events in the macrophages of lung. Then the silica contained by dying macro-phages is released and taken up a second time by previously healthy cells, a repeating cycle of uptake, cell death and release that stimulates the deposition of fibrous tissue charcteristic of the disease (Allison, 1969).

It has recently been suggested (Weissman and Rita, 1972) that an analagous mechanism operates in the inflammation seen in acute gout. Monosodium urate is phagocytized by polymorphonuclear leucocytes and disrupts lysosomes from within by hydrogen bonding with their membranes. In experiments using liposomes (artificial spherules with lipid membranes) testosterone enhanced the disruptive effect of urate where $17-\beta$–oestradiol afforded a considerable degree of protection. This could be the basis of an explanation of why gout is chiefly an affliction of males.

It is possible to envisage leakage of enzymes from lysosomes occur-ring to a lesser extent that was not lethal to the cell. It has been suggested that certain carcinogens that concentrate in lysosomes may act by causing release into the cytoplasm of DNAase which subsequently penetrates the nucleus and causes chromosome damage (Allison, 1969). Also that the cellular hyperplasia seen in the synovial lining in rheuma-toid arthritis might be due to lysosomal proteinases reaching the nucleus and degrading regulatory proteins (Weissman, 1971a). These proposals must be regarded as highly speculative.

Another, less direct, way in which lysosomal membranes can be rendered unstable is the accumulation of dyes and other substances that absorb electro-magnetic radiation of particular wavelengths. Thus a cell exposed to certain dyes concentrates them in its lysosomes. If the cell is irradiated, the lysosomes specifically absorb large amounts of energy and in consequence rupture. Photosensitization that occurs in certain human and animal diseases may have an analagous mechanism (Slater, 1969).

Early work revealed that the stability of lysosomes after their isolation from tissues was affected markedly by a variety of substances. Thus cortisone and other steroids stabilized lysosomes whereas vitamin A caused them to rupture more quickly. At the time these observations were made it was thought that intracellular rupture of lysosomes might be a widespread phenomenon in normal physiology and that modification of this process could explain the action of a number of drugs. More recently it has been shown that in the living cell most of these substances modify not lysosomal membrane stability but the rate of extrusion of lysosomal enzymes from cells. Their effects are therefore considered below.

C. MODIFIERS OF EXOCYTOSIS

The exocytosis of lysosomal enzymes was described in Section IV as being important in the remodelling of connective tissue and bone, and its occurrence to an excessive degree is probably the cause of many inflammatory conditions such as the local lesions in rheumatoid arthritis. Vitamin A increases the exocytosis of lysosomal enzymes and the depletion of cartilage matrix seen in hypervitaminosis A is the result of the extracellular action of released lysosomal proteinases, chiefly cathepsin D. Cortisone, in contrast, depresses exocytosis, and this is probably the basis of its therapeutic efficacy in some inflammatory lesions (Weissman, 1971b).

REFERENCES AND RECOMMENDED READING

Allison, A. C. (1969). Lysosomes and cancer. *In*: "Lysosomes in Biology and Pathology" (Dingle, J. T. and Fell, H. B. eds), Vol. 2, 178–204. North Holland Pub. Co. Amsterdam.

Allison, A. C. and Young, M. R. (1969). Vital staining and fluorescence microscopy of lysosomes. *In*: "Lysosomes in Biology and Pathology".(Dingle, J. T. and Fell, H. B. eds), Vol. 2, 600–628. North Holland Pub. Co., Amsterdam.

Allison, A. C. (1972). Role of membranes in effector systems for hormone and drug action. *Chem. Phys. Lipids* **8**, 374–385.

Bainton, D. F. and Farquhar, M. G. (1966). Origin of granules in polymorphonuclear leucocytes. Two types derived from opposite faces of the golgi complex in developing granulocytes. *J. Cell Biol.* **28**, 277–301.

Barrett, A. J. (1972). Lysosomal enzymes. *In*: "Lysosomes. A Laboratory Handbook" (Dingle, J. T. ed.), 46–135. North Holland Pub. Co., Amsterdam.

Beaufay, H. (1972). Methods for the isolation of lysosomes. *In*: "Lysosomes. A Laboratory Handbook" (Dingle, J. T. ed.), 1–45 North Holland Pub. Co., Amsterdam.

Beck, F., Lloyd, J. B. and Squier, C. A. (1972). Histochemistry. *In*: "Lyosomes. A Laboratory Handbook." (Dingle, J. T. ed.), 200–239, North Holland Pub. Co., Amsterdam.

Bowers, W. E. (1969). Lysosomes in lymphoid tissues: spleen, thymus, and lymph nodes. *In*: "Lysosomes in Biology and Pathology" (Dingle, J. T. and Fell, H. B. eds), Vol. 1, 167–191., North Holland Pub. Co., Amsterdam.

Brambell, F. W. R. (1970). "The transmission of passive immunity from mother to young". North Holland Publishing Company, Amsterdam.

Caro, L. G. and Palade, G. E. (1964). Protein synthesis, storage and discharge in the pancreatic exocrine cell. *J. Cell Biol.* **20**, 473–495.

Ciba Foundation (1963). Symposium on Lysosomes. J. and A. Churchill, London.

Cohn, Z. A. and Fedorko, M. A. (1969). The formation and fate of lysosomes. *In*: "Lysosomes in Biology and Pathology" (Dingle, J. T. and Fell, H. B. eds), Vol. 1, 43–63. North Holland Pub. Co. Amsterdam.

Daems, W. Th., Wisse, E. and Brederoo, P. (1969). Electron microscopy of the vacuolar system. *In*: "Lysosomes. A Laboratory Handbook" (Dingle, J. T. ed.), 150–199. North Holland Pub. Co., Amsterdam.

Davies, M., Lloyd, J. B. and Beck, F. (1971). The effect of trypan blue, suramin and aurothiomalate on the breakdown of ^{125}I-labelled albumin within rat liver lysosomes. *Biochem. J.* **121**, 21–26.

De Duve, C. (1959). Lysosomes, a new group of cytoplasmic particles. *In* "Subcellular Particles", (Hayashi, T. ed.), 128–159. Ronald Press, New York.

De Duve, C. (1964). From cytases to lysosomes. *Fed. Proc.* **23**, 1045–1049.

De Duve, C. (1965). The separation and characterization of subcellular particles. *Harvey Lect.* **59**, 49–87.

De Duve, C. (1967). General principles. *In*: "Enzyme Cytology", (Roodyn, D. B. ed.), 1–26. Academic Press, New York and London.

De Duve, C. (1968). Lysosomes as targets for drugs. *In*: "The Interaction of Drugs and Subcellular Components in Animal Cells", (Campbell P. N. ed.), 155–169. J. and A. Churchill, London.

De Duve, C. (1969). The lysosome in retrospect. *In*: "Lysosomes in Biology and Pathology" (Dingle, J. T. and Fell, H. B. eds), Vol. 1, 3–40, North Holland Pub. Co., Amsterdam.

De Duve, C. and Wattiaux, R. (1966). Functions of lysosomes. *Ann. Rev. Physiol.* **28**, 435–492.

Dingle, J. T. (1972), ed. "Lysosomes. A laboratory handbook", North Holland Publishing Co., Amsterdam.

Dingle, J. T. (1973), ed. "Lysosomes in Biology and Pathology", Volume 3. North Holland Publishing Co., Amsterdam.

Dingle, J. T., Poole, A. R., Lazarus, G. S. and Barrett, A. J. (1973). Immunoinhibition of intracellular protein digestion in macrophages. *J. Exp. Med.* **137**, 1124–1141

Dingle, J. T. and Fell, H. B. (1969), eds. "Lysosomes in Biology and Pathology", Volumes 1 and 2. North Holland Publishing Co., Amsterdam.

Farquhar, M. G. (1969). *In*: "Lysosomes in Biology and Pathology" (Dingle, J. T. and Fell, H. B. eds), Vol. 2, 462—482. North Holland Publishing Co., Amsterdam.

Hers, H. G. (1965). Inborn lysosomal diseases. *Gastroenterology* **48**, 625–633.

Hers, H. G. and Van Hoof, F. (1973), eds. "Lysosomes and Storage Diseases". Academic Press, New York and London.

Jacques, P. J. (1969). Endocytosis. *In*: "Lysosomes in Biology and Pathology". (Dingle, J.T. and Fell, H. B. eds), Vol. 2, 395–420. North Holland Publishing Co., Amsterdam.

Jacques, P. J. (1973). The endocytic uptake of macromolecules. *In*: "Pathologic Aspects of Cellular Membranes", (Trump, B. J. and Arstila, A. ed.) Academic Press, New York and London.

Jollie, W. P. and Triche, T. J. (1971). Ruthenium labelling of micropinocytotic activity in the rat visceral yolk-sac placenta. *J. Ultrastructure Res.* **35**, 541–553.

Leighton, F., Poole, B., Beaufay, H., Baudhuin, P., Coffey, J. W., Fowler, S., and De Duve, C. (1968). *J. Cell Biol.* **37**.

Lloyd, J. B. (1973). Experimental support for the concept of lysosomal storage disease. *In*: "Lysosomes and Storage Diseases". (Hers, H. G. and Van Hoof, F. eds), 173–195. Academic Press, New York and London.

Novikoff, P. M., Novikoff, A. B., Quintana, N. and Hann, J. J. (1971). Golgi apparatus, Gerl and lysosomes of neurones in rat dorsal root ganglia studied by thick section and thin section cytochemistry. *J. Cell Biol.* **50**, 859–886.

Nichols, B. A., Bainton, D. F. and Farquhar, M. G. (1971). Differentiation of monocytes. Origin, nature and fate of their azurophilic granules *J. Cell Biol.* **50**, 498–515.

Slater, T. F. (1969). Lysosomes and experimentally induced tissue injury. *In*: "Lysosomes in Biology and Pathology". (Dingle, J. T. and Fell, H. B. eds). Vol. 1, 469–492. North Holland Pub. Co., Amsterdam.

Smith, R. E. and Farquhar, M. G. (1966). Lysosome function in the regulation of the secretory process in cells of the anterior pituitary gland. *J. Cell Biol.* **31**, 319–347.

Straus, W. (1964). Cytochemical observations on the relationship between lysosomes and phagosomes in kidney and liver by combined staining for acid phosphatase and intravenously injected horseradish peroxidase. *J. Cell Biol.* **20**, 497–507.

Weissmann, G. (1971a). Lysosomes and the mediation of tissue injury in arthritis. *In*: "Rheumatoid Arthritis", (Müller, W., Harwerth, H.-G. and Fehr, K. eds), 141–154. Academic Press, London and New York.

Weissmann, G. (1971b). The effects of corticosteroids, especially on lysosomes and biomembranes. *In*: "Rheumatoid Arthritis", (Müller, W., Harwerth, H.-G. and Fehr, K. eds), 577–584. Academic Press, London and New York.

Weissmann, G. and Rita, G. A. (1972). Molecular basis of gouty inflammation: interaction of monosodium urate crystals with lysosomes and liposomes. *Nature, Lond.* **240**, 167–172.

Williams, K. E., Lloyd, J. B., Kidston, E. M. and Beck, F. (1973). Biphasic effect of trypan blue on pinocytosis. *Biochem. Soc. Proc.* **1**, 203–206.

Winterburn, P. J. and Phelps, C. F. (1972). The significance of glycosylated proteins. *Nature, Lond.* **236**, 147–151.

Wissig, S. L. and Graney, D. O. (1968). Membrane modifications in the apical endocytotic complex of ileal epithelial cells. *J. Cell Biol.* **39**, 564–579.

8.   Subcellular Pathology

A. C. ALLISON

Clinical Research Centre, London, England

The great German pathologist Rudolf Virchow (1821–1902) pointed out that disease should be considered in terms of the reactions of individual cells (Virchow 1858). Toxins affect more or less selectively certain cells in the liver, kidneys or other organs; inflammation—from whatever cause—is associated with infiltration of leucocytes; regeneration is due to controlled multiplication, and malignancy to uncontrolled multiplication, of certain cells; and so forth. Traditional morbid histology is soundly based on this type of cellular pathology. The next major step in understanding was largely due to R. Peters, who emphasized that pathological processes should be defined in terms of specific metabolic disturbances—biochemical lesions (Peters, 1953). Thus, the toxin fluoroacetate, present in the African plant *Dichapetalum cymosum* and ingested by livestock, is metabolized to fluorocitrate which blocks the enzyme aconitase, an essential step in the tricarboxylic cycle. This type of metabolic conversion into a damaging product Peters termed "lethal synthesis"; other examples will be discussed below. The metabolic disturbances in deficiency diseases could likewise be defined. In thiamine deficiency levels of co-carboxylase in the brain fall markedly; the reaction catalysed by carboxylase, pyruvate oxidation, requires thiamine pyrophosphate. Hence carbohydrate metabolism is impaired and this can lead to convulsions and other signs of neuronal dysfunction.

During the last three decades great progress has been made in cell biology. Methods have developed for fractionating cellular components, examining their ultrastructure by electron microscopy and their metabolic capabilities—especially by using radioactive isotopes. Even membranes, which for so long remained rather mysterious boundaries between cell compartments, are now becoming better understood. It has been possible to show that many pathological reactions affect primarily one cell organelle, and sometimes one metabolic step within that organelle. This information has accumulated to form a new science, subcellular pathology, which has as its aim the identification of the primary lesion in any pathological process in terms not only of the cells involved, but of the organelles and metabolic processes that are disturbed. However, changes in one system are often followed by changes in others, which may be equally important in the ultimate damage produced. Thus sequences of changes should also be considered. In this chapter examples will be given of disturbances primarily affecting different cellular organelles. Emphasis is placed on the application of principles of cell biology to explain causes and manifestations of disease. To illustrate the complexities of interaction, effects of major metabolic disturbances by agents depleting ATP or inhibiting synthesis of protein, RNA or DNA, are also considered.

I. The Plasma Membrane

Various methods have been developed for the isolation of plasma membranes, and it has been possible to show that their compositions are somewhat different from those of membranes elsewhere in the cell, e.g. smooth endoplasmic reticulum. Thus, plasma membranes have a relatively high proportion of sphingomyelin and cholesterol. Cholesterol "stiffens" membranes, and the plasma membrane is less flexible than other membranes (Bosmann *et al.*, 1969). About 30% of the plasma membrane mass is made up of proteins. According to the classical Danelli and Davson model (see Chapter 2), the membrane consists primarily of a bilayer of lipid with hydrophobic groups facing inward and hydrophilic groups facing outwards and interacting with protein. In electron micrographs of suitably fixed material, a characteristic trilaminar membrane structure is observed. Recent evidence suggests that this model is true for part of the membrane structure, but that in some regions proteins with hydrophobic side chains extend into the bilayer and interact with non-polar portions of lipid molecules (see also Chapter 2). Many of the functions of membranes, including selective transport across them, are thought to be due to protein carriers or enzymes within the membrane itself.

Many membrane proteins have attached carbohydrate moieties and the carbohydrate groups of the outer surface make an important contribution to the antigenicity of cells. The major blood group antigens, for example, are the terminal disaccharides of carbohydrate side chains which are attached to a peptide backbone. In electron micrographs the outer layer of membrane is seen to be covered by a glycopeptide cell coat the thickness of which varies. The coat can readily be sheared from the plasma membrane and remain attached to substrates.

A. EFFECTS ON ION PERMEABILITY

Several classes of compounds can affect the permeability of ions through plasma membranes. Only a few examples of selective effects can be considered here. Many plasma membranes have an adenosine triphosphatase, requiring Mg^{2+}, Na^+ and K^+ for activity, which links the active transport of Na^+ and K^+ with the energy liberated by splitting ATP (Whittam and Wheeler, 1970). This enzyme is inhibited by cardiac glycosides such as digoxin, digitoxin, strophanthin and ouabain. Valinomycin, a cyclic peptide derived from a fungus, can be inserted into membranes and increase the permeability of K^+ ions a thousandfold or more. In contrast, tetrodotoxin, a fish poison, applied to the outside of the membrane, blocks entry of Na^+, and with it the action potential in nerve. Procaine and other local anaesthetics are positively charged surface-active agents that become oriented at the membrane-aqueous interface and decrease permeability of Ca^{2+} and other cations across the membrane. These drugs affecting ion permeability are useful tools for the analysis of membrane physiology.

B. ENZYMES

Only a limited number of enzymes can attack the membranes of intact cells. Neuraminidase, from filtrates of *Vibrio cholerae* and other bacteria, can selectivity remove sialic acid-containing moieties from the surfaces of cells. Measurement of electrophoretic mobility of cells before and after neuraminidase treatment shows that the majority of the negative charges on the surface of various cells (which have a considerable net negative charge) are due to carboxyl groups of sialic acid. Trypsin and other proteases also remove sialic acid-containing peptides from the surfaces of cells. This again reduces the negative charge and facilitates dispersion of cells for subculture or other purposes.

Certain enyzmes enter membranes or other lipid-containing structures such as liposomes (spherules composed of concentric bilayers of lipid separated by an aqueous phase). Phospholipase A in snake venoms can convert phosphatides to highly surface-active lysophosphatides, and these can bring about lysis of the cells. Activity of snake venom phospholipase on intact cells is more effective if strongly basic polypeptides of the venom are also present. Phospholipases from bacteria, and also non-enzymic proteins such as streptolysin which can enter membranes and disrupt their structure, explain the well known haemolytic and cytolytic effects of the bacteria.

C. POSITIVELY CHARGED COMPOUNDS

Since the cell membrane has a net negative charge, it reacts with many positively charged substances, especially polyelectrolytes (Allison, 1968). Chemically synthesized basic polymers, such as polylysine or polyarginine, or the basic protein protamine, agglutinate cells and, in slightly higher concentrations, lyse them. Snake venoms contain basic peptides apart from those facilitating phospholipase-induced haemolysis. One polypeptide (crotamine) increases cation permeability through membranes, depolarizing muscle fibre membranes and producing the spasticity characteristic of rattlesnake venom. A dialysable basic polypeptide from cobra venom produces fibrillations. The physiologically active kinins, kallidin and bradykinin, which induce contraction of smooth muscle and increase capillary permeability, have positively charged C-terminal (arginine) and N-terminal (lysine-arginine or arginine) groups, respectively. The location of these positive charges may facilitate attachment at appropriate sites and increase membrane permeability to cations, which is the trigger for some smooth muscle contractions.

D. SPECIFIC INTERACTIONS OF SURFACE-ACTIVE CONSTITUENTS

Several groups of surface-active agents interact strongly with cholesterol and exert their effects on membranes only when they contain cholesterol. Examples are saponin, digitonin and the larger polyene antibiotics such as nystatin and amphotericin B (Kinsky, 1970). This is illustrated by the membranes of mycoplasma which do not normally contain cholesterol but incorporate the sterol if it is added to the culture medium; only when sterol is present are they sensitive to polyene antibiotics. Vitamin A is a surface-active agent with a wide range of effects on membranes, whether they contain cholesterol or not. In high concentrations vitamin A lyses cells; in lower concentrations it leads to release of lysosomal hydrolases from surviving cells.

E. LYSIS OF CELLS BY ANTIBODY AND COMPLEMENT

Among the most remarkable effects on plasma membranes is the capacity of antibody and complement to bring about lysis. A single molecule of IgM attached to a cell is sufficient for lysis, but in the case of IgG antibody more than one molecule is required; apparently two IgG molecules on the cell membrane must be sufficiently close together for complement components to form a bridge between them. Green et al. (1959) showed that after combination of antibody and complement with cells low-molecular weight cytoplasmic contents such as potassium and free amino acids leak out, but that macromolecules such as protein and RNA are lost only if osmotic effects are not counterbalanced by the presence of protein in the medium. Proteins are trapped within the plasma membrane, and an energy-driven pump offsets Na^+ leakage inwards. If the cation permeability of such a system increases beyond the capability of the pump, the non-diffusible proteins within will effect a Donnan redistribution of ions, and the concomitant rise in osmotic pressure and swelling will rupture the membrane and release its contents. Green and his colleagues concluded that antibody and complement produce holes in the cell membrane large enough to permit the free exchange of water and ions such as K^+ and Na^+ but too small to allow passage of macromolecules.

These holes have been visualized in electron micrographs of negatively stained membranes (Humphrey and Dourmashkin, 1969). When the surface of the membrane lies flat on the electron microscope grid (Fig. 1), a dark central part of the hole is observed, surrounded by a lighter ring. When a folded edge of the membrane is present, the clear ring can be seen projecting from the membrane surface, forming a hollow cylinder filled with negative stain. These holes are observed in the walls of bacteria, and formation of such holes, along with the action of the enzyme lysozyme (muramidase) contributes to bacterial lysis and host protection.

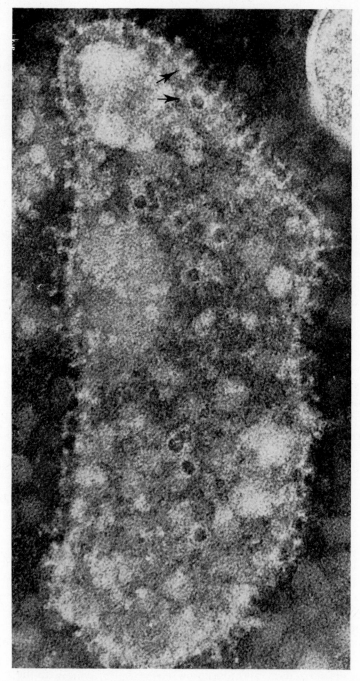

Fig. 1. Electron micrograph of a bacterial cell wall showing holes produced by serum complement from above and in profile (arrows). Negative staining, $\times 320\,000$ (Courtesy of R. Dourmashkin).

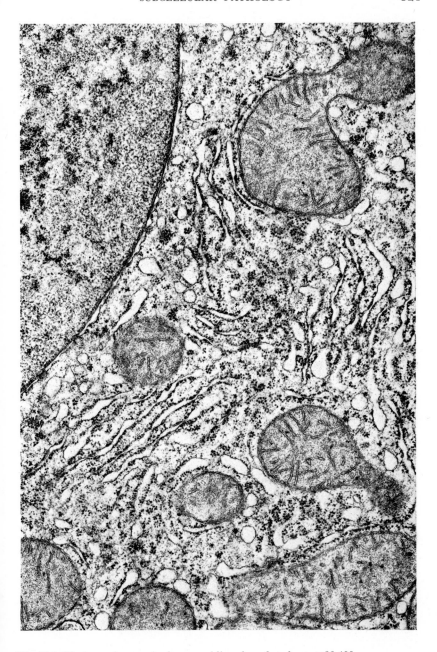

Fig. 2(a). Electron micrograph of a normal liver from fasted rat, ×23 400.

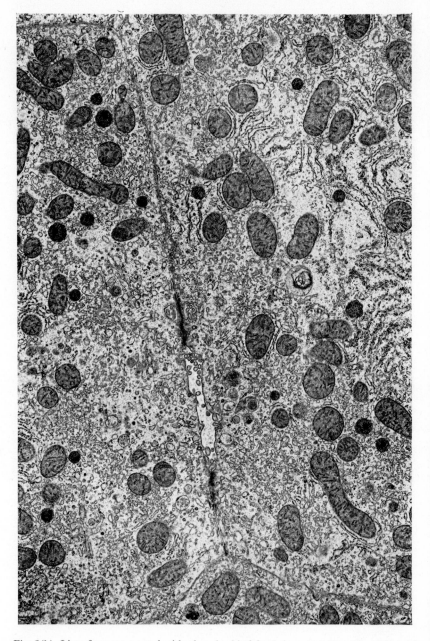

Fig. 2(b). Liver from rat treated with phenobarbital for 5 days, × 8 100. (Smuckler and Arcasoy, 1968, courtesy of the authors and publishers).

II. Smooth Membrane Endoplasmic Reticulum

One of the important functions of the smooth-membrane endo-plasmic reticulum, which is especially well developed in the liver, is the metabolism of hormones and drugs. Administration of barbiturates and other drugs causes a marked proliferation in smooth-membrane endo-plasmic reticulum observed in electron micrographs (Fig. 2) and in the activity of drug-metabolizing enzymes in the smooth microsome fraction of liver homogenates (Smuckler and Arcasoy, 1970). When the cell is homogenized, membranes of the endoplasmic reticulum break up and become resealed to form little spheres sedimented at high speed, the microsomes.

Evidence is accumulating that some toxic substances are meta-bolically converted into active derivatives, possibly free radicals, before they can damage cells (Slater, 1968). Thus carbon tetrachloride is probably metabolized into a chloroform radical (CCl_3^-) in smooth-membrane endoplasmic reticulum. This leads to selective destruction of important constituents of microsomal membranes such as cytochrome P-450; other constituents, such as NADP-cytochrome C reductase, cytochrome b_5 and neotetrazolium reductase, are unaffected. This is associated with structural changes in the endoplasmic reticulum illus-trated in Fig. 3. Fatty liver, inhibition of protein synthesis and release of lysosomal enzymes follows; these effects, but not destruction of cytochrome P-450, can be prevented by antioxidants.

Newborn rats or rats treated with drugs that inhibit mixed-function oxidases of smooth-membrane endoplasmic reticulum show reduced sensitivity to carbon tetrachloride hepatotoxicity, whereas pretreat-ment with phenobarbital increases sensitvity, which supports the inter-pretation that conversion to an active metabolite is involved. The same is true of carbon disulphide, which induces liver cell necrosis only when given with phenobarbital (Bond and de Matteis, 1969).

III. Rough Endoplasmic Reticulum and Free Polysomes

There is now substantial evidence that proteins are synthesized not on individual ribosomes but on polysomes—strings of ribosomes attached to messenger RNA. The number of ribosomes in polysomes is related to the length of the messengers involved. Proteins can be syn-thesized on polysomes which are free in the cytoplasm or polysomes which are attached to membranes, thereby forming rough endoplasmic reticulum.

The latter is believed to be the site of synthesis of proteins that are not released directly into the cytoplasm but are packaged for export in secretory granules, are included in lysosomes and other cytoplasmic

x

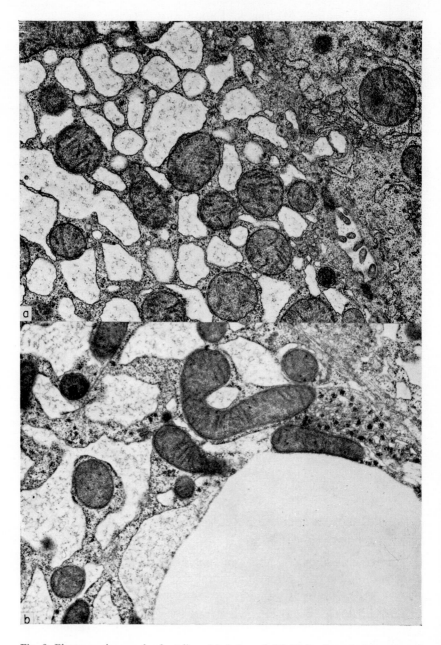

Fig. 3. Electron micrograph of rat liver (a) 3 h and (b) 12 h after administration of carbon tetrachloride, ×16 000 (Smuckler and Arcasoy, 1958, by courtesy of the authors and publishers).

organelles or are contributed directly to membranes. Membrane proteins are insoluble in aqueous media since many hydrophobic groups face outwards. Consequently such proteins must be transferred directly from polysomes to membranes, where they can interact with membrane lipid by hydrophobic bonding. At least some cytoplasmic constituents are synthesized on polysomes not bound to membranes, a well-studied example being synthesis of haemoglobin in red blood cell precursors. For these and other reasons, attachment of ribosomes to, and detachment from, membranes of the endoplasmic reticulum might exert an important controlling function on protein synthesis. It is therefore of interest that certain treatments can lead to detachment of ribosomes from membranes. Rabin and colleagues (1971) have found that when ribosomes are detached, sites on endoplasmic reticulum with unusual enzymic activity are exposed: these can rearrange the disulphide bonds in thermally inactivated ribonuclease and restore capacity of the latter to degrade RNA. This can be used as an assay for detachment, and results of such tests correspond quite well with electron microscopic observations. Carcinogenic chemicals with various structures, e.g. aflatoxin, aromatic amines and polycyclic hydrocarbons, all cause prolonged detachment of ribosomes from endoplasmic reticulum, and the resulting change in control of protein synthesis may play a role in cancer induction. In contrast, oestrogenic steroids in the male animal, and androgenic steroids in the female animal, tend to promote attachment of polysomes to membranes.

The effects of inhibitors of RNA synthesis on cytoplasmic polysomes, ribosomes and protein synthesis is still imperfectly understood. Actinomycin D produces marked inhibition of RNA synthesis and a slow, progressive inhibition of protein synthesis in rat liver, accompanied by breakdown of free, but not membrane-bound, polysomes within a 12 h period (Farber, 1971). No necrosis is produced, in contrast to the effects of aflatoxin.

As discussed in the section on protein synthesis below, the acute necrosis seen in many organs of animals treated with puromycin is probably not due simply to inhibition of protein synthesis. It is not observed in animals in which protein synthesis is depressed to a comparable degree with cycloheximide.

Exposure of rats to puromycin or ethionine gives rise to a fatty liver with a periportal distribution. The liver converts free fatty acids from the blood to triglycerides which are returned to the blood as lipoproteins. Inhibition of protein synthesis (or of synthesis of other required components such as phospholipid in choline deficiency) cuts off the supply of the protein, so that there is an accumulation of triglyceride. Evidently production of triglyceride is uncontrolled under these conditions. Farber (1971) suggests that, since the bulk of liver triglyceride is synthesized in endoplasmic reticulum, the maintenance of intact polysomes in close association with endoplasmic reticulum may allow regulatory control of triglyceride synthesis which is lost when the polysomes are no longer *in situ* in the membrane. With cycloheximide or actinomycin there is no displacement of polysomes from membrane or shift to polysome precursors (except in the case of actinomycin with free polysomes), and no fatty liver. With puromycin or with ethionine in female rats there is detachment and breakdown of polysomes, and triglyceride accumulation is marked.

IV. The Golgi Apparatus

The Golgi apparatus consists of stacks of flattened membranous lamellae extending outwards from the region of the centrosomes to form a network of channels closely related to the endoplasmic reticulum. The Golgi apparatus is thought to carry out two main functions. Proteins synthesized in rough endoplasmic reticulum are packaged for export or for incorporation into lysosomes at the periphery of the Golgi apparatus. Sequential autoradiographs at the electron microscopic level show that the most recently synthesized proteins (labelled with radioactive amino acids) are in the rough endoplasmic reticulum. They pass from there to vesicles of the Golgi system, condensing vacuoles and then to secretory vacuoles, lysosomes and other organelles. Such autoradiographic studies also show that some sugars and sulphate are incorporated into glycoproteins in the Golgi system. The Golgi system can now be isolated from homogenized cells in reasonably pure form, and the appropriate sugar transferase enzymes are found in it. No specific pathology of the Golgi system is known, but it can show swelling and other structural changes in a variety of situations, especially when the endoplasmic reticulum is damaged.

V. Lysosomes

Because of their content of hydrolytic, and therefore potentially damaging enzymes, their role in intracellular killing of micro-organisms and their participation in many cellular reactions to toxic substances, lysosomes are of special interest in subcellular pathology. Their involvement can be considered under seven main headings : autolysis of tissues that are injured or undergoing morphogenetic changes; formation of autophagic vacuoles; intracellular release of hydrolases producing cellular damage; extracellular release of hydrolases producing damage to connective tissue matrix; intracellular digestion of potentially pathogenic mico-organisms; "storage diseases" due to accumulation within lysosomes of materials which cannot be broken down because the required enzymes are congenitally lacking; and failure of intracellular digestion in histiotrophic nutrition, for example in the placenta, leading to fetal abnormalities. Chapter 7 is devoted to a discussion of lysosomes.

A. AUTOLYSIS

If liver or other tissue is examined at intervals after death, or after deprivation of blood supply, it is found that hydrolytic enzymes are released from the lysosomes and begin to digest cellular constituents. This occurs under sterile conditions. If the organs are homogenized and centrifuged, the redistribution of lysosomal enzymes can be shown by an increase in their activity in the supernatant and a decrease in the large-granule fraction. Histochemically, discrete staining for acid phosphatase and other marker enzymes is replaced by diffuse staining. However, lysosomal changes may often occur secondarily to damage in other systems, e.g. endoplasmic reticulum. Lysosomal digestion also takes place in the remodelling of tissues occuring during the course of normal development, a well-studied example being the disappearance of the tadpole's tail during metamorphosis (Weber, 1970).

B. FORMATION OF AUTOPHAGIC VACUOLES

When cells are exposed to certain stimuli, various cytoplasmic organelles (such as mitochondria and components of endoplasmic reticulum) are found to be enveloped by smooth membrane to form "autophagic vacuoles". Histochemical studies at the electron microscopic level show that lysosomal hydrolases are present in the autophagic vacuoles, and their constituents are digested. Among the agents that promote the formation of autophagic vacuoles are starvation (in amoeba and mammalian liver), glucagon in the liver and chloroquine in the liver and in malarial parasites. Some autophagy takes place in normal tissues such as the proximal convoluted tubules of the kidney (Fig. 4).

C. INTRACELLULAR RELEASE OF HYDROLASES

This occurs after tissue damage produced by a wide range of toxic substances, but in most situations it is difficult to establish whether the lysosomal changes precede or follow those occuring in the cell membrane, smooth-membrane endoplasmic reticulum or other organelles. A situation in which it can confidently be concluded that the initial damage is lysosomal is that in which silica and other toxic particles are ingested by macrophages. Usually inhaled carbon or other particles produce little disturbance, but silica or asbestos particles initiate a

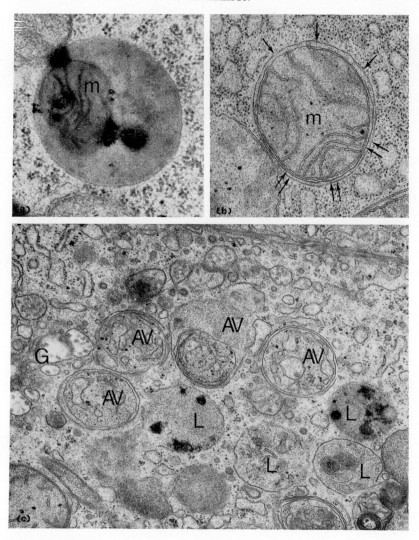

Fig. 4. Autophagy in proximal convoluted tubules of rat kidney. (a) Electron-dense phosphatase reaction product and a mitochondrion M in an autophagic vacuole, ×39 600. (b) Mitochondrion surrounded by two or three membranes, ×36 000. (c) Formation of autophagic vacuoles AV and secondary lysosomes L containing phosphatase reaction product, ×10 800 (Courtesy of J. L. E. Ericsson).

fibrous reaction in the lung known as silicosis or asbestosis. The inhaled particles are taken up by alveolar macrophages, which are killed; the particles are released and taken up by other macrophages which are in turn killed. The cytotoxicity can be reproduced with small quantities of silica particles in cultures of macrophages. Toxic and non-toxic

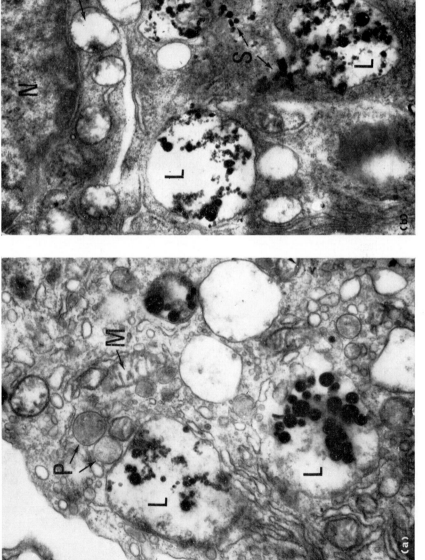

Fig. 5. a. Electron micrograph of the peripheral cytoplasm of a macrophage shortly after ingestion of silica particles which are seen in secondary lysosomes L. b. A macrophage 18 hours after uptake of silica particles S which have escaped into the cytoplasm. Mitochondria M are swollen and abnormal. The nucleus N is not yet involved, ×24 000.

particles (such as titanium dioxide or diamond dust) are ingested by macrophages and incorporated into secondary lysosomes: membrane-lined vacuoles into which hydrolases are discharged. Toxic particles interact with the membrane surrounding the secondary lysosome, releasing the hydrolases, whereas with non-toxic particles interaction with the membrane does not occur (Fig. 5). Evidence has been presented that the unusual hydrogen-bonding properties of the silica surface are responsible for the interaction with membranes, which can be illustrated also by the haemolytic effects of silica particles (Allison, 1971). Damage to macrophages stimulates collagen formation by fibroblasts by some mechanism which is not yet understood.

D. EXTRACELLULAR RELEASE OF HYDROLASES

An example of this phenomenon is provided by a series of investigations carried out by Fell, Dingle, Barrett and their colleagues on organ cultures of cartilaginous rudiments from chick embryos. This system is taken as a model of cartilage breakdown such as occurs in arthritic joints. When the cartilage cultures are exposed to high concentrations of vitamin A, to sucrose or other sugars which are accumulated in lysosomes but cannot be broken down because the necessary enzymes are lacking, or to anticellular antibody, the cells are not killed but continuously release large quantities of lysosomal hydrolases (Fig. 6).

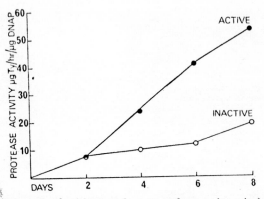

Fig. 6. Extracellular release of lysosomal protease from embryonic bone rudiments cultured in the presence of antibody and serum complement, and in the absence of serum complement (Dingle, Fell and Coombs, 1967, by courtesy of the authors and publishers).

This is due to increased synthesis of the lysosomal enzymes, and not just release of preformed enzymes. The considerable breakdown of cartilage matrix that follows (Fig. 7) is largely due to the lysosomal protease cathepsin D, as can be shown by inhibition with specific antibody against the enzyme. Release of hydrolases from the cells is inhibited by hydrocortisone (Dingle and Fell, 1969).

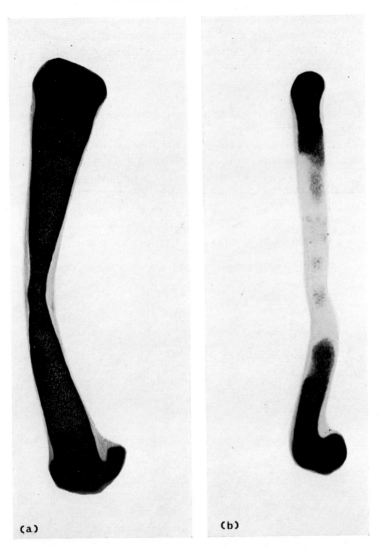

(a)   (b)

Fig. 7. Photomicrograph of embryonic bone rudiment cultured in normal medium (a) and in the presence of a high dose of vitamin A (b). In the latter the darkly staining cartilage matrix has been largely degraded (Courtesy of J. T. Dingle).

E. INTRACELLULAR KILLING OF MICRO-ORGANISMS

Bacteria, fungi, viruses and other micro-organisms are ingested by macrophages and neutrophil leucocytes, usually after "opsonization" by antibody, sometimes also with serum complement. In secondary lysosomes the organisms are exposed to lysosomal constituents includ-ing, in the case of neutrophils, specific basic proteins, lysozyme (muramidase) and myeloperoxidase. The basic proteins adhere to the bacterial cell walls and have an antibacterial effect. Lysozyme can break down the walls of damaged bacteria, and can act on bacteria in the presence of antibody and complement. Myeloperoxidase in the presence of halide can also kill certain bacteria, but it is not clear to what extent this is relevant to protection *in vivo* because most human subjects known to have a congenital absence of myeloperoxidase have not suffered from persistent or recurrent infections. Severe bacterial and fungal infections are, however, the rule in children with chronic granulomatous disease, of which two forms have been characterized. One, inherited as a sex-linked recessive, is manifested in young boys and is characterized by diminished activity of leucocyte NADH oxidase in phagocytic cells after ingestion of particles. It can be demonstrated conveniently by a histochemical test with nitroblue tetrazolium. The other is inherited as an autosomal recessive, manifested in both boys

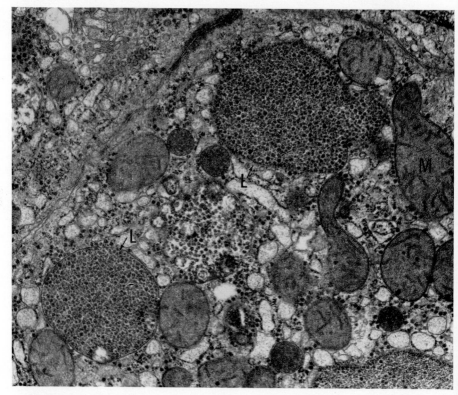

Fig. 8. Electron micrograph (× 36 000) of part of the cytoplasm of a liver cell from a patient with type II glycogenosis, showing accumulation of glycogen within lysosomes L. Mitochondria appear relatively normal (Hers and van Hoof, 1969, by courtesy of the authors and publishers).

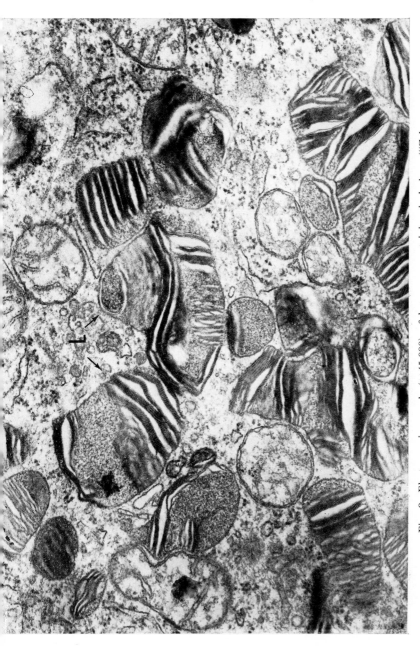

Fig. 9. Electron micrograph (×45 000) of lysosomes containing inclusions ("Zebra bodies") in a neuron of a patient with Hurler syndrome, lacking β-galactosidase (courtesy of A. Resbois).

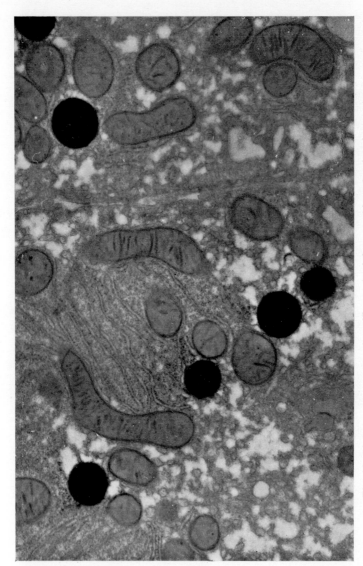

Fig. 10. Electron micrograph of rat liver stained for peroxidase showing dense micro-bodies 10 days after stopping injections of the drug allylisopropylacetamide (×15 000). (Courtesy of P. G. Legg and R. L. Wood).

and girls, and is associated with deficient glutathione perioxidase. Also inherited as an autosomal recessive is the Chediak-Higashi syndrome, in which some lysosomal and other intracellular granules are abnormally large. Affected subjects are unusually susceptible to infections.

The capacity of macrophages to kill organisms which can multiply in these cells, such as the bacteria *Listeria*, *Brucella*, *Salmonella* and *Mycobacterium* (causes of tuberculosis and leprosy) is increased when the macrophages are "activated" by a cell-mediated immune response (Mackaness, 1970). The content of lysosomal hydrolases in these cells is increased, but details of the mechanism of intracellular killing are not yet understood.

VI. Microbodies

These are a distinct group of cytoplasmic organelles, containing several oxidative enzymes, which yield hydrogen peroxide as a reaction product, and catalase, which reduces hydrogen peroxide to water by two distinct mechanisms, catalatic and peroxidatic. Proliferation of microbodies in hepatic cells is known to occur in three main circumstances: during foetal and early postnatal development, during the recovery phase following partial hepatectomy and following administration of certain chemical agents, notably salicylates and ethyl-α-p-chlorophenylisobutyrate (CPIB or Clofibrate, see Legg and Wood, 1970, and Fig. 10). Microbodies are further discussed in Chapter 6.

VII. Mitochondria

Mitochondria (see Chapter 6) have long attracted attention as the site of oxidative phosphorylation in eukaryotic cells. Recently interest has centred around the finding that mitochondria contain DNA and have an autonomous system for synthesizing RNA and proteins. In mammalian cells the active unit of the mitochondrial protein-synthesizing system is a 55S particle assembled into a 95S structure, producing six or seven polypeptides. Several specific inhibitors of mitochondrial RNA and protein synthesis are available. The trypanocidal drug ethidium bromide in low concentrations inhibits mitochondrial RNA synthesis but has little effect on RNA synthesis outside mitochondria. It also inhibits mitochondrial DNA replication and rapidly reduces mitochondrial protein synthesis to immeasurable levels. Chloramphenicol is a specific and complete inhibitor of mitochondrial protein synthesis but has little effect on protein synthesis in cytoplasmic polysomes. Thus the protein synthetic mechanism in mitochondria is remarkably like that in bacteria, which provides support for the view that mitochondria orginated as bacterial symbionts in primitive unicellular organisms.

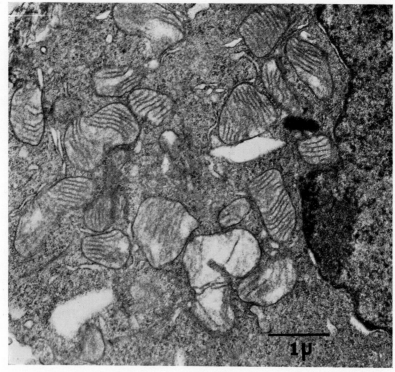

Fig. 11(a). Part of the cytoplasm of a normal HeLa cell ($\times$ 49 600).

Although only a few of the mitochondrial proteins are synthesized in the mitochondria, they appear to be necessary for the preservation of normal structure. HeLa cells growing in the presence of chloramphenical or ethidium bromide show distorted mitochondrial inner membranes and few cristae (Fig. 11). No specific cytopathology is associated with mitochondria, although these organelles frequently show non-specifically swelling and other structural changes when cellular metabolism is disturbed, e.g. after damage to smooth-membrane endoplasmic reticulum or lysosomes. Some normal and tumour cells, for example, in the thyroid gland, contain abundant mitochondria (Fig. 12).

VIII. MICROTUBULES

Electron micrographs of suitably fixed material show in the cytoplasm of most cells small tubular structures. The tubules are about 20–24 nm in diameter and of indefinite length. In cross section there are 13 subunits 4 nm in diameter, and these are regularly arranged

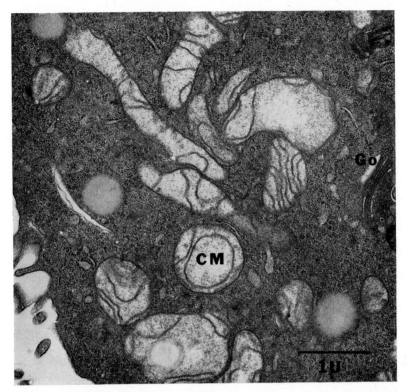

Fig. 11(b). Part of the cytoplasm of a HeLa cell cultivated in the presence of ethidium bromide, showing distortion of inner mitochondrial membranes, some of which are circular CM. (×41 600, courtesy of Dr R. Lenk).

as superimposed rings to make up the microtubule (Fig. 13). The protein subunits have a molecular weight of 60 000 with one mol of guanine nucleotide per mol of protein (Stephens, 1968). Microtubules are best known in cilia and flagella, in which they are characteristically arranged in the form of two central tubules surrounded by nine pairs of tubules (Fig. 14). This arrangement is common to the flagella and cilia of motile bacteria and plant and animal cells (with rare exceptions). As Behnke and Forer (1967) have pointed out, the microtubules of flagella and cilia are relatively stable to treatments that break down the structure of labile cytoplasmic microtubules. The latter include the numerous microtubules arranged along the long axis of the mitotic spindle. Microtubules are thought to provide rigidity and polarity in cytoplasm, and are often concentrated in regions where movement is taking place, as in the mitotic spindle, the contracting melanophore and lymphocyte uropod (Porter, 1966).

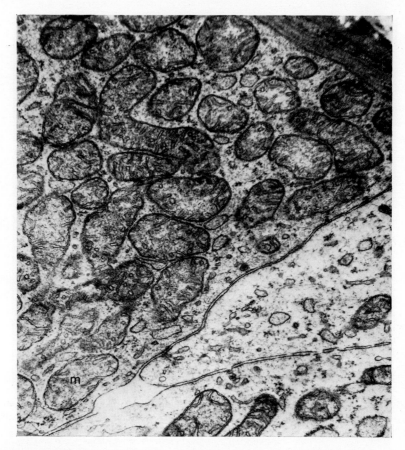

Fig. 12. Electron micrograph of normal thyroid showing numerous mitochondria M in an oxyphil cell (×27 600, courtesy of the Armed Forces Institute of Pathology; A.F.I.P. Atlas No. 67–3–5).

Labile microtubules, including those of the mitotic spindle, disappear when cells are exposed to cold (4°C) or high hydrostatic pressure; the disappearance is reversible on return to physiological temperature and pressure. Several drugs disperse labile microtubules, including colchicine, which is used in treatment of gout; vincristine and vinblastine (alkaloids from the periwinkle, *Vinca*) which are used to treat malignant disease; the anti-fungal agent griseofulvin; and podophyllotoxin. Colchcine and vincristine have been shown to bind to different sites on microtubular protein; in cells treated with vincristine, characteristic crystalline structures, thought to be microtubular protein, are observed. Colchicine arrests mitosis at metaphase because of interference with spindle function through dispersal of microtubules. Normal neurons show an

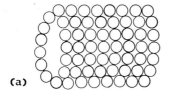

Fig. 13. Diagram of the arrangement of protein subunits in (a) microtubules and (b) actin filaments.

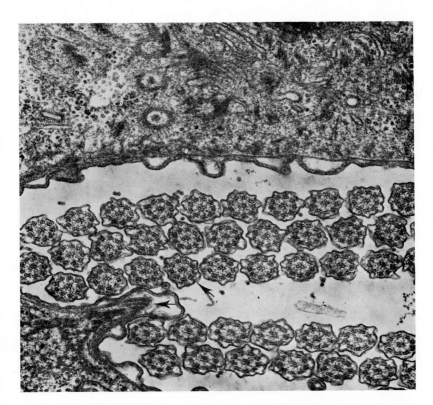

Fig. 14. Electron micrograph of the peripheral cytoplasm of *Tetrahymena*, showing microtubules (marked by arrows) in flagella and beneath the plasma membrane ($\times 35\,000$).

W*

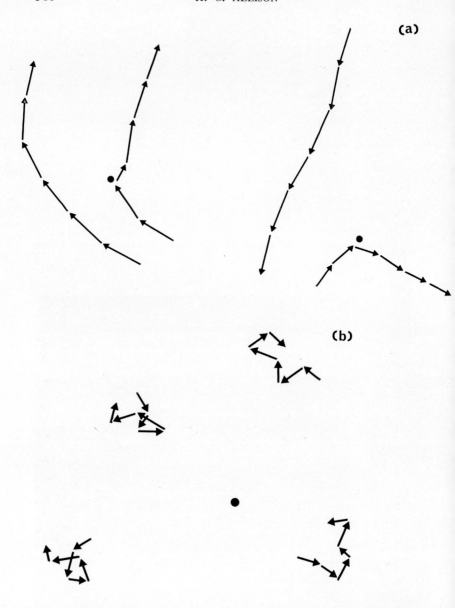

Fig. 15(a). Representation of the movement of mouse peritoneal macrophages cultivated on coverslips, as shown by time-lapse cine-micrography. Each arrow shows the displacement of a nucleus during five minutes, and the adjoining arrow the displacement during the next five minutes. The amoeboid movement of colchicine-treated cells (b) is nearly random, whereas the gliding movement of the normal cells is obviously directional. The movement during each time interval has a high probability of being in nearly the same direction as the previous time interval, unless the cells meet some obstruction. This is due to the intrinsic polarity within the cells.

ordered arrangement of microtubules, extending from the axon hillock down the axon to near its extremities. After vincristine treatment there is an irregular arrangement of microtubular protein. In neurons treated with colchicine or vincristine transport of protein down the axon is impaired (see Sjöstrand et al., 1970), and this may account for the peripheral neuropathy seen in patients and experimental animals receiving prolonged treatment with vincristine (Uy et al., 1967).

Allison and colleagues (1971) have found that inhalational anaesthetics in clinical concentrations bring about dispersal of labile cytoplasmic microtubules, including those of the mitotic spindle. This may account for the inhibition by anaesthetics of cell division; whether it plays any part in causing general anaesthesia is not yet known. The most obvious consequence of treating non-dividing cells with low concentrations of colchicine is that highly directional gliding movement is replaced by amoeboid randomly-directed movement (Fig. 15). From this observation it appears that microtubules play an important part in determining the polarity of the cytoplasm, which allows directional movement to take place (see Table 1).

Table 1. Effects of drugs on microtubules and actomyosin microfilaments

	Flagellar and ciliary microtubules	Labile cytoplasmic microtubules	Microfilaments
Colchicine	—*	+	—
Vinblastine	—*	+	—
Cytochalasin	—	—	+

+ Means inhibition of function, in the case of microtubules accompanied by breakdown of structure.
* No effect on the assembled microtubules, but their formation can be inhibited.

IX. Microfilaments

Many cells have in their peripheral cytoplasm a network of microfilaments about 7 nm in diameter. They are closely related to the actomyosin filaments in smooth muscle, and can be shown (Fig. 16) to bind a component of myosin (heavy meromyosin). The microfilaments are inserted into the plasma membrane, and appear to be responsible for movement of the membrane, and, with it, of the cell itself. Cells move by pseudopodial extension, attachment of the pseudopod and the contraction of the pseudopod with translocation of the cell body towards the point of attachment.

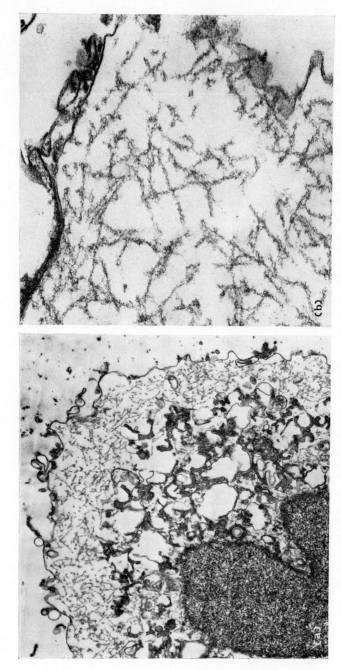

Fig. 16. Electron micrograph of mouse peritoneal macrophage after exposure to glycerol and heavy meromyosin. In the peripheral cytoplasm, beneath the plasma membrane, is a network of actin-like fibrils that has bound heavy meromyosin. The fibres are therefore thickened and show fuzzy outlines. (a) ×16 000; (b) ×59 000. From Allison et al., 1971.

Analysis of the role of contractile microfilaments in cell function has been greatly advanced by the discovery that the microfilaments are paralysed by the fungal product cytochalasin. Several cytochalasins have been isolated; they combine with actomyosin from skeletal muscle, smooth muscle and blood platelets, and may inhibit the magnesium-activated ATPase of myosin which is required for contraction. Further evidence that cytochalasin interacts directly with the contractile system is the inhibition by the drug of contraction of glycerolated smooth muscle brought about by ATP in the presence of Mg^{2+} and Ca^{2+}.

Cytochalasin stops the movement and contraction of a variety of cells including fibroblastic and epithelial cells, leucocytes and smooth muscle cells. It has no effect on spontaneous contractions of skeletal muscle cells in culture. Movement of ruffled membranes and the invagination of plasma membranes to form large endocytic vacuoles is inhibited by cytochalasin. The phagocytosis of bacteria by leucocytes is reversibly stopped by the drug. The release of granules from cells by exocytosis is also prevented by cytochalasin. This has been demonstrated for the release from mast cells, exposed to cell-bound antibody and antigen, of granules containing histamine and other mediators of acute allergic reactions. Cytochalasin also inhibits the release which is induced by acetylcholine of granules containing catecholamines from the adrenal medulla. Release of hormones from other cells, and of neurotransmitters from nerve terminals, appears to take place through a similar mechanism.

Allison (1973) has suggested that this is one of the two major ways in which hormones and drugs act on cells. Each reactive cell has on its membrane specific receptors for the agents to which it is responsive: some cells have receptors for catecholamines, others for acetylcholine, others for histamine, others for antibody + antigen, and so on. Attachment to the receptor of the appropriate agonist triggers one or both of two processes: A—contraction of microfilaments, which results in contraction of smooth muscle (in blood vessels, abdominal viscera, etc.) or in discharge from the cell of granules containing pharmacologically active agents; and B—activation of the membrane-associated enzyme adenyl cyclase and formation of 3′,5′-adenosine monophosphate (cyclic AMP). The latter in turn stimulates kinases that phosphorylate proteins, thereby activating enzymes and so changing the metabolism of target cells. The type A response is thus a short-term effect involving some sort of movement of the peripheral cytoplasm, whereas the type B response occurs over a longer period and involves cellular metabolism. Type A responses can be blocked either by interfering with the increased membrane permeability to cations which triggers the responses, e.g. by procaine and related local anaesthetics, or by inhibition of the contractile process itself, e.g. by cytochalasin or general anaesthetics administered by inhalation.

The distinction between agents acting on microtubules and on microfilaments is nicely illustrated by their effects on cell division. Microtubule poisons such as colchicine result in metaphase arrest, whereas in cells treated with cytochalasin movement of chromosomes at anaphase and reformation of daughter nuclei is normal; in contrast, cleavage, which depends on contraction of a ring of microfilaments beneath the cytoplasm, does not occur, so that the cells remain binucleate. Another common feature in cells treated with cytochalasin is that the nucleus comes to lie immediately beneath the plasma membrane. Bulging and extrusion of the nucleus quite often follows. It seems that the nucleus is normally kept away from the plasma membrane by the peripheral network of microfilaments, but that if this "cage" is paralysed the nucleus can be closely enveloped by the plasma membrane. Disappearance of contractile microfilaments in the maturing normoblast may account for nuclear extrusion from these cells.

In addition to actin-like microfilaments about 7 nm in diameter, which (except in smooth muscle cells) are mainly found in the periphery of the cytoplasm, cells have other microfilaments of unknown function. In many cells 10 nm of microfilaments are seen, and these are especially numerous after colchicine treatment. The existence in most cells of a cytochalasin-insensitive contractile system is shown by the lack of effect of the drug on the activity of the mitotic spindle, and a similar system may be active in the inner cytoplasm of interphase cells.

X. Nucleus

In cells infected with viruses or exposed to a variety of toxic agents, especially carcinogens, non-specific nuclear changes are seen. These include increase in size, structural distortions of various kinds, clumping and margination of chromatin and invagination of the nuclear membrane, sometimes producing intranuclear inclusions containing cytoplasmic organelles. An example is the changes in the livers of rats treated with pyrollizidine alkaloids (Allen et al., 1970), in which the liver cells and nuclei become unusually large.

Many different agents can bring about chromosome breaks and other aberrations, including ultraviolet and ionizing radiation, viruses and a variety of alkylating agents and other toxic substances.

XI. Nucleolus

Changes in nucleolar structure are seen in cells treated with inhibitors of RNA synthesis. The biochemical effects of these substances are

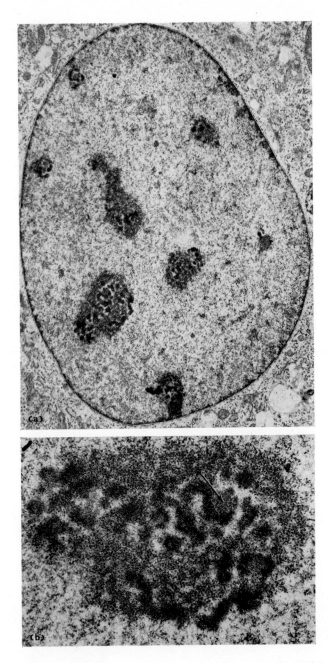

Fig. 17. Untreated Chinese hamster cells, showing nucleoli, one of which (b) is enlarged. Nucleoli are composed of a granular region which is often separated by electron-transparent areas from a dense fibrous region. Small, light fibrous regions (arrow) can sometimes be discerned juxtaposed to dense fibrous regions. (a) ×6 800; (b) ×25 200; Phillips and Phillips, 1971, by courtesy of the authors and publishers).

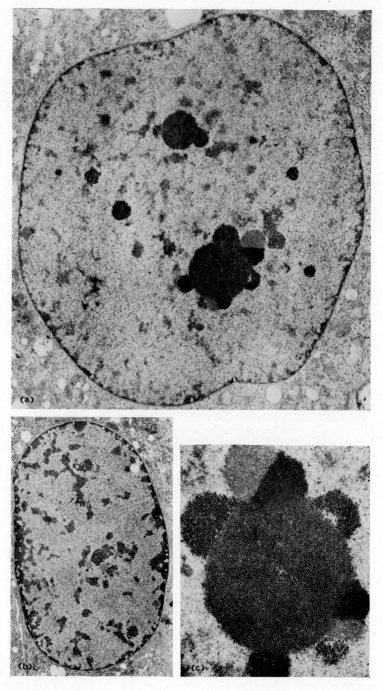

Fig. 18(a). L. cell treated with actinomycin D for 2 h Nucleoli of cells are segregated
into different components (a) ×4 200; (b) Chinese hamster cell treated for 3 h with
actinomycin D, showing nucleolar disintegration (×4 400; (c) segregation ×12 000;
Phillips and Phillips, 1971, by courtesy of the authors and publishers).

discussed below. Two main types of morphological change are produced. The commoner is *nucleolar segregation* : the granular and fibrillar components, which are intermingled in the normal nucleolus, become separated into distinct zones (Fig. 18). This is found after administration of many compounds that interfere with RNA synthesis by combining with DNA, including actinomycin D, 4-nitroquinoline-N-oxide, proflavine, daunomycin, nogalamycin, ethidium bromide, aflatoxin and pyrrolizidine alkaloids (Farber, 1971). It is also seen after ultraviolet irradiation or infection with mycoplasma and several different viruses. The second type of change is *nucleolar fragmentation*, with scattering of the pieces around the nucleus (Fig. 18). This is seen after exposure to ethionine, alpha-amanitin or toyocamycin, which inhibit RNA synthesis directly without combining with the DNA template, and also actinomycin D treatment under certain conditions (Phillips and Phillips, 1971).

XII. ATP DEPLETION

A single metabolic lesion may affect several organelles and functions. An example is the depletion of ATP in liver that follows administration of ethionine, the ethyl analogue of methionine, to the rat (Farber, 1971). Reaction of ATP with ethionine to give S-adenosylethionine occurs more rapidly than recovery of the adenosine component through trans-ethylation and *de novo* synthesis of adenine nucleotides from available precursors, so that ATP deficiency is rapidly produced. This is followed by fatty liver, depletion of liver glycogen, hypoglycaemia, a major shift in the polysome-ribosome equilibrium towards ribosomes and ribosome subunits, concomitant marked inhibition of protein synthesis, inhibition of RNA synthesis and the accompanying characteristic abnormalities of the nucleolus. All these effects of ethionine can be prevented or reversed by supplying adenine or adenine nucleotide precursors, so that the biochemical lesion in the animal can be turned on or off at will. A more transient depletion of ATP, with similar consequences, follows intravenous administration of D-fructose to rats. Phosphorylation of fructose to fructose-1-phosphate occurs more rapidly than splitting of fructose-1-phosphate into glyceraldehyde and dihydroxyacetone phosphate; the latter step, catalysed by aldolase, is rate limiting.

XIII. INHIBITION OF OXIDATIVE PHOSPHORYLATION

Triethyltin strongly inhibits the coupling of mitochondrial respiration to phosphorylation, with generation of ATP, in the brain and other organs (Aldridge and Street, 1970). Triethyltin binds strongly to mitochondria, probably to histidine groups of the inner mitochondrial membrane, and disorganizes the major mitochondrial energy-generating system. Triethyltin is metabolically stable and induces extensive cerebral oedema in the white matter of the brain and spinal cord, associated with progressive weakness, paralysis and convulsions. The oedema is intracellular in glial cells.

XIV. Inhibition of Protein Synthesis

Many subcellular abnormalities produced by toxic substances have been attributed to inhibition of protein synthesis. However, some of these substances have multiple effects : for instance, ethionine not only inhibits protein synthesis but also depletes ATP. One of the first compounds shown to inhibit protein synthesis was puromycin, which can be activated and substitutes for charged transfer RNA at the P site on the ribosome. Polypeptide chain growth is terminated prematurely and incomplete peptides are released containing a terminal puromycin molecule. Puromycin in whole animals inhibits protein synthesis in all cells so far studied, with acute, widespread, severe damage.

However, it seems unlikely that this is due to inhibition of protein synthesis *per se* because cycloheximide, given to rats in doses sufficient to produce almost complete inhibition of protein synthesis for many hours leads to only a few cellular changes in selected organs (Farber, 1971). Unlike puromycin, cycloheximide acts only in eukaryotic cells and not in prokaryotes, and in the former it inhibits general cytoplasmic ribosomal but not mitochondrial protein synthesis. In doses comparable to those used *in vivo* cycloheximide acts almost exclusively as an inhibitor of peptide chain termination or release. *In vivo* administration of cycloheximide leads to death of many cells in the germinal centres of lymphoid organs. There is also a disappearance of mitotic cells in intestinal crypt epithelium and other organs, including regenerating liver. Protein synthesis is required for the late G1, S (DNA synthetic) and early G2 phases of the cell cycle, but not for late G2 and M (mitosis); hence inhibition of protein synthesis soon leads to a fall in the number of cells observed in mitosis. The relationship of inhibition of protein synthesis to fatty liver is discussed in the section on rough endoplasmic reticulum.

XV. Inhibition of RNA Synthesis

Many inhibitors of RNA synthesis also inhibit DNA synthesis and other reactions, so that interpretation of their effects is complicated. Even compounds that are relatively selective inhibitors of RNA synthesis may do so in different ways and may not affect all classes of

RNA equally. There are at least three classes of ribosomal RNA (28S, 18S and 5S), more than 50 varieties of *t*RNA, unique messengers for each polypeptide synthesized and some species of RNA that are degraded in the nucleus and do not reach the cytoplasm.

A. ACTIONMYCIN D

The most widely used inhibitor of RNA synthesis has been actino-mycin D, which combines non-covalently with guanine residues in DNA. At low doses the drug inhibits predominantly ribosomal RNA synthesis and consequently the incorporation of uridine or other pre-cursors which label the nucleolus. This may be attributable to the high cytosine or cytosine plus guanine content of ribosomal RNA (Perry, 1964). As the dose of actinomycin is increased, more species of RNA are affected, until at concentrations of the order of 1 mg per kg body weight 95% of RNA synthesis is inhibited. Several other effects follow, including accelerated breakdown of some fractions of recently synthesized nuclear RNA, nucleolar segregation and inhibition of the normal passage of ribosomal and transfer RNA from the nucleus to the cytoplasm.

B. ALPHA AMINITIN

This is one of a group of cyclic peptide toxins, collectively known as amatoxins, found in the poisonous mushroom, *Amanita phalloides*. Alpha aminitin, an octapeptide, powerfully inhibits RNA polymerase II, a manganese-activated enzyme in the nucleoplasm, but not RNA polymerase I, which is associated with the nucleolus. Alpha aminitin produces nucleolar fragmentation but not death of liver cells (Farber, 1971).

C. ETHIONINE

This amino acid analogue strongly depresses RNA synthesis *in vivo*, partly by depleting ATP which is required for the reaction. It has been found to inhibit RNA polymerase I and II (Jacob, Sajdel and Munro, 1970), and leads to nucleolar fragmentation. The *in vivo* effects are reversible by adenine administration.

D. TOYOCAMYCIN

This antibiotic produces highly selective inhibition of the normal transformation of 45S nuclear RNA to ribosomal RNA, presumably by being incorporated into the 45S fraction. It leads to fragmentation of the nucleolus.

XVI. Inhibition of DNA Synthesis

It has been widely believed that inhibition of DNA metabolism leads to death of proliferating cells. However, certain inhibitors of protein synthesis, such as cycloheximide, also impair DNA synthesis in rapidly proliferating cells but do not decrease cell viability. In contrast, agents such as 1-β-D-arabinofuranosyl cytosine (cytosine arabinoside), alkylating agents and X-rays, which are known to affect DNA, kill cells such as germinal centre cells and the crypt cells of the small intestine. This type of cell killing can in fact be prevented by inhibitors of protein synthesis administered shortly after cytosine arabinoside, nitrogen mustard of X-radiation (Farber, 1971). Protein synthesis may therefore be required for manifestation of metabolic perturbations following administration of agents that have DNA as important targets.

XVII. Defective Cross-Linking of Connective Tissue Structural Proteins

Copper deficiency in young growing mammals and chickens leads to weakness and malformation of bones and arteries, often terminating in rupture of the heart or major blood vessels when they are three or four months old. There is a reduction in the tensile strength of the aorta of copper-deficient animals, a reduction in elastin content and a corresponding increase in the content of cellular material and mucopolysaccharide. There is also irregular fragmentation of the internal elastic lamina and dissecting haemorrhages are observed. There is an increase in the proportion of collagen extractable with saline at neutral pH. Similar changes are seen in pyridoxine deficiency.

A constituent of the sweet pea (*Lathyrus odoratus*), β-(N-γ-glutamyl)-aminopropionitrile, was found to produce lathyrism, a severe and crippling disease of the bones and connective tissue (see Levine, 1970). The toxic properties were later shown to be due to the β-aminopropionitrile component. The tissues of lathyrogen-treated animals have no change in total collagen content but a large increase in cold saline-soluble collagen. The aorta and large blood vessels show abnormalities in elastin similar to those seen in copper deficiency (Fig. 19). Several nitriles and other classes of carbonyl blocking agents also produce lathyrism. They can be arranged according to activity, nitriles being most potent, then ureides, hydrazides and hydrazines. It therefore appeared likely that lathyrogenic compounds block carbonyl groups which participate in the normal cross-linking of collagen.

With both collagen and elastin, fibrogenesis is brought about by synthesis in fibroblasts of a soluble precursor protein which is later polymerized at the site of fibre formation to produce, after a considerable lapse of time, a stable three-dimensional cross-linked structure by the formation of covalent bonds. Normally collagen consists of three

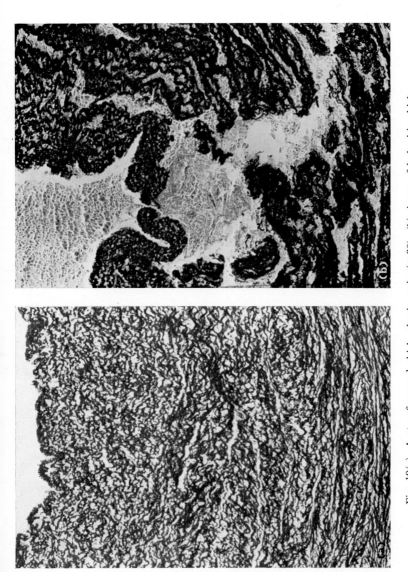

Fig. 19(a), Aorta of normal chick, elastin stain (×80). (b) Aorta of lathyritic chick, elastin stain (×80), showing disorganization of elastin and rupture of wall (courtesy of S. M. Partridge).

supercoiled peptide chains with cross-linking, whereas collagen from lathyrogen-treated animals consists of immature α-chains, suggesting that cross-linking is defective.

The chemistry of cross-linking has been more fully studied in elastin than in collagen. When soluble proelastin has arrived at the site of fibre formation, the side chains of certain lysine residues are oxidised to aldehyde. The fact that copper is involved in the formation of cross-links and that the processes is inhibited by aldehyde reagents such as semicarbazide suggests that a diamine oxidase (with pyridoxal phosphate as co-factor) is involved. During a period of time before the formation of permanent irreversible crosslinks, there is a dynamic equilibrium of unstable bonds (Schiff bases) and more stable aldol condensations. Eventually stable bonds are formed. Lysine residues are converted to two amino acids uniquely present in elastin, desmosine and isodesmosine. Thus during the growth of animals (or in wound healing) formation of new elastin has such a requirement. This explains why the effects of copper or pyridoxine deficiency, and of lathyrogenic agents, are observed with greatest intensity in young animals (Partridge, 1969).

XVIII. Comment

Examples are now known of disorders which affect more or less selectively most of the major cell constituents. Sometimes the underlying biochemical abnormalities have also been characterized. It is nevertheless becoming clear that cells can tolerate marked perturbations of metabolism without necrosis. These include ATP depletion, severe inhibition of protein, RNA or DNA synthesis and considerable loss of K^+ and accumulation of Na^+. Some metabolic lesions, especially those affecting plasma and lysosomal membranes, appear rather rapidly to produce irreversible damage. Several toxins produce sequences of changes affecting many organelles. Although many problems remain unresolved, enough information is available to provide a sound basis for the science of subcellular pathology.

References

Aldridge, W. N. and Street, B. W. (1970). *Biochem. J.* **118**, 171.

Allen, J. R., Carstens, R. A. and Norback, D. H. (1970). *Cancer Res.* **30**, 1857.

Allison, A. C. (1968). *Brit. Med. Bull.* **24**, 135.

Allison, A. C. (1971). *Arch. of Intern. Med.* **128**, 131.

Allison, A. C. (1973). *Ciba Foundation Symposium on Locomotion in Tissue Cells.* Amsterdam, Associated Scientific Publishers

Allison, A. C., Hulands, G. H., Nunn, J. F., Kitching, J. A. and Macdonald, A. C. (1971). *J. Cell Sci.* **7**, 483.

Behnke, O. and Forer, A. (1967). *J. Cell Sci.* **2**, 169.

Bond, E. J. and de Matteis, F. (1969). *Biochem. Pharmacol.* **18**, 2531.

Bosmann, H. B., Hagopian, A. and Eylar, E. H. (1969). *Arch. Biochem. Biophys.* **150**, 573.

Dingle, J. T. and Fell, H. B. (1969). "Lysosomes in Biology and Pathology". North Holland Press, Amsterdam.

Farber, G. (1971). *Ann. Rev. Pharmacol.* **11**, 71.

Green, H., Barrow, P. and Goldberg, B. (1959). *J. Exp. Med.* **110**, 699.

Humphrey, J. H. and Dourmashkin, R. R. (1969). *Advan. Immunol.* **11**, 75.

Jacob, S. T., Sajdel, E. M. and Munro, H. N. (1970). *Biochim. Biophys. Res. Commun.* **38**, 765.

Kinsky, S. C. (1970). *Ann. Rev. Pharmacol.* **10**, 119.

Legg, P. G. and Wood, R. L. (1970). *J. Cell Biol.* **45**, 118.

Levine, C. I. (1970). "Mechanisms of Toxicity". (W. N. Aldridge ed.), p. 67. Macmillan, London.

Partridge, S. M. (1969). *Gerontologia* **18**, 85.

Perry, R. P. (1964). *Nat. Cancer Inst. Monogr.* **14**, 73.

Peters, R. (1953). "Biochemical Lesions and Lethal Synthesis". Macmillan, London.

Phillips, D. M. and Phillips, S. G. (1971). *J. Cell Biol.* **49**, 803.

Porter, K. R. (1966). *In* "Principles of Biomolecular Organization". (G. E. W. Wolstenholme and M. O'Connor, eds.), p. 308. Churchill, London.

Rabin, B. R., Blyth, C. A., Doherty, D., Freedman, R. B., Roobol, A., Sunshine, G. and Williams, D. J. (1971). *In* "Effects of Drugs on Cellular Control Mechanisms". (B. R. Rabin and R. B. Freedman, eds.). Churchill, London.

Sjöstrand, J., Frizell, M. and Hasselgren, P. O. (1970). *J. Neurochem.* **17**, 1573.

Slater, T. F. (1968). *In* "The Biological Basis of Medicine", (C. E. and N. Bittar, eds.). Vol. 1, p. 369. Academic Press, London and New York.

Smuckler, E. A. and Arcasoy, M. (1968). *Int. Rev. Exp. Pathol.* **7**, 411.

Stephens, R. (1968). *J. Mol. Biol.* **33**, 517.

Uy, P. L., Moen, T. H., Johns, R. J. and Owens, A. J., Jr. (1967). *Johns Hopkins Med. J.*, **121**, 349.

Virchow, R. (1958). "Disease, Life and Man". Translated by C. J. Pather. Stanford University Press, 273 pp.

Weber, R. (1969). *In* "Lysosomes in Biology and Pathology". (J. T. Dingle and H. B. Fell, eds.). Vol. 2, p. 437. North Holland Press, Amsterdam.

Whittam, R. and Wheeler, K. P. (1970). *Ann. Rev. Physiol.* **32**, 21.

Author Index

Subject Index